Lecture Notes in Artificial Intelligence 1692

Subseries of Lecture Notes in Computer Science
Edited by J. G. Carbonell and J. Siekmann

Lecture Notes in Computer Science

Edited by G. Goos, J. Hartmanis and J. van Leeuwen

Springer

Berlin
Heidelberg
New York
Barcelona
Hong Kong
London
Milan
Paris
Singapore
Tokyo

Václav Matoušek   Pavel Mautner
Jana Ocelíková   Petr Sojka (Eds.)

Text, Speech and Dialogue

Second International Workshop, TSD'99
Plzen, Czech Republic, September 13-17, 1999
Proceedings

Springer

Series Editors

Jaime G. Carbonell, Carnegie Mellon University, Pittsburgh, PA, USA
Jörg Siekmann, University of Saarland, Saarbrücken, Germany

Volume Editors

Václav Matoušek
Pavel Mautner
Jana Ocelíková
Faculty of Applied Sciences, University of West Bohemia in Plzeň
Department of Computer Science and Engineerig
Universitní 22, 306 14 Plzeň, Czech Republic
E-mail: {matousek/mautner/jnetrval}@kiv.zcu.cz

Petr Sojka
Faculty of Informatics, Masaryk University Brno
Department of Programming Systems and Communication
Botanická 68a, 602 00 Brno, Czech Republic
E-mail: sojka@informatics.muni.cz

Cataloging-in-Publication data applied for

Die Deutsche Bibliothek - CIP-Einheitsaufnahme

Text, speech and dialogue : second international workshop ; proceedings / TSD
'99, Plzen, Czech Republic, September 1999. Václav Matousek ... (ed.). -
Berlin ; Heidelberg ; New York ; Barcelona ; Hong Kong ; London ; Milan ; Paris
; Singapore ; Tokyo : Springer, 1999
(Lecture notes in computer science ; Vol. 1692 : Lecture notes in artificial
intelligence)
ISBN 3-540-66494-7

CR Subject Classification (1998): I.27, H.3, H.4, I.7

ISBN 3-540-66494-7 Springer-Verlag Berlin Heidelberg New York

© Springer-Verlag Berlin Heidelberg 1999
Printed in Germany

Typesetting: Camera-ready by author
SPIN 10704486   06/3142 – 5 4 3 2 1 0   Printed on acid-free paper

Preface

This book contains the collection of papers presented at the *Second Workshop on Text, Speech and Dialogue — TSD'99* held in Plzeň and Mariánské Lázně (Czech Republic) on 13–17 September 1999. The general objective of the workshop was to present state–of–the–art technology and recent achievements in the field of natural language processing.

A total of 57 papers and 19 posters contributed by 128 authors (63 from Central Europe, 11 from Eastern Europe, 33 from Western Europe, 2 from Africa, 13 from America, and 6 from Asia) were included in the workshop proceedings.

The workshop is an interdisciplinary forum, which brings together research in speech and language processing as well as research in the Eastern and Western hemisphere. We feel that the mixture of different approaches and applications gives all of us a great opportunity to benefit and learn from each other.

We would like to gratefully thank the invited speakers and the authors of the papers for their valuable contributions, the Medav GmbH (Uttenreuth, GER) and the SpeechWorks (Boston, USA) for their financial support, and Prof. Vostracký for greeting the workshop on behalf of the University of West Bohemia.

Last but not least, we would like to express our gratitude to the authors for their efforts to provide the papers on time, to members of the program committee for their careful reviews, to the editors for their hard work in preparing these proceedings, and to the members of the local committee for their efforts in organising the workshop.

Baltimore and Erlangen, June 1999   F. Jelinek (General Chair)
                                                E. Nöth (Executive Chair)

Organization

TSD'99 was organized by the Faculty of Applied Sciences, University of West Bohemia in Plzeň in cooperation with Masaryk University in Brno. The workshop's Web page is http://www-kiv.zcu.cz/events/tsd99.

Program Committee

Jelinek, Frederick (USA) *General Chair,*
Nöth, Elmar (Germany) *Executive Chair,*
Baudoin, Geneviève (France),
Čermák, František (Czech Republic),
Ferencz, Attila (Romania),
Hajičová, Eva (Czech Republic),
Hanks, Patrick (GB),
Hynek, Hermansky (USA),
Hovy, Eduard (USA),
Kilgariff, Adam (GB),
Kopeček, Ivan (Czech Republic),
Krauwer, Steven (Netherland),
Kučera, Karel (Czech Reublic),
Matoušek, Václav (Czech Republic),
Moon, Rosamund (GB),
Norling-Christensen, Ole (Danmark),
Pala, Karel (Czech Republic),
Pavešić, Nikola (Slovenia),
Petkevič, Vladimír (Czech Republic),
Psutka, Josef (Czech Republic),
Schukat-Talamazzini, E. Günter (Germany),
Skrelin, Pavel (Russia),
Vintsiuk, Taras (Ukraine),
Wilks, Yorick (GB).

Organizing Committee

Matoušek Václav *(Chair),*
Benešová Helena *(Secretary),*
Černý Martin,
Dudáček Karel,
Hesová Jana *(Secretary),*
Kindlová Svatava *(Secretary),*
Klečková Jana,

Kopeček Ivan,
Krutišová Jana,
Mautner Pavel,
Ocelíková Jana,
Pala Karel,
Sojka Petr,
Zuzák František.

Table of Contents

Invited Talks

Research Issues for the Next Generation Spoken Dialogue Systems 1
  *E. Nöth, F. Gallwitz, M. Aretoulaki, J. Haas, S. Harbeck, R. Huber,
  H. Niemann*

Data-Driven Analysis of Speech 10
  Hynek Hermansky

Towards a Road Map for Machine Translation Research 19
  Steven Krauwer

The Prague Dependency Treebank: Crossing the Sentence Boundary 20
  Eva Hajičová

Text

Tiered Tagging and Combined Language Models Classifiers 28
  Dan Tufiş

Syntactic Tagging 34
  Alena Böhmová, Jarmila Panevová, Petr Sgall

Information, Language, Corpus and Linguistics 39
  František Čermák

Prague Dependency Treebank: Restoration of Deletions 44
  Eva Hajičová, Ivana Kruijff-Korbayová, Petr Sgall

Some Types of Syntactic Ambiguity 50
  Jarmila Panevová, Markéta Straňáková

Semantic Annotation of (Czech) Corpus Texts 56
  Karel Pala

The General Principles of the Diachronic Part of the Czech National Corpus 62
  Karel Kučera

Performing Adaptive Morphological Analysis Using Internet Resources ... 66
  Marek Trabalka, Mária Bieliková

Automatic Text-to-Speech Alignment: Aspects of Robustification 72
  R. Schmidt, R. Neumann

Czech Translation of G. Orwell's '1984': Morphology and
Syntactic Patterns in the Corpus 77
  Vladimír Petkevič

Handling Word Order in a Multilingual System for Generation of Instructions 83
Ivana Kruijff-Korbayová, Geert-Jan M. Kruijff

Text Structuring in a Multilingual System for Generation of Instructions .. 89
Ivana Kruijff-Korbayová, Geert-Jan M. Kruijff

Leveraging Syntactic Information for Text Normalization 95
Deborah A. Coughlin

Automatic Structuring of Written Texts 101
Marek Veber, Aleš Horák, Rostislav Julinek, Pavel Smrž

Implementation of Efficient and Portable Parser for Czech 105
Pavel Smrž, Aleš Horák

Word Sense Disambiguation of Czech Texts 109
Ondřej Cikhart, Jan Hajič

The Acquisition of Some Lexical Constraints from Corpora............. 115
Goran Nenadić, Irena Spasić

Run-Time Extensible (Semi-)Top-Down Parser 121
Michal Žemlička, Jaroslav Král

Enhancing Readability of Automatic Summaries by Using Schemas....... 127
Mariem Ellouze, Abdelmajid Ben Hamadou

Use of a Weighted Topic Hierarchy for Document Classification 133
Alexander Gelbukh, Grigori Sidorov, Adolfo Guzman-Arénas

Speech

Remarks on Sentence Prosody and Topic-Focus Articulation............. 139
Petr Sgall

Speech Recognition Using Elman Neural Networks 146
L.J.M. Rothkrantz, D. Nollen

Use of Hidden Markov Models for Evaluation of Russian Digits
Pronunciation by the Foreigners 152
Alexei Machovikov, Iliana Machovikova

Allophone-Based Concatenative Speech Synthesis System for Russian 156
Pavel A. Skrelin

Intonation Questions in English and Armenian:
Results of the Perceptual Study.................................... 160
Nina B. Volskaya, Anna S. Grigoryan

Methods of Sentences Selection for Read-Speech Corpus Design 165
Vlasta Radová, Petr Vopálka

Speaker Identification Using Discriminative Centroids Weighting –
A Growing Cell Structure Approach 171
  Bogdan Sabac, Inge Gavat

Speech Analysis and Recognition Synchronised by One-Quasiperiodical
Segmentation.. 175
  Taras K. Vintsiuk, Mykola M. Sazhok

Spanish Phoneme Classification by Means of a Hierarchy of Kohonen
Self-Organizing Maps... 181
  A. Postigo Gardón, C. Ruiz Vázquez, A. Arruti Illarramendi

Information Theoretic Based Segments for Language Identification 187
  Stefan Harbeck, Uwe Ohler, Elmar Nöth, Heinrich Niemann

Fast and Robust Features for Prosodic Classification................... 193
  Jan Buckow, Volker Warnke, Richard Huber, Anton Batliner,
  Elmar Nöth, Heinrich Niemann

A Segment Based Approach for Prosodic Boundary Detection 199
  Volker Warnke, Elmar Nöth, Heinrich Niemann, Georg Stemmer

Speech Recognition and Syllable Segments 203
  Ivan Kopeček

Text Preprocessing for Czech Speech Synthesis 209
  Robert Batůšek, Jan Dvořák

MLPs and Mixture Models for the Estimation of the Posterior
Probabilities of Class Membership.................................. 215
  Alexei V. Ivanov, Alexander A. Petrovsky

A Simple Spanish Part of Speech Tagger for Detection and Correction of
Accentuation Error.. 219
  S. N. Galicia-Haro, I. A. Bolshakov, A. F. Gelbukh

Slovene Interactive Text-to-Speech Evaluation Site – SITES 223
  Jerneja Gros, France Mihelič, Nikola Pavešić

Developing HMM-Based Recognizers with ESMERALDA 229
  Gernot A. Fink

Large Vocabulary Speech Recognition for Read and Broadcast Czech 235
  W. Byrne, J. Hajič, P. Ircing, F. Jelinek, S. Khudanpur, J. McDonough,
  N. Peterek, J. Psutka

Rules for Automatic Grapheme-to-Allophone Transcription in Slovene 241
  Jerneja Gros, France Mihelič, Nikola Pavešić

Speech Segmentation Aspects of Phone Transition Acoustical Modelling .. 248
  Simon Dobrišek, France Mihelič, Nikola Pavešić

Context Dependent Phoneme Recognition............................ 252
  Dušan Krokavec

State-Space Model Based Labeling of Speech Signals 258
  Dušan Krokavec, Anna Filasová

Very Low Bit Rate Speech Coding: Comparison of Data-Driven Units with
Syllable Segments 262
  Jan Černocký, Ivan Kopeček, Geneviève Baudoin, Gérard Chollet

Storing Prosody Attributes of Spontaneous Speech 268
  Jana Klečková

Dialogue

An Overview of the State of the Art of Coding Schemes for Dialogue Act
Annotation.. 274
  Marion Klein

Structural and Semantic Dialogue Filters 280
  Zdeněk Mikovec, Martin Klíma, Pavel Slavík

A Retrieval System of Broadcast News Speech Documents through
Keyboard and Voice... 286
  Hiromitsu Nishizaki, Seiichi Nakagawa

Situations in Dialogs 290
  Petr Hejda

Components for Building an Automatic Voice-Dialogue Telephone System 296
  Miroslav Holada

Modeling of the Information Retrieval Dialogue Systems 302
  Ivan Kopeček

Improvement of the Recognition Rate of Spoken Queries to the Dialogue
System 308
  Václav Matoušek, Jana Ocelíková

Analysis of Different Dialog Strategies in the Slovenian Spoken Dialog System 315
  Ivo Ipšić, France Mihelič, Nikola Pavešić

Posters

Dispersion of Words in a Language Corpus 321
  Jaroslava Hlaváčová, Pavel Rychlý

Corpus-Based Rules for Czech Verb Discontinuous Constituents......... 325
  Eva Žáčková, Karel Pala

Automatic Modelling of Regional Pronunciation Variation for Russian 329
  Kseniya B. Shalonova

Experiments Regarding the Superposition of Emotional Features on
Neutral Korean Speech 333
  Cheol-Woo Jo, Attila Ferencz, Dae-Hyun Kim

Modeling Cue Phrases in Turkish: A Case Study 337
  Bilge Say

Speaker Identification Based on Vector Quantization 341
  Vlasta Radová, Zdeněk Švenda

Robustness in Tabular Deduction for Multimodal Logical Grammar - Part 1 345
  Geert-Jan M. Kruijff

Classifying Visemes for Automatic Lipreading 349
  Michiel Visser, Mannes Poel, Anton Nijholt

Semantic Inference in the Human–Machine Communication 353
  Leo Hadacz

Playing with RST: Two Algorithms for the Automated Manipulation of
Discourse Trees 357
  Floriana Grasso

Another Step in the Modeling of Basque Intonation: Bermeo 361
  *Gorka Elordieta, Iñaki Gaminde, Inma Hernáez, Jasone Salaberria,
  Igor Martin de Vidales*

Electronic Dictionaries: For Both Humans and Computers 365
  Igor A. Bolshakov, Alexander F. Gelbukh, Sofia N. Galicia-Haro

Statistical Evaluation of Similarity Measures on Multi-lingual
Text Corpora... 369
  R. Neumann, R. Schmidt

Document Title Patterns in Information Retrieval..................... 372
  Manuel Montes-y-Gómez, Alexander F. Gelbukh, Aurelio López-López

Statistical Approach to the Automatic Synthesis of Czech Speech 376
  Jindřich Matoušek, Josef Psutka, Zbyněk Tychtl

Language Model Representations for the GOPOLIS Database 380
  Janez Žibert, Jerneja Gros, Simon Dobrišek, France Mihelič

Recognition of Alkohol Influence on Speech......................... 384
  Richard Menšík

Recording of Czech and Slovak Telephone Databases within SpeechDat-E 388
  Jan Černocký, Petr Pollák, Milan Rusko, Václav Hanžl, Marián Trnka

Pragmatic Features of the Electronic Discourse in the Internet 392
  Irina Potashova

Author Index 395

Research Issues for the Next Generation Spoken Dialogue Systems*

E. Nöth[1], F. Gallwitz[1], M. Aretoulaki[2], J. Haas[1], S. Harbeck[1], R. Huber[1], and H. Niemann[1]

[1] University of Erlangen-Nuremberg, Chair for Pattern Recognition
Martensstr. 3, D-91058 Erlangen-Nuremberg,
[2] Bavarian Research Centre for Knowledge-Based Systems (FORWISS)
Am Weichselgarten 7, D-91058 Erlangen, Germany
noeth@informatik.uni-erlangen.de

Abstract. In this paper we present extensions to the spoken dialogue system EVAR which are crucial issues for the next generation dialogue systems. EVAR was developed at the University of Erlangen. In 1994, it became accessible over telephone line and could answer inquiries in the German language about German InterCity train connections. It has since been continuously improved and extended, including some unique features, such as the processing of out-of-vocabulary words and a flexible dialogue strategy that adapts to the quality of the recognition of the user input.

1 Introduction

The spoken dialogue system EVAR was developed at the University of Erlangen over a period of almost 20 years. Different system architectures have been implemented and evaluated, and intensive research has been performed in the areas of word recognition, linguistic analysis, knowledge representation, dialogue management, and prosodic analysis. To our knowledge, EVAR was the first spoken dialogue system in the German language that was made available to the general public when it was connected to the telephone line in January 1994 (EVAR's phone number: ++49 9131 16287). Since that time, a corpus of approximately 3000 spontaneous human-machine dialogues has been compiled, most of them in the train timetable information domain, involving mainly naive users who are not familiar with the speech recognition and understanding technology. These dialogues are constantly transcribed and used for retraining, improving, and evaluating the various system components.

EVAR ("*Erkennen, Verstehen, Antworten, Rückfragen*", or "Recognize, Understand, Answer, Ask back") was designed as a research platform, where current

* This research was funded by the DFG (German Research Foundation) under contract number 810 939-9, by the EC in the framework of the Copernicus project COP-1634 (SQEL), and by the German Federal Ministry of Education, Science, Research and Technology (*BMBF*) in the framework of the VERBMOBIL Project under Grant 01 IV 701 K5. The responsibility for the contents of this article lies with the authors.

F. Jelinek, E. Nöth (Eds.): TSD'99, LNCS 1692, pp. 1–9, 1999.

developments can be implemented and evaluated with real users, and speech data can be collected. The dialogue strategy adopted is less strict than in the case of systems that are intended for commercial applications, because to us, the collection of actual spontaneous speech data is more important than an optimal dialogue success rate. Nevertheless, word accuracy, semantic accuracy, and dialogue success rates have all considerably increased over the last few years, partly due to improved algorithms and dialogue strategies, and partly to the increasing availability of training data.

An overview of the system architecture is given in Section 2. A number of the unique features of EVAR is then outlined. In Section 3, the new WWW database interface, which is a generic information interface, is presented. In Section 4, we explain how the EVAR dialogue strategy adapts to the acoustic channel and the cooperativeness of the user. In Section 5, we explain how out-of-vocabulary words are detected and classified by the word recognizer and how this information is used to generate an appropriate system reaction. In Section 6, we briefly report on the evolution of EVAR into a multilingual and multifunctional system that automatically detects the appropriate language and domain. Stochastic methods for semantic analysis are discussed in Section 7, which should complement and enhance the traditional linguistic methods. A new approach involving the integrated recognition of words and prosodic boundaries is presented in Section 8, which allows the implicit detection of syntactic structure during the word recognition process. Finally, we outline our current research on the detection of user emotion, which is an essential issue to be dealt with if spoken dialogue systems are to be used in real–world applications.

2   System Architecture

The architecture of EVAR is depicted in Figure 1. User utterances are first digitalized by an AD/DA converter. Then word recognition is performed and the best word chain (e.g. *"I would like to go to Frankfurt"*), or alternatively a word graph, is handed on to the linguistic processor. The linguistic processor extracts a set of semantic concepts (semantic attribute–value pairs) from the word recognizer result (e.g. [goalcity:frankfurt]) and forwards them to the dialogue manager. The dialogue manager checks whether all necessary parameters are available and, if so, sends a query to the application database. Depending on the dialogue history and the current dialogue strategy, the user is asked to confirm the parameter (e.g. *"You want to go to Frankfurt?"*) and/or another parameter is requested (e.g. *"What time would you like to leave?"*); otherwise the result of the database search is verbalized. The generated message is then synthesized by a text–to–speech module and played to the user over the AD/DA converter that is connected to the telephone line.

3   WWW Database Access

The long–term vision regarding EVAR is its evolution into a multimodal environment, where text, speech, and even image processing are integrated over

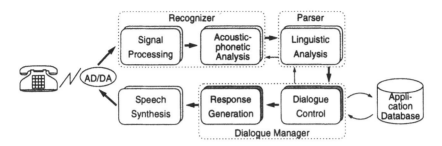

Fig. 1. The basic architecture of EVAR

the phone and the internet. A step in this direction has been the development of a search engine that poses the user's query to multiple travel information databases on the WWW. The search engine constitutes the interface between the dialogue system and the WWW databases. To date, the following databases are accessed: German Railways (*DB*), *Lufthansa*, and Swiss Railways (*SBB*). During the search, a number of dynamic HTML documents are created and accessed holding the intermediate results collected. These are temporarily saved in a local cache which is continually updated until the search is ended. The engine also provides the facility for a number of local databases to be set–up for regularly–accessed data. Although the retrieved entries are filtered according to the parameters specified by the user in the course of the dialogue, constraints can be relaxed, when the initial query does not match any of the stored data.

4 Flexible and Adaptive Dialogue Strategy

One of the distinctive traits of EVAR is the adoption therein of 'open' dialogue strategies, allowing the user to freely formulate their queries and carry out the transaction quite flexibly. The user is allowed to take the initiative regarding the order in which task parameter specification takes place and is also usually able to change the current subgoal of the interaction; e.g. in correcting a parameter that has already been dealt with, at a time when the system is expecting information about another parameter. This contrasts to the more common approach of presenting the user with menus to which they have to comply and answer with yes or no. In this sense, EVAR is quite intelligent. As a result, however, there are more possibilities regarding the content of the next user utterance, thus increasing the probability of misrecognitions and misunderstandings. To remedy this, the system tries to always confirm each task parameter as they are specified by the user. This defensive strategy safeguards that the correct database entry will be retrieved at the expense of interaction speed.

Confirmation can be both explicit and implicit, depending on whether or not there has been a history of failures in the current dialogue. A potential correction by the user initiates a clarification subdialogue, which can contribute to the repair of the most crucial errors and, hence, to the successful completion of the task in most cases. This is effected by asking the user to repeat or confirm a parameter on its own, or to spell it in the worst case, thereby restricting the user's usual freedom. Normally, however, the user is expected to confirm a parameter indirectly in conjunction with the specification of other parameters, which contributes to greater user satisfaction [4]. Consequently, EVAR can dynamically modify its communicative and repair strategies employing the user's reactions in the course of the dialogue as a guide and the frequency of conflict between the system's beliefs and the user's goals.

5   Robustness towards Out–of–Vocabulary Words

One of the most important causes of failure in spoken dialogue systems is usually neglected: the problem of words that are not covered by the system's vocabulary (Out–Of–Vocabulary or OOV words). In such a case, the word recognizer usually recognizes one or more different words with a similar acoustic profile to the unknown. These misrecognitions often result in possibly irreparable misunderstandings between the user and the system, as exemplified in Figure 2 (a). This is due to the fact that users rarely realise that they have crossed the boundaries of the system's knowledge, but just notice its suddenly weird behaviour. Therefore, it is desireable to have the system detect unknown words and inform the user about them so that they might correct the error. This will increase not only the dialogue success rates but also the acceptability of the system to the user.

In [7] we presented an approach to directly detect OOV words during the speech recognition process and, at the same time, to classify the word with respect to a set of semantic classes. The semantic tag to the OOV word is of considerable importance for initiating an appropriate system reaction. For example, most OOV words in the train timetable domain, such as the name of the caller or word fragments, should be ignored by the dialogue manager. However, when an unknown departure or destination city is requested, the caller should be informed about the limitations of the system vocabulary, as demonstrated by the example dialogue in Figure 2 (b). EVAR was extended in this manner which, to our knowledge, makes it the first end–to–end speech understanding system with this capability [2].

In other domains, such as the VERBMOBIL speech translation task, more complex system reactions are desirable. The OOV word should be included into the recognition vocabulary, which — as in human-human communication — may involve a clarification dialogue. For example, the system might request the users to repeat their name to obtain additional acoustic evidence on its pronunciation, or to spell it if its orthography is task–relevant.

6   Multilinguality and Multifunctionality

Within the EC Copernicus-project SQEL, EVAR was extended with respect to multilinguality and multifunctionality [1]. The current SQEL–demonstrator

S:	*Good morning, this is the train timetable information system EVAR.*
	How can I help you?
U:	*Well, I need a train connection from Munich to St. Petersburg.*
S:	*You want to go from Munich to Heidelberg. At what time?*
U:	*No, to St. Petersburg!*
S:	*You want to go to Stuttgart?*
U:	*No! St. Pee – ters – burg!*
S:	*You want to leave around three thirty?*

(a) Baseline system

S:	*Good morning, this is the train timetable information system EVAR.*
	How can I help you?
U:	*Well, I need a train connection from Munich to St. Petersburg.*
S:	*I believe your destination is not covered by our database. We currently only provide information on German InterCity train connections. Do you need any other information?*

(b) System extended by OOV word classification and processing

Fig. 2. Example dialogues between a human user (**U**) and an automatic train timetable information system (**S**) that involves an out–of–vocabulary word (St. Petersburg).

can handle four different languages and domains: Czech (Czech train connections), German (German train connections), Slovak (Slovak train connections), and Slovenian (European flights). The system starts up in German with a German opening phrase, but the user is free to use any of the implemented languages. The language is identified implicitly by a multilingual word recognizer [1]. During the recognizer beam search, search paths related to other languages are normally eliminated in the first few seconds, thus keeping the overhead low compared to running four monolingual recognizers in parallel.

After completing the recognition process, the recognizer passes to the linguistic processor both the identity of the language spoken and the best matching word chain. The recognition results for the multilingual recognizer are almost as good as when the monolingual version is run for the language that was spoken. Once the language has been identified by the word recognizer, it is associated with the corresponding domain, which calls the appropriate database and task parameters. All domains are already accommodated for in the generic concept ontology used in EVAR. This ontology covers concepts such as source and goal location, departure and arrival time and date of travel. The existence of such language–independent semantic units has meant that porting the system to a new language involves mainly the development of new lexica and grammars for the analysis and the generation phases (apart from the word recognizers) and not an extensive restructuring of the interpretation process within the dialogue manager. This is because the dialogue manager of EVAR is sufficiently flexible to switch between the different domains, i.e. to the appropriate parser, generator, and application database for each language and domain.

7   Stochastic Methods for Semantic Analysis

The motivation behind the use of statistical methods in speech processing tasks has been the improved and sometimes more efficient analysis of the data in hand, while also reducing the amount of expert knowledge that needs to be incorporated in the corresponding system. The result is the fast and straightforward adaptation to new tasks and domains. This is why one of the major research areas we currently focus on is the introduction of statistical methods at the level of semantic and linguistic analysis.

The basic assumption behind using stochastic methods in semantic analysis is that the analysis effort can be greatly reduced when operating in a restricted domain such as train timetable information. It is not necessary to carry out a complete linguistic analysis, because it is sufficent for the system to locate and understand those parts of an utterance which exercise influence on the database query to be performed, as well as those parts influencing the dialogue structure. For example, it is essential to know the city of departure but it is completely irrelevant to know who wants to travel or why someone needs a connection. Those parts that are important to the dialogue system are called semantic attributes and semantic/pragmatic analysis could be limited to them.

In [3], a new concept was introduced of using stochastic models for semantic analysis. The idea is to avoid using a grammar that describes all important parts at once, and to write instead small partial grammars, each of them covering a special attribute; e.g. a grammar that is able to analyse time expressions. Given that not every utterance includes all semantic attributes, we use stochastic methods to predict the occurence of attributes in the current utterance, only using partial grammars for the semantic analysis of the predicted attributes. The advantage is that partial grammars are easy to maintain and reuseable in other systems for different tasks. In addition, the analysis is carried out faster without a decrease in accuracy.

8   Integrated Recognition of Words and Boundaries

It is widely recognized that prosodic information is a useful source of information in speech understanding. Nevertheless, most existing systems do not make use of it. The most probable reason is, that these systems can usually only handle fairly short and simple utterances in a very restricted domain, where prosodic functions - such as the marking of *sentence mood, accentuation,* and *phrasing* - are believed to be of little importance. In such applications, efficient linguistic processing is possible without any prosodic information about the structure of the utterance, and the prosodic marking of the sentence mood and accentuation are mostly redundant. An acceptable system performance can also be achieved, if prosodic information is totally ignored. This is also the case in the EVAR domain.

When moving to more complex tasks, such as in VERBMOBIL human–to–human spontaneous speech translation [12], prosody becomes an important issue. In VERBMOBIL, utterances tend to be considerably longer than those in EVAR;

the average number of words per utterance in EVAR dialogues is only 3, with an average of 7 words in the first utterance, whereas an average VERBMOBIL utterance is made up of 22 words. This makes it particularly important to identify prosodic phrase boundaries. In spontaneous speech, prosodic boundaries are even more crucial in understanding an utterance than punctuation marks are in written language. Words which "belong together" from the viewpoint of meaning are grouped into *prosodic phrases*, and it is widely agreed upon that there is a close correspondence between prosodic and syntactic phrase boundaries [10,5,13,9].

In [6], we propose the direct integration of the classification of phrase boundaries into the word recognition process. HMMs are used to model phrase boundaries, which are also integrated into the stochastic language model. The word recognizer then determines the optimal sequence of words and boundaries. In the VERBMOBIL domain, even without additional prosodic features, we obtain phrase boundary recognition rates that are comparable to those achieved with a separate prosodic classifier based on the word recognition result and a large set of prosodic features. At the same time, a word error rate reduction of 4 % is attained without any increase in computational effort [6].

9 User Emotion

Just as people kick soda vending machines when these don't work, it can be observed in the EVAR speech corpus that users get angry at spoken dialogue systems, when a dialogue with such a system goes wrong. In the context of call–center applications, it is important to identify such a situation and to initiate an appropriate reaction, such as referring the customer to a human operator or starting a clarification sub–dialogue, if one does not want to lose a potential customer forever. The detection of emotion and an adequate reaction to an angry user will certainly lead to a higher degree of acceptance of the system.

Even though there are many different emotions that can be expressed in human speech – such as sadness, joy, or fear – we are currently only interested in the distinction between anger and normal speaking. Other types of emotion are most probably not relevant to the implementation of the emotion detector in spoken dialogue systems, such as EVAR. Besides, it is more difficult to distinguish between emotions, such as joy and anger [11].

Emotion can be expressed, at least, in two verbal ways (in addition to non–verbal cues like body language): by the lexical information carried by certain words of an utterance, e.g. swear words; and by way of acoustic prosodic cues, such as sharp changes in the loudness and/or the fundamental frequency (F_0) of the speech signal, as well as changes of the duration values. All these changes do not have to occur in a single utterance. They can also take place in conjunction with earlier utterances of a dialogue. Furthermore, emotion can be expressed by means of a combination of the above parameters. We are currently concentrating on the use of acoustic prosodic cues to locate emotional utterances, without using the lexical information of words.

For the classification of emotion 276 prosodic features are used in our approach, based mainly on durational cues and fundamental frequency and energy contours. An artificial neural network is used as classifier. On a training set of 1530 utterances and a test set of 300 utterances, we attained a precision rate of 87 % and a recall rate of 92 % for the detection of angry vs. normal speaking style (cf. [8]).

10 Conclusion and Future Work

Although the first commercial spoken dialogue systems for train timetable information are beginning to emerge, running EVAR on the public telephone line still gives us the valuable opportunity of assessing newly developed techniques on real–world users. In this paper we presented EVAR and some of its unique features, as well as the main research areas that are directly related to the further development of EVAR.

One of our most ambitious goals is to merge our approach to the integrated recognition of words and phrase boundaries with that of stochastic semantic analysis. The classification of words, phrase boundaries, and semantic attributes will then be subject to a single integrated search procedure, which employs many different sources of information, such as acoustics, prosody, statistical language models for word and boundary sequences, statistical models for semantic attributes, and statistical language models for sequences thereof. By replacing the second pass of the word recognizer with this search procedure, all these sources of information can be made available at an early stage in the recognition process. The result of this search will be an '*interpretation graph*' which contains not only words and syntactic boundaries, but also semantic interpretations of word sequences.

References

1. M. Aretoulaki, S. Harbeck, F. Gallwitz, E. Nöth, H. Niemann, J. Ivanecký, I. Ipšić, N. Pavešić, and V. Matoušek. SQEL: A Multilingual and Multifunctional Dialogue System. In *Int. Conf. on Spoken Language Processing*, Sydney, 1998.
2. M. Boros, M. Aretoulaki, F. Gallwitz, E. Nöth, and H. Niemann. Semantic Processing of Out-Of-Vocabulary Words in a Spoken Dialogue System. In *Proc. European Conf. on Speech Communication and Technology*, volume 4, pages 1887–1890, Rhodes, 1997.
3. M. Boros, J. Haas, V. Warnke, E. Nöth, and H. Niemann. How Statistics and Prosody can guide a Chunky Parser. In *Proc. of the AIII Workshop on Artificial Intelligence in Industry*, pages 388–398, Stara Lesna, Slovakia, 1998.
4. W. Eckert. *Gesprochener Mensch–Maschine–Dialog*. Berichte aus der Informatik. Shaker Verlag, Aachen, 1996.
5. C. Féry. *German Intonational Patterns*. Niemeyer, Tübingen, 1993.
6. F. Gallwitz, A. Batliner, J. Buckow, R. Huber, H. Niemann, and E. Nöth. Integrated Recognition of Words and Phrase Boundaries. In *Int. Conf. on Spoken Language Processing*, Sydney, 1998.

7. F. Gallwitz, E. Nöth, and H. Niemann. A Category Based Approach for Recognition of Out-of-Vocabulary Words. In *Int. Conf. on Spoken Language Processing*, volume 1, pages 228–231, Philadelphia, 1996.
8. R. Huber, E. Nöth, A. Batliner, J. Buckow, V. Warnke, and H. Niemann. You BEEP Machine — Emotion in Automatic Speech Understanding Systems. In *Proc. of the Workshop on TEXT, SPEECH and DIALOG (TSD'98)*, pages 223–228, Brno, 1998. Masaryk University.
9. R. Kompe. *Prosody in Speech Understanding Systems*. Lecture Notes for Artificial Intelligence. Springer-Verlag, Berlin, 1997.
10. M. Steedman. Grammar, Intonation and Discourse Information. In G. Görz, editor, *KONVENS 92*, Informatik aktuell, pages 21–28. Springer–Verlag, Berlin, 1992.
11. B. Tischer. *Die vokale Kommunikation von Gefühlen*, volume 18 of *Fortschritte der psychologischen Forschung*. Psychologie Verlags Union, Weinheim, 1993.
12. W. Wahlster. Verbmobil — Translation of Face-To-Face Dialogs. In *Proc. European Conf. on Speech Communication and Technology*, volume "Opening and Plenary Sessions", pages 29–38, Berlin, 1993.
13. C. Wightman, S. Shattuck-Hufnagel, M. Ostendorf, and P. Price. Segmental Durations in the Vicinity of Prosodic Boundaries. *Journal of the Acoustic Society of America*, 91:1707–1717, 1992.

Data-Driven Analysis of Speech

Hynek Hermansky

Oregon Graduate Institute, Portland, Oregon,
International Computer Science Institute, Berkeley, California
hynek@ece.ogi.edu

Abstract. We show on results taken from recent studies from our laboratory that conventional speech analysis techniques for ASR (such as Mel cepstrum or PLP) in combination with dynamic features (such as estimates of derivatives of cepstral feature trajectories) are sub-optimal and could be improved. The improvements can be derived by employing large labeled databases which allow for studying how is the linguistic information distributed in time and in frequency as well as for a design of discrimitative spectral basis and temporal RASTA filters.

1 Current Automatic Recognition of Speech

1.1 Main Components of a Typical ASR System

A typical automatic speech recognition (ASR) system consists of three main modules. The speech analysis module converts the speech signal into vectors of features. The pattern classification module classifies the feature vectors into selected linguistic units (phonemes, words,...). The decoder with language modeling finds the optimal sequence of linguistic units using language constraints, specific for the given application. One of the major advances in ASR over the past two decades was replacing the pre-wired prior knowledge in pattern classification and in the language model by knowledge derived from large training databases.

1.2 The Roles of Analysis and of Pattern Classification in ASR

The boundary between the speech analysis and the pattern classification modules is not well defined. The speech analysis module typically contains general knowledge, which is independent of the particular application domain. This knowledge is still mostly derived from intuitions and beliefs. One such belief is that the ASR analysis should emulate human hearing [4].

The feature extraction (analysis) is important in any pattern classification task and it is true for ASR too. Speech analysis could help in alleviating non-linguistic variability and to improve reliability of ASR in realistic environments. Apart from some advances in the auditory-like modification of the short-term spectrum and in employing some simple measures of spectral dynamics, the

F. Jelinek, E. Nöth (Eds.): TSD'99, LNCS 1692, pp. 10–18, 1999.
© Springer-Verlag Berlin Heidelberg 1999

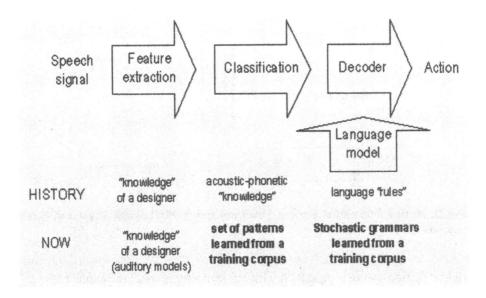

Fig. 1. To reduce the need for extensive training of the classifier in ASR, the speech feature extraction (analysis) module should provide as much general speech knowledge as possible. This general knowledge may be acquired from a separate training data, not necessarily specific for a given recognition task.

speech analysis module is still essentially based on short-term spectral techniques used in early vocoding.

The goals of speech analysis in speech coding (preserving perceptually dominant components of the signal) and in ASR (alleviating non-linguistic components of the signal) are very different. It is important to remember that there is no need to reconstruct the original speech from ASR features and that a significant loss of harmful information (e.g. the information about a communication environment or information about a health or emotional state of a particular speaker) could (and probably should) be achieved in the speech analysis module for ASR.

The stochastic part (pattern classification module) of the ASR system should need to learn only the task-specific knowledge. Then, the current excessive need for re-training of the ASR systems on each new task could be somehow alleviated.

2   Towards Alternative Approaches to Acoustic Modeling

One of problems of the current ASR is the need for re-training of the system on every new task. The acoustic analysis should provide the task-independent speech-specific knowledge which could to some extent alleviate this excessive training. In that respect, the adaptive approaches to ASR in which the general knowledge is contained in the un-adapted stochastic models, are steps in the

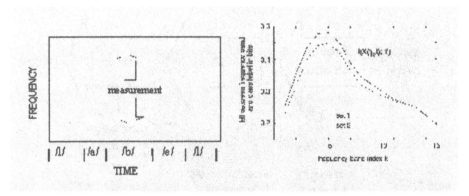

Fig. 2. Mutual information between phonemes and spectral points alligned with the phonemes.

right direction and in competition with our strive for more of the explicit speech-specific knowledge in the system. However, as indicated in this paper, we still believe that improvements in speech analysis are needed and possible.

Such improvement could come from good use of large labeled speech databases which are available today. A significant progress has been achieved in ASR over past two decades by using stochastic approaches together with large speech databases.. The progress, however, was mainly made in the pattern classification and in the language modeling. It appears that efforts for data-driven techniques in speech analysis for ASR began only recently and that some dramatically different techniques then the ones in use today may emerge as we progress. Some of our efforts in this direction are described below.

2.1 Where is the Information about Phonemes in the Short-Term Spectrum of Speech?

Yang et al. [7] considered phoneme labels a categorical target variable and measurements in the time-frequency plane as variables carying the information about the target variable. The phoneme-labeled database was used to investigate where in the critical-band time-frequency plane is the information about underlying phonemes. Since the targert variable is categorical and the time-frequency measurements are continuous variables with generally non-gaussian distributions) they evaluated the mutual information between the two. To include more than one time-frequency variable, the concept of joint mutual information was used [7]. The main conclusions from their study are summarized below.

Investigating along the frequency axis (see Fig. 2) indicates the non-equal distribution of information-bearing elements in frequency. The single measurement yields the most important components in the vicinity of 5 Bark (compare to the shape of the first discriminant spectral base in the work on the LDA-derived spectral basis described later in this paper).

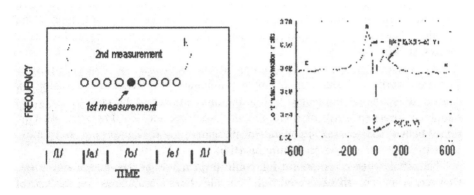

Fig. 3. Additional information provided by a second measurement at the same frequency but at different time.

The second measurement at the same time but at different frequency (especially when taken from more reliable frequency channels avoiding the low frequencies) provides an additional information [7].

So far up to three measurements at different frequencies (and at the same time) were investigated. These provide about 34 % of total information carried by phoneme labels [6].

Misaligning measurement in time gradually decreases the amount of the information about underlying phonemes (the lower limit is obtained by random scrambling of the labels). However, only severe misalignment of more than 200 ms provide no measurable information about the underlying phonemes [7], thus indicating that the information about phonemes is spread in the signal over a considerable time.

The second measurement at the same frequency but at a different time yields an interesting result (shown in Fig. 3) which suggests that the most important information in time can be obtained by looking at the previous 50 ms of the signal. The amount of the additional temporal information becomes constant (but still considerable) only when the time span between the two measurements in time exceeds 200 ms.

Three measurements at the same frequency (5 Bark) but at different times could provide about 28 % of the total information about the underlying phoneme labels [6].

2.2 Data-Driven Feature Extraction

Providing that speech evolved to be heard, properties of hearing should reflect in the speech signal. Thus, finding a way of optimizing speech analysis directly from the signal could provide an alternative to the direct emulation of textbook properties of hearing in speech analysis module and should yield results consistent with human hearing.

The goal of optimization is the classification to sub-word units. In our work we mostly aim at context independent phonemes as the sub-word units to be classified. However, our techniques are general and can be also adapted to other choices of sub-words units.

Our current justification for aiming at the phonemes as subword units is that two words can differ in a single phoneme. The reason for focusing on context-independent phonemes is that humans appear to be able to recognize phonemes independently of their phonetic environment [8]. Therefore we currently believe that context-independent phonemes are reasonable sub-word units to be derived during speech communication.

The initial speech representation could in principle be the speech waveform. However, it is generally accepted that a cochlea does a crude spectral analysis of the incoming sound and hair-cells do a one-way rectification of activity on the cochlea. Thus, let us assume that short-term power spectrum could represent the initial speech representation for deriving features for speech recognition.

The optimization is done by using a linear discriminant analysis (LDA). The LDA is a technique attepting to find such linear projection of the feature vector space which would maximize linear separability between target classes. In speech, it requires avalability of labeled speech data. We us phoneme hand-labeled OGI Stories data in our optimizations. The data contain about 3 hours of telephone speech from 200 adult male and female speakers, each producing about 1 minute of free fluent speech.

Depending on a way the vector space for the LDA is obtained, the LDA can yield either spectral basis functions (if the feature vectors are spectral vectors) or FIR RASTA filters if the feature vectors are cut out of time trajectories of spectral energies (Fig. 4)

Spectral Basis Hermansky and Malayath found that the LDA on labeled short-term fft-derived spectral vectors yields spectral basis which examine lower frequencies with more detail [5]. This is consistent with Mel or Bark spectral warping employed in Mel cepstrum or PLP.

LDA applied to critical-band spectrum from PLP analysis yield spectral basis shown in Fig. 5. On the Bark scale, spectral basis have more uniform resolution. Some of the discriminants are rather frequency-selective, thus employing only parts of the available spectral range. The most important discriminant evaluates spectral energy in the lower half of the auditory-like spectrum (peaking at about 5 Bark-compare with the outcome of the mutual information experiments described above), most likely an indication of voicing. The second discriminant evaluates whole spectrum with resolution of about 5 Bark, the third and fourth have somehow higher resolution of about 3 Bark, the third one again also evaluating the spectral slope (voicing), the fourth one focusing on the spectrum above 5 Bark. More details of the experiment can be found in [5].

Temporal RASTA Filters Hermansky and his colleagues show that LDA on 1 s long temporal vectors of critical band spectral energies (labeled with respect

LDA gives basis for
projection of spectral space

LDA gives FIR filters for
processing time trajectories of
spectral energies

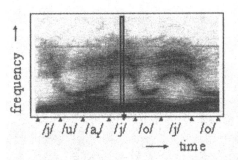
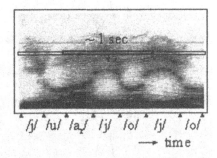

Fig. 4. Two ways of using LDA for data-driven design of speech extraction module. When the feature vector space is formed by spectral vectors, LDA yields spectral basis functions. When the feature vectors are cut from temporal trajectories of spectral energies at a given frequency, the LDA yields FIR RASTA filters.

to their centers) yields FIR RASTA filters to be applied to time trajectories of critical band energies [1,9]. Impulse and frequency responses of the first three discriminant filters derived from about 6 hours of phoneme-labeled Switchboard data [10] are shown in Fig. 6.

The filters are band-pass filters, suppressing modulation frequency components below 1 Hz and above 10 Hz. The first discriminant filter can be approximated by a difference of two Gaussians, the higher ones approximately represent derivatives of the first one [2]. Active parts of their impulse responses span about 200 ms. More details about the data-driven design of RASTA filters can be found in [9,10,2,3].

3   Conclusions

The paper summarizes results of work on acoustic processing in ASR from our Anthropic Signal Processing laboratory. Several themes are emphasized:

- Use of large labeled speech database for optimizing speech analysis.
- Exploitation of relatively large (up to 1 s) temporal spans of the signal in phoneme-based ASR.

Our experience from this work so far can be summarized as follows:

- Evaluation of the mutual information between phoneme labels and points in the time-frequency plane derived by a short-term critical-band analysis indicates the non-uniform distribution of information in frequency. It also

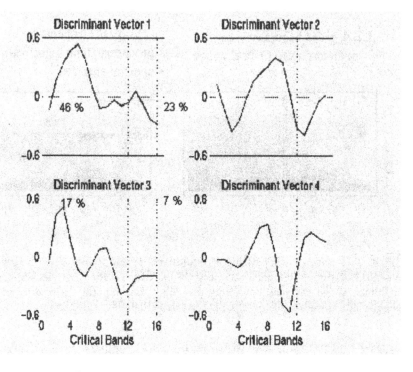

Fig. 5. Spectral basis derived by LDA.

shows that additional measurements spread in time can provide some additional information with a maximum of information coming from about 200 ms neigborhood of the current time instant.

– The auditory-like spectral resolution of Mel cepstrum or PLP, which is decreasing with increasing frequency, appears to be desirable. However, the cosine spectral basis used in coputing the cepstrum, as well as conventional techniques for exploiting speech dynamics such as delta features, are suboptimal in several aspects.

  – The cosine spectral basis functions merely project on a direction of maximum overall variability. It is more desirable to project on a direction of maximum linguistic variability. This can be achieved by discriminant techniques such as LDA applied with a phoneme-labeled speech database. The LDA-derived spectral basis are not cosines and yield improvements in ASR performance.

  – Impulse responses of FIR filters implied by delta and double-delta features appear to be too short and their frequency responses too selective. The lengths of impulse responses of discriminant RASTA filters derived by LDA (which yield improvement in ASR performance) are of the order of 200 ms, their shapes are consistent with difference of Gaussians and its derivatives, and the frequency response of the first discriminant filter is a bandpass with pass-band between about 1 Hz and 15 Hz.

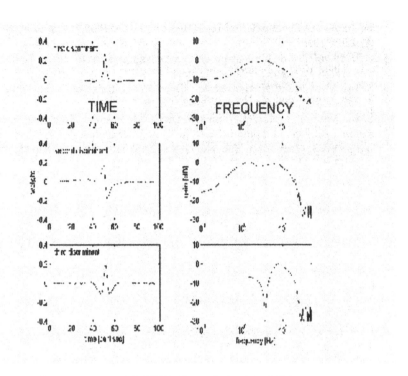

Fig. 6. Temporal RASTA filters derived by LDA.

Acknowledgement. The paper presents an overview of works done jointly with Naren Malayath, Sangita Sharma, Sarel van Vuuren, and Howard Yang. The research was supported by DoD under MDA904-98-1-0521, NSF under IRI-9712579, and by industrial grants from Texas Instruments and Intel Inc.

References

1. C. Avendano, S. van Vuuren and H. Hermansky. Data-based RASTA-like filter design for channel normalization in ASR. In *ICSLP'96*, volume 4, pages 2087–2090, Philadelphia, PA, USA, October 1996.
2. H. Hermansky. The modulation spectrum in automatic recognition of speech. In S. Furui, B.-H. Juang and W, Chou, editor, *1997 IEEE Workshop on Automatic Speech Recognition and Understanding*. IEEE Signal Processing Society, 1997.
3. H. Hermansky. Should recognizers have ears? In *Tutorial and Research Workshop on Robust speech recognition for unknown communication channels*, pages 1–10, Pont-a-Mousson, France, April 1997. ESCA-NATO.
4. H. Hermansky. Should recognizers have ears? *Speech Communication*, 25(1-3):3–27, 1998.
5. H. Hermansky, N. Malayath. Spectral basis functions from discriminant analysis. In *ICSLP'98*, Sydney, Australia, 1998.
6. H. Yang. Personal communications.

7. H. Yang, S. van Vuuren and H. Hermansky. Relevancy of time-frequency features for phonetic classification of phonemes. In *ICASSP'99*, pages 225–229, Phoenix, Arizona, 1999.

8. J.B. Allen. How do humans process and recognize speech? *IEEE Trans. on Speech and Audio Processing*, 2:567–577, 1994.

9. S. van Vuuren and H. Hermansky. Data-driven design of RASTA-like filters. In *Eurospeech'97*, Rhodes, Greece, 1997. ESCA.

10. S. van Vuuren, T. Kamm, J. Luettin and H. Hermansky. Presentation of the 1997 summer workshop on innovative techniques for continuous speech asr. In *available on the* http://www.clsp.jhu.edu. Johns Hopkins University, August 1997.

Towards a Road Map for Machine Translation Research

Steven Krauwer

ELSNET / Utrecht Institute of Linguistics OTS,
Trans 10, 3512 JK Utrecht, The Netherlands
steven.krauwer@let.uu.nl

Abstract. A recent phenomenon in European information technology is the so called technological Road Map: a longer term view on developments in a specific technology area, which identifies longer term goals and priorities, as well as intermediate milestones along the road. Such a Road Map may help the community to focus their research and development work, in order to achieve a maximal impact, and the milestones can be used to measure progress made.

The field of Machine Translation (MT) is typically an area for which such a road map can and should be made, and as you may know ELSNET, the European Network of Excellence in Language and Speech will start preparing one in the course of next year.

In my presentation I will try to take the first few steps into this direction. To this end I will give an overview and an analysis of recent activities in MT, as reported on at three main conferences in Natural Language Processing and MT organised summer 1999:

- EACL'99: The Ninth Conference of the European Chapter of the Association for Computational Linguistics, Bergen (Norway), June 8–12;
- ACL'99: The 37th Annual Meeting of the Association for Computational Linguistics, University of Maryland (USA), June 23–26;
- TMI'99: The 8th International Conference on Theoretical and Methodological Issues in Machine Translation, University College, Chester (UK), August 23–25.

I will try to distill from this a first outline road map. It should be pointed out that this analysis can only be tentative and partial. First of all it is based on only three conferences, and secondly it does not cover an aspect of MT that is at least as important as developments in research: the MT market.

F. Jelinek, E. Nöth (Eds.): TSD'99, LNCS 1692, p. 19–19, 1999.

The Prague Dependency Treebank: Crossing the Sentence Boundary

Eva Hajičová

Faculty of Mathematics and Physics, Charles University, Prague, Czech Republic
E-mail: hajicova@ufal.mff.cuni.cz

Abstract. The units processed by tagging procedures - both automatic and manual - are sentences (as occurring in the texts in the corpus), but the human annotators are instructed to assign (disambiguated) structures according to the meaning of the sentence in its environment, taking contextual (and factual) information into account. We focus in the paper on two issues: how to capture (i) the topic-focus articulation as one of the fundamental properties of sentence structure, which is related to the use of the sentence in a broader context, be it a suprasentential or a situational one, and (ii) the coreferential links in the text.

1 Introduction: The Prague Dependency Treebank

The Prague Dependency Treebank (PDT) project is conceived of as a collection of tree structures representing sentences of (a part of) the Czech National Corpus (CNC) in the shape of syntactic trees (tagged both on the analytical and the tectogrammatical levels, in addition to the morphological tags). The tagging on the tectogrammatical layer is based on the theoretical framework of Functional Generative Description (FGD, see [3]). The units processed by tagging procedures - both automatic and manual - are sentences (as occurring in the texts in the corpus) but the human annotators are instructed to assign (disambiguated) structures according to the meaning of the sentence in its environment, taking contextual (and factual) information into account. Another aspect that has led us to think about the context in which the given sentence occurs, is the regard to the use of the PDT as a resource for linguistic research not only within the limits of the sentence. These two considerations have their consequences for several points in the specification of the tectogrammatical tree structures (TGTSs), of which we would like to concentrate in our paper on the attribute TFA and on the order of nodes (Sect. 2), adding some preliminary remarks on some special attributes reflecting the linking of sentences in the text (Sect. 3).

2 Representing Topic-Focus Articulation in TGTSs

2.1.0   One of the basic claims of the theoretical framework of Functional Generative Description (FGD) concerns the relevance of topic-focus articulation (TFA) for the meaning of the sentence; the representations of meaning

F. Jelinek, E. Nöth (Eds.): TSD'99, LNCS 1692, pp. 20–27, 1999.
© Springer-Verlag Berlin Heidelberg 1999

(tectogrammatical representations, TR's) thus capture both the syntactic (dependency) relations and the TFA. TFA reflects the communicative function of the sentence: topic can be informally paraphrased as what the sentence is about' and focus as the information that is 'asserted' about the topic. As a matter of fact, this dichotomy is derived from the primary distinction of contextually bound and non-bound nodes in the syntactic tree and from the underlying order of nodes (corresponding to the so-called communicative dynamism). A detailed empirical analysis and a formal account of TFA is given in the writings quoted below;

2.1.1   The relevance of TFA for the meaning of the sentence, and thus also for annotations on the underlying level, can be illustrated by the following cases:

(i)   The semantic relevance of TFA has already been pointed out by [2] and is exemplified in (1); in Czech no special constructions to change the TFA are needed in this case. (The capitals denote the placement of the intonation centre.)

(1)   a.   English is spoken in the SHETLANDS.

b.   In the Shetlands, one speaks ENGLISH.

The sentence (1)(b) is certainly true (at least in the actual world); this is not the case with (a), which in a prototypical situation implies that the Shetlands are the only (or, maybe, the most important, prevailing etc.) country where *English* is spoken. In (a), we speak 'about' English (*English* is in the topic), and inform the hearer in which countries it is spoken; in (b), we speak 'about' the Shetlands (*the Shetlands* is in the topic) and state which language is spoken there.

(ii)   The semantic relevance of TFA is also supported by the semantics of negation, which shows a close relation to the position of negation in topic or focus:

(2)   John didn't come because he was ill.

a.   The reason for John's not-coming was his illness.

b.   The reason for John's coming (e.g. to the doctor) was not his illness but something else (e.g. he wanted to invite the doctor for a party).

If (2) is uttered with the meaning of (b), John might have been ill but not necessarily so, and it is implied that John did come, while (a) implies that John was ill and he did not come. This difference in meaning is a result of different TFA's of (a) and (b): (a) is 'about' John's not-coming (*John didn't come* is in the topic part of the sentence); when the operator of negation is in the topic, the end of the scope of negation coincides with the boundary between topic and focus and the elements in the focus trigger, in a prototypical case, a presupposition. In (b), the sentence is about John's coming, and what is negated is that the reason was not his illness; the operator of negation together with the *because*-clause is in the focus and as such triggers allegation rather than a presupposition.

(iii)   Similar situation obtains in case of operators called in the recent linguistic literature focalizers; there belong such particles as *only, also, even*, etc.

A detailed analysis of the meaning of constructions with such focalizers is given in [1]; in the present paper, we will reserve ourselves to one example:

(3)   John only introduced Sue to BILL.

The meaning of the focalizer *only* indicates that an alternative is being chosen from a set of alternatives: the statement can be understood as being 'about' John's introducing Sue (topic), who could have chosen several people to whom he could introduce her; it is said that it was Bill (focus) and no other person, to whom Sue was introduced by John. The focus of the focalizer *only* prototypically coincides with the focus of the sentence.

It should be noticed in this connection that an important role is played by the position of the intonation center ; a change of the position of the intonation centre indicates a different TFA of the sentence:

(4)   John only introduced SUE to Bill.

Now the statement can be understood as being 'about' John's introducing to Bill (topic), and it is said that it was Sue (focus) and no other person, who was introduced to Bill by John. Here again the focus of the focalizer *only* is equal to the focus of the sentence.

2.1.2   Paying due respect to TFA offers a good support for the assignment of reference, as illustrated by examples (5) and (6).

(5)   a.   The chair stood in front of a TABLE. This was old and shabby.
      b.   The chair stood in front of a TABLE. It was old and shabby.

(6)   The chair stood in front of a TABLE. It was small, round, with three legs.

The strong pronoun *this* in the second sentence in (5)a refers to the item displaying the highest activation, i.e. *a table*; prototypically, it is the item constituting the focus proper and as such carrying the intonation center of the sentence. The reference by a strong pronoun in such cases is unambiguous, though it should be kept in mind that the reference to the item with the highest activation can be overshadowed by inferencing, based on world knowledge.With a weak pronoun in the second sentence in (5)b, the preferable reference is to the subject of the preceding sentence, i.e. *the chair*. The reference by a weak pronoun is ambiguous, though there is a preferred reading keeping the syntactic symmetry (there is a tendency to preserve the subjects in the successive sentences, if possible).

2.2   Three values of the TFA attribute are distinguished:
(i)   T (a non-contrastive contextually bound node, with a lower degree of communicative dynamism, CD, than its governor),
(ii)   F (a contextually non-bound node, "new" piece of information),
(iii)   C (a contrastive (part of) topic; in the present stage, this value is assigned only in cases in which the node concerned is in a non-projective position).

It is assumed that an F node is always more dynamic and a C node is less dynamic than a sister or parent T node.

The following examples illustrate these three values:

(7) (Nadpis: Volby v Izraeli.)
Po volbách(T) si Izraelci(T) zvykají(F) na nového(F) premiéra(F).
(Headline in the newspapers: Elections in Israel.)
After the elections(T), the Israelis(T) get used(F) to a new(F) Prime Minister(F).

(8) Sportovec(C) on(T) je(F) dobrý(F), ale jako politik(C) nevyniká(F).
Sportsman(C) he(T) is(F) good(F), but as a politician(C) he does not excel(F).

2.3   The instructions for the assignment of these values are formulated as follows:

2.3.1   In prototypical cases, i.e. cases of projective ATSs:

(i)   left side dependents on the verb get T (except for cases in which this dependent would clearly carry the intonation center, IC),

(ii)   the rightmost dependent of the verb gets F (under the assumption that it carries the IC; if the IC is placed more to the left, then every item dependent on the verb and placed after IC gets T),

(iii)   the verb and those of its dependents that stand between the verb and the node assigned F and are ordered (without an intervening sister node) according to the systemic ordering (for Czech the systemic ordering (SO) of the main types of dependency is Actor < Temporal < Location < Instrument < Addressee < Patient < Effect; for the notion of SO see [3]), get F, unless they are repeated (perhaps coreferential, associated with or included in the meaning of their antecedent) from the previous sentence or context; the nodes between the verb and the node assigned F and the repeated nodes get T, as well as those placed more to the left than what would correspond to SO,

(iv)   embedded attributes get F, unless they are only repeated or restored in the TGTS,

(v)   indexical expressions such as 'já' [I], 'ty' [you], 'teď' [now], 'tady' [here], weak forms of pronouns, as well as pronominal expressions with a general meaning 'někdo' [somebody], 'jednou' [once upon a time] get T, except in clear cases of contrast or as bearers of IC,

(vi)   strong forms of pronouns get the value F; after prepositions, the assignment of T or F these forms is guided by the general rules (i) – (iii),

(vii)   restored nodes (i.e. those that are absent in ATSs but are added in the corresponding TGTSs) are always assigned T (and as such depend on their governors from the left.

2.3.2   An application of the above instructions leads to the following assignments of the values of the TFA attribute in sentences (9) through (14).

(9)   Některé(T) ekologické(F) iniciativy(T) označily(F) informaci(F) o chys-
      taném(F) teroristickém(F) útoku(F) za provokaci(F).
      Some(T) ecological(F) initiatives(T) denoted(F) the information(F) about
      a prepared(F) terroristic(F) attack(F) as a provocation(F).

The node for *iniciativy* gets T according to (i), the node standing for *za pro-
vokaci* gets F according to (ii), the node for the verb *označily* and the node for
informaci, which carries the functor Patient and as such stands in the hierarchy
of systemic ordering before Effectum (i.e. the order is in accordance with the
systemic ordering) get F according to (iii), and the nodes representing the attributes *některé, ekologické, chystaném, teroristickém, útoku* receive the value F
according to (iv). According to the definition of topic and focus in the Functional
Generative Description, this assignment of TFA values results in the following
bipartition of the sentence into topic and focus:

(9') topic: některé ekologické iniciativy
      focus: označily informaci o chystaném teroristickém útoku za provokaci

Even though we work with written texts, it is sometimes evident that the author
of the text assumed the sentence to be 'read' with a non-prototypical placement
of the intonation centre, see (10):

(10) (Většina ministrů Stěpašinovy nové vlády patří k věrným druhům nej-
      známějšího ruského intrikána Berezovského.) I Aksjoněnko(F) udržuje(T)
      s Berezovským(T) blízké(F) styky(T).
      (The majority of the ministers of Stěpašinov's new government belongs to
      faithful friends of the best known Russian intriguer Berezovskij.) Even(F)
      Aksjoněnko(F) keeps(T) with Berezovskij(T) close(F) contacts(T).

The value (F) with the node for *Aksjoněnko* is assigned according to (i) because
in the given context this word would be a bearer of the intonation centre; the
node for *contacts* gets T inspite of the fact that *contacts* is the last word of the
sentence; this is in accordance to the instruction (ii).
In a prototypical case, the embedded attributes are more dynamic than their
head words and thus receive F; in specific cases of repetitions or restoration
of the respective node (as in (11)) they get T (the restored nodes in (11') are
enclosed in square brackets):

(11) (Tento týden se opět sešla poslanecká sněmovna.) Včera zasedaly parla-
      mentní komise pro bezpečnost a pro zahraniční styky.
      (This week again the parliament is in session.) Yesterday there was a
      meeting of the parliament committee for security and for international
      relations.

(11') Včera(T) zasedaly(F) parlamentní(F) komise(F) pro bezpečnost(F) a
      [parlamentní(T)] [komise(T)] pro zahraniční(F) styky(F).

The instructions (v) and (vi) hold for the nodes for *tady* (here) and for the strong
form of pronoun *jeho* (him) in (12), respectively:

(12) (Pro českou hudbu je Charles Mackerras jedinečnou osobností.) Tady(T)
      je(F) doma(F), a proto si organizátoři(T) Pražského(F) jara(F) pro in-
      terpretaci(T) Smetany(F) vybrali(F) právě(F) jeho(F).
      (For Czech music Charles Mackerras is an unequalled personality.) Here(T)
      he-is(F) at home(F), and therefore the organizers(T) of the Prague(F)
      Spring(F) for the interpretation(T) of-Smetana(F) have-chosen(F) just(F)
      him(F).

The value of the TFA attribute with nodes that are added in the TGTSs (i.e.
those that are deleted in the ATSs and restored in TGTSs) is T; this concerns e.g.
all nodes with the lexical value Gen(eral), as indicated in (13'), or contextually
licensed deletions as in (14'):

(13) V Českém Krumlově byl zahájen kulturní program seznamující se stře-
      dověkými zvyky.
      Lit.: In Český Krumlov (there) was opened a cultural programme ac-
      quainting with medieval customs.

(13') V Českém(F) Krumlově(T) [Gen.Actor(T)] byl zahájen(F) kulturní(F)
      program(F) seznamující(F) se středověkými(F) zvyky(F).

(14) (Kam uprchlíci nejčastěji směřují?) Do Makedonie.
      (Where the refugees most frequently head for?) To Macedonia.

(14') [uprchlíci(T)] [nejčastěji(T)] [směřují(T)] do Makedonie(F)

2.3.3   For non-projective ATSs specific rules are formulated; a node N dependent to the left in a way not meeting the condition of projectivity will be
assigned C and will be placed more to the right, to meet that condition. The
nodes depending on N (directly or indirectly) will move together with N and will
get the value T or F according to 2.3.1 above. Thus, e.g. the sentence (15) will
have a TGTS in (15'), in which *jásot* depends on *důvod*, has the index C and is
placed to the right of the verb.

(15) K jásotu(C) není(F) nejmenší(F) důvod(F).
      lit. For triumphing(C) is-not(F) the-least(F) reason(F)

(15') (neg.F) být.F (důvod.F (jásot.C) (nejmenší.F))
      (neg.F) be.F (reason.F (triumphing.C) (least.F))

3 Attributes Capturing Coreferential and Other Links

In Sect. 2 we have presented an outline of a possibility how to capture in a
annotated corpus (the PDT, in our concrete case) a fundamental property of the
sentence structure that is related to the use of the sentence in a broader context,
be it a suprasentential or a situational one, namely its topic-focus articulation.
Another important property of sentences that links them to each other and to

the context of situation are the coreferential links. This issue actually reaches beyond the system of language, but we are believe that its treatment, even if a rather preliminary and tentative way, is a necessary ingredient of annotation schemata. Let us illustrate the matter on ex. (16) with two successive sentences (a) and (b).

(16)   a.   Rakouská vláda se rozhodla, že bude vyvíjet tlak na Prahu ve věci stavby jaderné elektrárny v Temelíně.

The Austrian government decided that it will execute a pressure on Prague in the matter of building the nuclear power station in Temelin.

       b.   Rakouští představitelé dali jasně najevo, že otázku Temelína spojují s přijetím Veska do Unie.

Austrian representatives have made it clear that they connect the issue of Temelin with the acceptance of Czechia to the Union.

The expression *představitelé* (representatives) in (b) refers back to *vláda* (government) in (a); in other words, the expression *vláda* is an antecedent of *představitelé*. The relation between the two expressions can be captured by a special attribute attached to each expression, the value of which would be the lexical value of its antecedent. However, it can be easily shown that this is not enough, see a slight modification of (16) in (17):

(17)   a.   Rakouská vláda se rozhodla, že bude vyvíjet tlak na pražskou vládu ve věci stavby jaderné elektrárny v Temelíně.

The Austrian government decided that it will execute a pressure on Prague government in the matter of building the nuclear power station in Temelin.

       b.   Rakouští představitelé dali jasně najevo, že otázku Temelína spojují s přijetím Česka do Unie.

Austrian representatives have made it clear that they connect the issue of Temelin with the acceptance of Czechia to the Union.

In (17)(b) the expression *představitelé* (representatives) again refers back to *vláda* (government) in (a), but to the first rather than to the most recent occurrence of this expression. Therefore, in addition to the attribute capturing the lexical value of the antecedent we need also to register which occurrence of the antecedent is referred to; this can be ensured e.g. by putting the serial number of the antecedent as the value of another attribute attached to each node.

Sentence (16) brings about a still another problem: in TGTSs, new nodes are added in case of deletion of elements in the surface shapes of sentences.

Thus, in the TGTS of (16)(a) the node [on.Fem] has to be restored as the Actor of the second clause (a similar situation obtains for the Actor of the second clause in (16)(b)), as indicated in (16'):

(16')   a.   Rakouská vláda se rozhodla, že [on.ELID.Fem.Sg.Actor] bude vyvíjet tlak na pražskou vládu ve věci stavby jaderné elektrárny v Temelíně.

b.   Rakouští představitelé dali jasně najevo, že otázku
Temelína [on.ELID.Anim.Pl.Actor] spojují s přijetím Česka do
Unie.

To indicate whether the antecedent is in the same sentence or in the preceding context, we have added a third attribute, with the value 0 for the former case and the value PREV for the latter.

To sum up, three attributes are introduced in the TGTSs to account for the three ingredients sketched above: the attribute COREF with a value equal to the lexical value of the antecedent, the attribute CORNUM with a value equal to the (serial) number of the antecedent, and the attribute CORSTC with two values, namely PREV (obtained if the antecedent is in the previous sentence(s)) and 0 (in case the antecedent is in the same sentence). Thus, the node *představitelé* in (16)(b) and the two restored nodes in (16)(a) and (b) will have the following values in these three attributes:

(16")   a.   on.ELID.Fem.Sg.Actor: COREF [vláda]

CORNUM [2]

CORSTC [0]

   b.   představitelé: COREF [vláda]

CORNUM [2]

CORSTC [PREV]

on.ELID.Anim.Pl.Actor: COREF [představitel]

CORNUM [2]

CORSTC [0]

Anaphoric relations crossing sentence boundaries are captured only in the so-called 'exemplary' set of TGTSs and they will be registered in the further stages of the project, in which also the distribution of degrees of salience in the stock of shared knowledge will be taken due account of.

Acknowledgements Research for this paper was supported by a grant of the Czech Ministry of Education VS 96/151 and by Czech Grant Agency GAČR 405/96/K214.

References

1. Hajičová E., Partee B. H., Sgall P.: Topic-Focus Articulation, tripartite Structures, and Semantic Content. Dordrecht: Kluwer Academic Publishers (1998)
2. Sgall P.: Functional sentence perspective in a generative description. In: Prague Studies in Mathematical Linguistics 2, Prague: Academia (1967) 203–225
3. Sgall P., Hajičová E., Panevová J.: The Meaning of the Sentence in Its Semantic and Pragmatic Aspects. Dordrecht: Reidel (1986)

Tiered Tagging and Combined
Language Models Classifiers

Dan Tufiş

RACAI-Romanian Academy, 13, '13 Septembrie', Ro-74311, Bucharest
tufis@valhalla.racai.ro

Abstract. We address the problem of morpho-syntactic disambiguation
of arbitrary texts in a highly inflectional natural language. We use a large
tagset (615 tags), EAGLES and MULTEXT compliant [5]. The large
tagset is internally mapped onto a reduced one (82 tags), serving statistical disambiguation, and a text disambiguated in terms of this tagset
is subsequently subject to a recovery process of all the information left
out from the large tagset. This two step process is called *tiered tagging*.
To further improve the tagging accuracy we use a combined language
models classifier, a procedure that interpolates the results of tagging the
same text with several register-specific language models.

1 Introduction

One issue recurrent in the tagging literature refers to the tagset dimension vs.
tagging accuracy dichotomy. In general, it is believed that the larger the tagset,
the poorer the accuracy of the tagging process, although some experiments [4]
show that this does not always hold provided enough training data is available
and the tagset cardinality varies within reasonable limits (say 100–200 tags).
However, when the target tagset gets larger (600–1000 tags or even more), the
problem becomes the current tagging technology. We describe tiered tagging, a
two-step process, as a possible solution for reconciling the tagging accuracy with
the large number of tags in the target tagset (as many highly inflectional languages require). The two levels of the tiered tagging have two different tagsets:
a reduced one, used for training and producing a language model (LM) which
a proper tagging needs, and a large tagset containing the same information
as the small one plus supplementary lexicon information. To further improve
the tiered tagging accuracy we developed a combined language models classifier
which tags the input text using different register-specific LMs and, interpolating the differences, produces a final more accurate tagged text. Although the
so far experiments limit to Romanian texts, our methodology, called tiered tagging with combined language models (TT-CLAM), we believe, is not language
dependent.

2 Tiered Tagging

For highly inflectional languages, traditional linguistics distinguishes a large
number of morpho-syntactic features and associated values. For Romanian, we

F. Jelinek, E. Nöth (Eds.): TSD'99, LNCS 1692, pp. 28–33, 1999.
© Springer-Verlag Berlin Heidelberg 1999

constructed a word-form lexicon [9], the items of which (427983 word-forms, 38807 lemmas) were described by a set of 615 morpho-syntactic descriptors (MSDs), EAGLES and MULTEXT compliant [5]. However, not all attributes or values present in these descriptors are distributionally sensitive or equally good contextual predictors/restrictors either. Moreover, some attribute values may depend on other attribute-values of a given wordform. Based on this set of morpho-syntactic descriptors (MSDs), and for tagging purposes, we designed a reduced tagset (RT) containing 82 tags (plus 10 punctuation tags) [8]. The reduced tagset, obtained by a trial&error process, eliminated attributes or merged attribute values which were either distributionally irrelevant or fully lexicon recoverable based on the remaining attribute values. Yet some attributes and values, although fully recoverable, were preserved in the reduced tagset, because they help disambiguate the surrounding words. The main property of this reduced tagset is what we call *recoverability*, to be described as follows.

Let MAP: RT→ LT^m be a function that maps a tag from RT onto an ordered set of tags from the large tagset (LT), AMB: W→ LT^n, a function that maps a word onto its ambiguity class (from the lexicon) and TAG: W→RT, a selector that returns for a word the tag assigned by a tagger (in a specific context). Then, recoverability (as achieved in our tagset design) means:

$$CARD(AMB(w) \bigcap MAP(TAG(w))) = \begin{cases} 1 \text{ in more than } 90\% \text{ cases} \\ \geq 2 \text{ for less than } 10\% \text{ cases} \end{cases}$$

The reduced tagset has the property that one tag assigned to a given word w can be deterministically mapped back onto the appropriate MSD in the large tagset in more than 90 % of the cases. Note that although this mapping is almost deterministic, one tag may be mapped differently, depending on the context and the word it is assigned to. The underlying idea of the tiered tagging is the recoverability property of the reduced tagset. Having a training corpus annotated in terms of the reduced tagset, we can build an LM that is to be used to tag new texts. Then, thanks to the recoverability property, the tags are mapped onto MSDs (the large tagset).

For the rare cases of the mapping ambiguities (when a coarse-grained tag is not mapped onto a unique MSD but onto a list of MSDs), we use 14 very simple contextual rules (regular expressions). They specify, by means of relative offsets, the local restrictions made on the current tag assignment. Our rules inspect the left, the right or both contexts with a maximum span of 4 words. Such a rule, headed by a list representing the still there ambiguity, is a sequence of pairs (*MSD: conditions*) where *conditions* is a disjunction of regular expressions which, if applied to the surrounding tokens (defined as positive or negative offsets), returns a truth-value. If *true*, then the current token is assigned the *MSD*, otherwise the next pair is tried. If no one of the conditions returns a *true*value, the mapping ambiguity remains unsolved. This happens very rarely (for less than 1 % of the whole text). For instance, the following rule considers a tag class DS corresponding to two merged MSD classes (possessive pronouns and possessive determiners/adjectives).

Ps|Ds

Ds.$\alpha\beta\gamma$: (-1 Nc$\alpha\beta\gamma$y)||(-1 Af.$\alpha\beta\gamma$y)|||(-1 Mo.$\alpha\beta\gamma$y)|| (-2 Af.$\alpha\beta\gamma$n and -1 Ts)||
    (-2 Nc$\alpha\beta\gamma$n and -1 Ts)|||(-2 Np and -1 Ts)|||(-2 D..$\alpha\beta\gamma$ and -1 Ts)

Ps.$\alpha\beta\gamma$: *true*

The rule reads as follows (α, β, γ represent shared attribute values, "." represent an "any" value):

IF any of the conditions a) to g) is true

a) previous word is a definite common noun

b) previous word is a definite adjective

c) previous word is a definite ordinal numeral

d) previous words are an indefinite adjective followed by a possesive article

e) previous words are an indefinite common noun followed by a possesive article

f) previous words are an indefinite proper noun followed by a possesive article

g) previous words are a determiner followed by a possesive article

THEN choose the determiner MSD; shared attribute values set by the context

ELSE choose the pronominal MSD; shared attribute values set by the context.

The second phase of the tiered-tagging is practically error-free, so in order to improve the overall accuracy of the output, the proper statistical tagging done at the first step has to be as accurate as possible. To this end, we developed the combined language model classifier, to be described in the next section.

3   Combined Language Models Classifiers

In general terms, a classifier is a function that, given an input example, assigns it to one of the K classes the classifier is knowledgeable of. Recent work on combined classifier methods ([1,2,3,6] etc.) has shown one effective way to speed up the process of building high quality training-corpora with a corresponding cost cut-down. The combined classifier methods in POS-tagging naturally derived from the work done on the taggers evaluation. In combining classifiers, one would certainly prefer classifiers of which errors would not coincide. The basic idea in combining classifiers is that they complement each other's decisions so that the number of errors is minimized. Of different statistical tests for checking error complementarity, we used McNemar's [2] and Brill&Wu's [1].

The combined classifiers methods [1,2] are based on the combination of the output from different taggers trained on the same data. Such an approach considerably improves single tagger performance, and its applicability relies on the assumption that the errors made by one tagger are not a subset or superset of those of another tagger. This conjecture, which we called *error complementarity*, is supported by all the experiments we know of (e.g. [1,2] also our own tests). The difference in taggers performance is mainly explained by the tagging methods, and, to a lesser extent, by the very linguistic nature of the training data. The linguistic relevance of the training text is not easily measurable.

The proposed methodology, even though similar to the one above at first sight, is actually different: instead of using several taggers and the same training corpus, we use one tagger (ideally, this should be the best available) but train it on various register corpora. For the work reported here, we used a modified version of Oliver Mason's QTAG http://www-clg.bham.ac.uk/oliver/java/qtag). Each training session, based on comparable-size corpora, results in a register specific LM. A new text is tagged with all LMs and their outputs are combined for the final result. The *combined classifier* is based on static data structures (*credibility profiles*) constructed during tagger training on various corpora. In our experiments, none of the training corpora contained less than 110,000 hand-tagged items. The credibility profile (LM dependent) encodes, among other things, the probability of correct assignment for each tag, its confusion-probabilities and the overall accuracy of the LM. Our experiments and intensive tests and evaluations with various classifiers (simple majority voting and three types of weighted voting – out of which the one based on the credibility profiles performed the best) brought evidence for several challenging hypotheses which we believe are language independent:

- the *error-complementarity* conjecture holds true for the LMs combination. We tested this conjecture with 18 LMs combinations on various texts (about 20.000 words each) in three different registers (fiction, philosophy and journalism) and no experiment contradicted it;
- a text T_i belonging to a specific register R_i is more accurately tagged with the LM_i learnt for that register than if using any other LM_j. As a consequence, by tracking which one of the (LM-dependent) classifiers came closer to the final tag assignment, one could get strong evidence for text-type/register identification (with the traditional methods further applicable)[1];
- the combined LMs classifier method does not depend on a specific tagger. The better the tagger, the better the final results.

4   Evaluation, Availability and Conclusions

Based on George Orwell's *'1984'*, Plato's *'The Republic'* and several issues from *'România Liberă'* and *'Adevărul'* (the daily newspapers with the largest distribution in Romania), we constructed three different register training corpora (fiction, philosophy and journalism). They cover all the MSDs and more than 94 % of the MSD-ambiguity-classes defined in the lexicon. The three training corpora were concatenated (the *Global* corpus) and used in the generation of another LM, to be referred in the following as LM_{Global}.

For testing, we specially hand-tagged about 60,000 words from different texts in the same registers: **Fiction**, **Philosophy** and **Newspapers** (articles extracted from newspapers others than those used for training).

Table 1 shows the results of McNemar's test on various LM combinations applied to the three test corpora. Our interest was in evaluating whether the paired classifiers were likely to make similar errors on new texts in the given register.

---

[1] According to an anonymous reviewer, the idea of register identification by seeing which LM models the text the best is also strongly supported by Beeferman, Berger, Lafferty: Statistical Methods for Text Segmentation, to appear in *Machine Learning, Special Issue on Natural Language Learning* vols. 1/2/3, 1999

The threshold for the null hypothesis with a 0.95 % confidence is $\chi^2_{0.95} = 3.84146$. Accepting the null hypothesis here implies that the two classifiers are expected to make similar mistakes on texts in the given register (e.g. Rep&News for Fiction, 1984&Rep, 1984&News News&Global for PHILOSOPHY and News&Global for NEWSPAPERS)[2]. For instance, in tagging the FICTION test corpus, the classifiers based respectively on *Rep* and *News* LMs performed equally well (or better said, equally bad) with a McNemar coefficient of 1.04. Rejecting the null hypothesis means that the two classifiers would make quite different errors and one of them is expected to make fewer mistakes than the other.

Table 1. McNemar's test for pairs of classifiers

FICTION				PHILOSOPHY				NEWSPAPERS			
LM	Rep	News	Global	LM	Rep	News	Global	LM	Rep	News	Global
1984	9.28	15.35	7.56	1984	**1.41**	**1.32**	8.14	1984	14.86	66.98	59.03
Rep	*	**1.04**	35.81	Rep	*	5.63	16.24	Rep	*	20.57	13.12
News	**1.04**	*	39.58	News	5.63	*	**1.47**	News	20.57	*	**3.46**

The results of tagging with combined LMs classifiers on the test texts (not included in the training corpora) are shown in Table 2. The classifiers based on single LMs (1984, Rep, News and Global) show a high level of correct agreement with less than 1% of wrong agreement. The bottom lines in the table display the accuracy of two combined classifiers: MAJORITY (MAJ) and CREDIBILITY (CRED).

Table 2. Evaluation results

LM	1984	Rep	News	Global	1984	Rep	News	Global	1984	Rep	News	Global
Test Texts	Fiction (20109w)				Philosophy (20136w)				Newspapers (20038w)			
(%) single classifiers	98.51	98.15	98.16	98.67	98.31	98.21	98.41	98.50	97.63	97.97	98.37	98.24
(%) right agreement	97.09				96.70				97.15			
(%) wrong agreement	0.59				0.83				0.72			
(%) MAJ. combiner	98.66				98.52				98.41			
(%) CRED. combiner	98.78				98.57				98.45			

The evaluation results point out at least two important things:

a) splitting a balanced training corpus (*Global* in our case) into specialised register training corpora is worth considering: although LM_{Global} generally provides better results than a model based on a subcorpus, even the simplest combiner – *majority*, scores in most cases better;

---

[2] One may note that similarity is not transitive, as Rep&News on PHILOSOPHY are shown not to behave similarly, in spite of the pairs 1984&Rep and 1984&News.

b) the high level of correct agreement and the negligible percentage of false agreement allow the human expert annotator to concentrate quite safely on the cases of disagreement only. With less than 2.5 % of the tagged text requiring human validation (see Table 2), the hand disambiguation of large training corpora becomes a less costly task.

The combined LMs classifier tagging system works in a client-server architecture with individual classifiers running on different machines. On a Pentium-II/300, under Linux, and with most of the programs written in Java, Perl and TCL, the individual classifiers' speed was about 15,000 words/min (most of the time being spent in accessing dictionaries). A new version of the entire system (using Oracle 8^{TM} and most TCL and Perl code rewritten in C and Java) is expected to improve the speed for at least 3-4 times. If the speed factor is critical, using the single LM_{Global}-based classifier is the option of choice. The tagging system described in this paper is used in a program for automatic diacritics insertion for Romanian language texts [7] with a very high level of accuracy (more than 98.5 %). The platform for tiered tagging with combined LMs classifier (containing the tokenizer, QTAG* tagger, the tag-to-MSD mapping and the combined LMs classifier) is designed as a public service on the web and, along with the required language resources for Romanian, is free (license-based) for research purposes.

Acknowledgements: The work reported here built on the main results of the Multext-East (COP106/1995) and TELRI (COP200/1995) European projects and was partly funded by a grant of the Romanian Academy (GAR188/1998).

References

1. Brill, E., and Wu, J. (1998): Classifier Combination for Improved Lexical Disambiguation *In Proceedings of COLING-ACL98* Montreal, Canada, 191–195
2. Dietterich, T. (1998) Approximate   Statistical Tests for Comparing Supervised Classification Learning Algorithms, 1998, http://www.cs.orst.edu/~tgd/cv/pubs.html.
3. Dietterich, T. (1997): Machine Learning Research: Four Current Directions, *In AI Magazine*, Winter, 97–136
4. Elworthy, D. (1995): Tagset Design and Inflected Languages, *In Proceedings of the ACL SIGDAT Workshop*, Dublin, Ireland (also available as cmp-lg archive 9504002)
5. Erjavec, T., Monachini., M. eds. (1997): Specifications and Notation for Lexicon Encoding of Eastern Languages. *Deliverable 1.1F Multext-East* http://nl.ijs.si/ME
6. v. Halteren, H., Zavrel, J., and Daelemans, W. (1998): Improving Data Driven Wordclass Tagging by System Combination *In Proceedings of COLING-ACL98*, Montreal, Canada, 491–497
7. Tufiş, D., Chiţu, A. (1999): Automatic insertion of diacritics in Romanian Texts, *In Proceedings of COMPLEX 99*, Pecs, Hungary
8. Tufiş, D., Mason O. (1998): Tagging Romanian Texts: a Case Study for QTAG, a Language Independent Probabilistic Tagger *In Proceedings of First International Conference on Language Resources and Evaluation*, Granada, Spain, 589-596
9. Tufiş, D., Barbu, A. M., Pătraşcu, V., Rotariu, G., Popescu C. (1997). "Corpora and Corpus-Based Morpho-Lexical Processing" in Dan Tufiş, Poul Andersen (eds.) *Recent Advances in Romanian Language Technology*, Editura Academiei, 35–56 (also available at http://www.racai.ro/books)

Syntactic Tagging: Procedure for the Transition from the Analytic to the Tectogrammatical Tree Structures

Alena Böhmová, Jarmila Panevová, and Petr Sgall

Faculty of Mathematics and Physics, Charles University, Prague, Czech Republic
E-mail: {bohmova,panevova,sgall}@ufal.mff.cuni.cz

Abstract. The syntactic tagging of the Prague Dependency Treebank (PDT) is divide into two steps, the first resulting in analytic tree structures (ATS) and the second in tectogrammatical tree structures (TGTS). The present paper describes the transition procedures, automatic and manual, from ATS to TGTS and illustrates these procedures on two Czech sentences.

Syntactic tagging in The Prague Dependency Treebank Project is conceived of in two steps: (i) analytic tree structures (ATS), in which every word form and punctuation mark is explicitly represented as a node of a rooted tree, with no additional nodes added (except for the root of the tree of every sentence) and with edges of the tree corresponding to (surface) syntactic dependency relations, (ii) tectogrammatical tree structures (TGTS) corresponding to the underlying sentence representations; TGTSs have the shape of dependency trees with the verb as the root of the tree and its daughter nodes representing nodes depending on the governor (on each layer of the tree). The two dimensions of the tree represent the syntactic structure of the sentence (the vertical dimension) and the topic-focus articulation of the sentence, based on the underlying word order (the horizontal dimension). In contrast to the ATSs, functional words (such as prepositions, auxiliaries, subordinating conjunctions etc.) as well as punctuation marks principally are not represented by nodes of their own; their functions are captured as parts of the labels (tags) of the nodes standing for autosemantic words. For technical reasons, the coordinating conjuntions are represented as specific nodes, which have the positions of the head nodes of coordinated constructions.

The transition from the ATSs to the TGTSs is conceived of as a transduction procedure (see [1]), consisting of two phases: (A) an automatic 'pre-processing' module, and (B) a manual tagging with the help of a 'user-friendly' software.

We want to illustrate here the automatic module, the input of which are the ATSs (with the accessibility of both the morphological and the analytical syntactic tags). The task of the module is then to process the ATSs in view of two aspects:

(a) to prune the tree structures, i.e. to devoid them of nodes that are counterparts to auxiliary forms in the surface structure of the sentence, without losing any important pieces of information these auxiliary forms carry;

F. Jelinek, E. Nöth (Eds.): TSD'99, LNCS 1692, pp. 34–38, 1999.

(b) to translate (by means of linguistically substantiated transduction rules) the semantically relevant information given in the ATSs into the terms of the underlying structure.

The task under (a) concerns e.g. cancellation of the auxiliary node for the sentence and other "technical" nodes, transduction of the nodes standing for the final sentence boundary to the modality grammatemes with the governing verb, putting analytical forms together (and placing them in the position of the 'highest' of their parts), and adding the information they convey in the form of indices, grammatemes and other parts of the TGTS complex tags. The part (b) includes first of all the assignment of 'grammatemes' (i.e. for the values of morphological categories such as number, tense, modality etc.) in those cases in which they can be derived from ATS.

The procedure under (b) mainly concerns transduction of the analytic functions (such as Subject, Object, Adverbial, Attribute) into their tectogrammatical counterparts, i.e. Actor, Patient, Addressee, different kinds of Free Modification using the information on their form and their immediate context (e.g. the information encoded in the prepositions).

Example (1)

Iniciátoři   dosud   nesehnali potřebných třicet podpisů   poslanců
The initiators not yet collected needed   thirty signatures deputies

z   každé sněmovny, aby schůze   obou
from each   Chamber, for the sessions two

komor   FS   mohly být předčasně svolány
Chambers of the Federal Parliament to   be specially summoned.

'The initiators have not yet collected the needed thirty signatures of deputies from each Chamber, for the sessions of the two Chambers of the Federal Parliament to be specially summoned.'

AuxS is for an auxiliary node for sentence, Pred: predicate, Sb: subject, Atr: attribute (Czech: 'atribut'), Adv: adverbial, Obj: object, AuxP is for preposition, AuxC is for auxiliary node for conjunction, AuxX is for auxiliary node for comma, AuxV: verb, AuxK is for final interpunction. For more detailed explanation of analytic functions see [2].

ACT is for actor, PAT is for patient, TTILL: time adverbial 'till', TWHEN: time adverbial 'when', DIR1 is for direction, RESTR: a restrictive adjunct, APP: appurtenance, AIM: aim adverbial. ENUNC: ennunciation, IND is for the indicative verb mode, ANT and SIM are for verb tense (anterior and simultaneous, respectively), POSS and DECL are for deontic mode (possibilitive and declarative), CPL is for complex aspect.

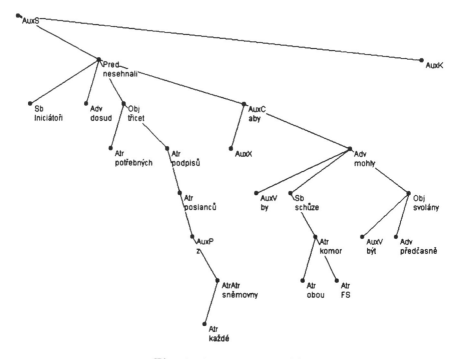

Fig. 1. ATS of sentence (1)

Example (1')

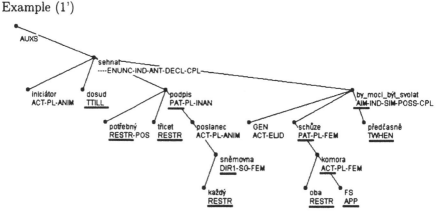

Fig. 2. TGTS of sentence (1)

Example (2)

Slovo "elita" se ovšem v Československu stále ještě chápe   trochu
word "élite", however, in Czechoslovakia still   is understood a little

pejorativně, jako podezřelá kategorie samozvaně   privilegovaných.
pejoratively as a suspicious category of self-appointed privileged.

'The word "élite", however, in Czechoslovakia still is understood a little pejoratively, as a suspicious category of self-appointed privileged people.'

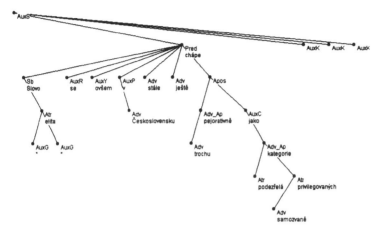

Fig. 3. ATS of sentence (2)

AuxG is for graphical symbols, AuxR is for a reflexive particle.

Example (2')

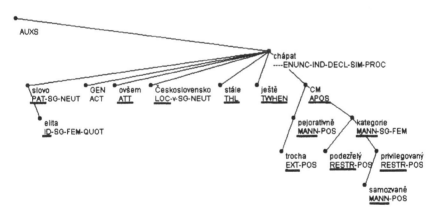

Fig. 4. TGTS of sentence (2)

LOC is for location, THL: time adverbial 'how long', APOS: apposition, MANN: manner.

By the ex. (1), (1') and (2), (2') we exemplify the subdivisions of tasks between the procedures (A)(a),(b) and (B); examples (1) and (2) are outputs from the ATS, their counterparts (1') and (2') correspond to the (simplified) output from TGTS. The tags left out for processing in the manual (B) procedure are underlined.

In ex. (1') we recieve as a result of the application of the automatic procedure (A) the following information:

(i)   the orientation of the relation between head and its modifier in the construction *třicet podpisů* gets changed,

(ii)   the nodes in ATS corresponding to the analytical forms of the verb and the modal verb with its infinitive complement (with the analytical function Obj) are combined in a single node (*by mohly být svolány*); the preposition (AuxP – 'z') is deleted as a node in the tree and it is stored in the attribute for the "future" value of the syntactic grammateme of the noun *sněmovna*),

(iii) all morphological grammatemes expressing the meaning of verbal categories (Verbmod, Deontmod, Tense and Aspect), gender and number with nouns and degrees of comparision with adjectives and adverbs are filled on the basis of their morphological tags (some asymmetries between forms and their respective functions will be solved later during the manual procedure),

(iv)   the grammatemes of Sentmod with the root of the tree is specified automatically (this attribute is assigned to all heads of main clauses on the basis of the data present in the analytical tree),

(v)   the analytical function Subject with the verb in active voice is converted into the tectogrammatical functor ACT (actor). The rest of functors will be determined on the basis of the "user-friendly" software manually by the procedure (B), which also includes the addition of a new node for a general actor (in the embedded clause with the verb in passive voice).

In ex. (2'), the same steps as in (1'), i.e. (ii), (iv) were applied; in this example an automatic procedure adds

(vi)   the "special" grammateme QUOT for the quote word in quotation marks,

(vii) the analytical function AuxR denoting the reflexive passive is converted into a node with lexical value "general" (actor).

In the ex. (2) and (2') the representation of the apposition (which is analogical to the coordination) in ATS and in TGTS are illustrated.

Neither the automatic nor the manual part of the tagging can achieve a complete formulation of tectogrammatical representations. Several types of grammatical information will be specified only after further empirical investigations. Thus, e.g., the disambiguation of the functions of prepositions and conjunctios can only be completed after lists of nouns and verbs with specific syntactic properties are established. However, the annotated corpus will offer a suitable starting point for monographic analysis of the problem concerned. Whenever possible, also statistical methods will be used.

References

1. Hajičová E.: Prague Dependency Treebank: From analytic to tectogrammatical annotations. Proceedings of the Conference TSD '98, Brno (1998)
2. Hajič J.: Building a syntactically annotated corpus: The Prague Dependency Treebank. In: Issues of Valency and Meaning. Studies in Honour of Jarmila Panevová (ed. by E. Hajičová), Prague: Karolinum (1998) 106–132

Information, Language, Corpus and Linguistics*

František Čermák

The Institute of the Czech National Corpus
Faculty of Philosophy and Arts, Charles University, Prague
Frantisek.Cermak@ff.cuni.cz

Abstract. At the turn of the century, linguistics is more and more dependent on corpora; at the same time it is evident that corpora become a primary source of information. Yet the original enthousiasm of corpus linguists has to be confromted with a number of new problems which have to be faced. Some aspects of corpora and language, applications and linguistics proper are discussed and a number of open questions is mentioned.

1 Language and Linguistics at the Turn of the Century

Most of information about anything is to be found in language; there are, in fact, very few areas of human life based to a higher degree on non-verbal symbols and there seem to be none at all which could do without language completely. Language seems to be something which is self-evident and quite basic.

The forthcoming twenty-first century and its people are now usually labeled *Information Society*, where the attribute *information* conveys the substance and basic nature of that new era.

In a global view, it is not language information, or, rather, information in language about a given time period that is of consequence, but also information across different time periods as well as across languages, for example, by means of gradually improved possibilities of automated machine translation which is important.

It is still very much true that there are, perhaps, inconveniently too many approaches to linguistics and too many languages to be covered. The underlying conviction of Chomskyan linguists has always been that linguists are language users, too, and their made-up sentences are as good as any other and that context and representative sources do not matter. The point made in their criticism was that in view of the complexity and richness of language it is justified to be skeptical to such specimens of language which have no reference to context and to the possibility of ever be certain that the phenomena thus illustrated cover the whole of the language without leaving out something of substance. Equally, it is doubtful, how justified any generalization of the phenomena and relationships thus illustrated could be.

---

* This research was supported by the GACR, Grant Nr. 405/96/K214.

F. Jelinek, E. Nöth (Eds.): TSD'99, LNCS 1692, pp. 39–43, 1999.

With the arrival of computers into science the phenomenon of language has proved to be amenable to a thorough documentation and exhuastive coverage exceedingly better than ever before. It was also shown that manual maintenance and upkeep of the traditional language archives of citation slips (which were dictionary archives only) was no longer possible for obvious reasons. In a sense, the historically given supply of available data, necessary for any science, makes much of the past of language studies and of the quality of their judgements problematic and dated: there has never been enough data. This is being radically changed and fundamentally improved; so much so that linguists are now starting to fight the overflow of data and look for ways how to avoid getting swamped with too much data. This is basically true of synchronic linguistics. And this is also why linguists, who have taken up this computer *corpus* challenge and possibility, feel bemused by the case of Chomsky, who has, rather early, refused corpus data as being too skewed to be of any real use.

Problems facing corpus linguistics seem to lie elsewhere, at the opposite end of the scale. It is really quite possible, expecially for a small language like Czech, to imagine a point in time, when – in a purely static and straightforward view – a corpus of the given period would be finally recognized as having come to its numeric limits in the community. The impression of inexhaustibility of resources is patently false.

Thus, an obvious problem, which may prove extremely difficult to solve and which might be tackled by small languages first, is to establish further and consistent links with past and future. While one can hardly expect more texts from the past than the number we already have, it is certainly possible to imagine that whatever we might have one day could be put to other uses, too. Specifically, since language is in a constant development and change, any synchronic corpus will have to become, in a matter of a few years, part of history, of a diachronic corpus, too. At the same time, however, it could also become a filter for monitoring any new items and changes in the future of the language. Linking thus contemporary corpus to past would eventually make it possible to have a comparatively continuous coverage of the language. This might be the only reliable way how to really understand continuity and discontinuity of the language.

Certainly, one can object that this is only the written, by far the smaller, part of language. It is also true that computers are still far from being able to record, to any substantial degree, the spoken language and hence we find ourselves in a position of wishful thinking only. If this were possible, our data would dramatically improve further. To come back to the written language, our extrapolation of conclusions from the bulk of data we have at our disposal today is highly reliable, considerably more s o, that we tend to forget to be cautious. Although the degree of the coverage of the language by a large corpus is considerable, it by no means true that today's corpora reflect language as a whole.

To be able to do at least some of this, the support and care of no-one less than the State which I have started with would be necessary. This fact is stressed in a rather harsh wording by A. Zampolli, president of ELRA, in a speech he gave at the opening of a conference in Granada in spring 1998.

On the worst hypothesis, citizens who are not able to communicate in the languages implemented in the global network would be denied full participation in the Information Society... A recent EUROMAP draft survey shows that the support of language technology is extremely uneven across Europe at the national level. Several member states have no policy on the support of their national language within Information Society, "a situation which threatens the survival of those languages in the mainstream." (pp. XXI–XXII, Introduction of the General Chairman, Antonio Zampolli, Rubio, A., LREC 1998)

2 Corpus and Corpus Linguistics

It is now clear that a language without a computer corpus is rare today, at least in Europe. Most European languages have some sort of a corpus already, although it may be small, and the growing awareness that there is an obvious place for them to be filled makes also their importance grow (there are now corpora also for Chinese, Japanese, Korean etc.). Despite initial disapprovals voiced by some linguists doubts are disspelled by obvious and undisputable facts: nobody has ever been able to manually collect and process so much data in his or her lifetime as the computer in a matter of a very short time. Also the project of the *Czech National Corpus* has now entered its fifth year of implementation and its second hundred million of words (see also http://ucnk.ff.cuni.cz/cnc). It seems obvious now that highest and justified expectations in linguistics are closely tied to corpus linguistics, both in this and the next decade at least.

It is *corpus linguistics*, aptly named after its major preoccupation, that has been prominent for and linked to this developement, which is concerned with the research of corpora of national languages, followed later by research across the boundaries of one language only. Albeit a young branch, it is undisputedly one of the fast developing ones.

3 Corpus Data and Information

Strictly speaking, corpus data may not be and, many people would say, should not be just texts taken from, for example, newspapers in any straigtforward way. An intricate complex of data features is due to their linear, sequential character which is, however, often not strictly sequential as many informational and linguistic units are made up of parts which are not adjacent. Much of this disadvantage may be reduced if an additional textual and linguistic annotation and tagging is introduced into the corpus. There are three obvious aspects in which *corpus data* is *related to other entities and areas* reflecting the relationship (1) between the corpus and language, (2) between the corpus and its applications, and (3) between the corpus and linguistics proper.

A basic prerequisite is the nature and size of the data used; this is an issue of the still developing corpus methodology. The *first* of the three aspects is also largely influenced by the degree to which new insights and knowledge are taken over from non-linguistic disciplines, such as mathematics, computer science etc., and effectively applied. The *second* aspect reflects such considerations which, on the one hand, include human resources and feasibility of the task at hand. On the other hand, one has to ask what the needs of the future corpus users are, i.e. both of those who are aware that they might need the corpus and also those who do not know yet. It is no witticism to say that many people really need the corpus and do not know they need it yet.

In an extended sense of what is sometimes called a monitor corpus, it is evident that compilation and upkeep of a corpus is an open-ended business and any really large corpus, once proclaimed finished, should not be viewed as finished at all. It should be developed further. To stop developing it would mean both to resign to the fact that our society is in constant development and change and to cherish a rather short-sighted view that we have got information enough for ever. Due to data becoming gradually da ted, it would also mean that, after a while, the initial large investment is wasted, basically.

The *third* aspect seems to be governed by the rule of proportion: problems and attitudes taken by future professionals in linguistics will be increasingly influenced by the growth of corpus data and their quality.

In contradistinction to tradition and to what linguists were used to in their fields, the data offered by corpora of today, is rather different, and no doubt better, both in their quantity and quality. The difference, which might be sometimes radical, may be summed up in a number of aspects.

Thus, as far as their relation to the language as a whole is concerned, corpus data is *typical, synchronic, up-to-date and topical, non-selective, objective and realistic, generally sufficient, not due to chance,* and *available and obtainable fast and easily*, not to speak of their fast processing.

Objections against most of these seemingly obvious features can be easily raised, such as its relative typicality achieved at the cost of a serious lack of spoken data, or selectivity due to their retrieval based on still precious few and imperfect statistical tools, but things are rapidly developing even here.

It has been shown many times that the quality and quantity of information is directly related to the quantity and kind of data fed into the computer corpus. To illuminate this, there are several important points which should be raised. em First point is often formulated like this: one can get only that kind of special information which has been really input into the data, i.e. through annotation. In fact, one can get much more than that, since also the information which has not been directly fed int o the corpus is available, even in plain text, namely thanks to configurations of words and patterns of these configurations found through statistical methods.

The *second* point is, that data and additional information (through annotation) can be inserted into the corpus in an evolutionary way only. It is a very slow, gradual and painstaking process which is, in principle, enabled through automatic approaches only.

The *third* point to be made wisely points to past and sobering experience of linguists with a limited benefit drawn from results of their predecessors. It simply means that any annotation has to be stored separately from the raw language data. Theories change, but data remains and as a sort of historical record of its time, the data should remain sacrosanct, also for those to come yet.

The *fourth* point, related to the first one, is concerned with the nature of entities the corpus linguist is dealing with. In contradistinction to traditional approaches, one has to come to grips with the following scales:

1. There are both traditional discrete *units* which are different from *combinations*, combinatorial phenomena. A particular challenge is the delimitation of *multi-word units*, whether they belong to the domain of *idioms* or *terms*.
2. There are forms that are *rare*, even occasional, and forms that are highly *typical* and central in the corpus. However, linguists are far from agreeing on where to draw the dividing line between the two.
3. The third scale, which may not appear to be a scale at all at first, is one between such terms as *(textual) word form, lemma and multi-word unit*; the traditional semiotic distinction of *token* and *type* is particularly shifted to mean something else in corpus and computational linguistics, unfortunately.

The last, *fifth* point to be stressed has already been implied above and concerns the general scalar quality of most types of information in language. What huge corpora show is rather different: most of the information is scalar, obtainable in stepwise batches with hazy edges only, where the best help available is statistics and fuzzy approaches and no longer black-and-white truths and clear-cut classification boxes. To put it differently, instead of insisting on getting straightforward answers of the type *yes-no* we have to elicit answers of the type *rather this than that*, or *more of this and less of that*. Needless to say this has and will still have its impact on the methodology of linguistics and attitudes of the bulk of linguists.

References

1. Biber D., S. Conrad, R. Reppen, 1998. Corpus Linguistics. Investigating Language Structure and Use. Cambridge U.P.: Cambridge.
2. Čermák F. 1997. Czech National Corpus: A Case in Many Contexts. International Journal of Corpus Linguistics, 2(2): 181–197.
3. Hlaváčová J.. 1998. Technical Insights into the Birth of a Corpus. A Few Words about Building the Czech National Corpus. In Text, Speech, Dialog, eds. P. Sojka, V. Matoušek, K. Pala, I, Kopeček. Masaryk University: Brno, pp. 55–60.
4. McEnery, T. & A. Wilson, 1996. Corpus Linguistics. Edinburgh University Press: Edinburgh.
5. Rubio A. et al., 1998, First International Conference on Language Resources and Evaluation, Vol. I, II, ELRA: Granada.
6. Svartvik Jan, ed., 1992, Directions in Corpus Linguistics: Proceedings of Nobel Symposium 82, Mouton: Berlin.

Prague Dependency Treebank: Restoration of Deletions*

Eva Hajičová, Ivana Kruijff-Korbayová, and Petr Sgall

Institute of Formal and Applied Linguistics, Faculty of Mathematics and Physics,
Charles University, Prague, Czech Republic,
{hajicova,korbay,sgall}@ufal.mff.cuni.cz,
WWW home page: http://ufal.mff.cuni.cz

Abstract. The use of the treebank as a resource for linguistic research has led us to look for an annotation scheme representing not only surface syntactic information (in 'analytic trees', ATS) but also the underlying syntactic structure of sentences and at least some aspects of intersentential links (in 'tectogrammatical tree structures', TGTS). We focus in this paper on some of the issues of the transduction of ATSs into TGTSs.

1 Two Steps of Syntactic Tagging in PDT

In the Prague Dependency Treebank (PDT) project, the structure of sentences is made explicit by means of two steps of syntactic tagging resulting in:

1. 'analytic' tree structures (ATSs), in which every word form and punctuation mark is represented as a node of the tree, and the edges of the tree correspond to (surface) syntactic dependency relations; and,
2. tectogrammatical tree structures (TGTSs) corresponding to underlying sentence representations and having the shape of dependency trees with the verb as the root of the tree.[1] In TGTSs the functional (synsemantic) words (such as prepositions, auxiliaries, subordinating conjunctions) as well as punctuation marks are principally not represented by nodes of their own; their functions are captured as parts of complex tags of the nodes standing for autosemantic (content) words. Surface deletions are 'restored' in TGTSs.

The syntactic information which is absent in the surface (morphemic) shape of the sentence is introduced – at least for the time being – in the manual phase of the transduction procedure ([1]), translating (in a 'user-friendly' environment) ATSs to TGTSs. Every added (restored) node gets the index ELEX (if its antecedent is an expanded head node) or ELID (if this is not so). The added nodes always depend on their governors from the left-hand side, except for certain cases in coordinated constructions (cf. (2) below).

---

* The work reported on in this paper has been supported by the grant of the Czech Ministry of Education VS 96/151 and by the Czech Grant Agency GAČR 405/96/K214.

[1] With the exception of TGTSs for coordinated constructions.

F. Jelinek, E. Nöth (Eds.): TSD'99, LNCS 1692, pp. 44–49, 1999.
© Springer-Verlag Berlin Heidelberg 1999

A specific case concerns coordinating conjunctions: although they belong to function words, they retain their status as nodes (labeled as CONJ, DISJ, etc.) in the TGTSs, which in this point differ from the theoretically substantiated form of tectogrammatical representations. This exception makes it technically possible to work with rooted trees, rather than with networks of more dimensions. One-to-one linearization of ATSs and TGTSs has been defined, which will be applied below, when presenting our examples of TGTSs.

2   Types of Lexical Labels of the Added Nodes

Two cases of node restoration according to the character of the lexical labels of the restored nodes can be distinguished: (a) restoration of full lexical information (i.e. adding a node with a particular lexeme in its label), and (b) restoration of a pronominal (anaphoric) element.

2.1   Restoration of Full Lexical Information

The lexical part of the complex label of the 'restored' (added) node consists in a particular lexeme, including a lexeme with a 'general' meaning, in the following situations:

(i) In Coordination: The restored node (included in square brackets in our examples) can be either a dependent node, as in (1), or a governor, as in (2).[2]

(1)   nové knihy a   časopisy ⇒ nové knihy a   [nové] časopisy
      new   books and journals ⇒ new   books and [new] journals

(2)   červená a   modrá barva ⇒ červená [barva] a   modrá barva
      red   and blue   paint ⇒ red   [paint] and blue   paint

We give precedence to a "constituent" coordination before a "sentential" one, whenever possible. Thus in the TGTS for (3) neither the Actor *Jirka* nor the Objective *Marii* will be 'doubled' because the coordination of the two verbs *potkal* and *pozdravil* will be treated as a coordination of two verbs that have a single Actor and a single Objective in common.

(3)   Jirka   potkal a   pozdravil Marii.
      George met   and greeted   Mary.

The complex labels for the coordinated nodes include a special symbol CO to distinguish them from nodes that modify the coordination as a whole. Thus, a simplified linearized representation (only with the lexical labels representing the respective nodes and with every dependent enclosed in a pair of parentheses) for (3) is given in (3').

---

[2] It should be noted that we give here only one of the possible interpretations of (1); (1) can be also understood as '(nové knihy) a (časopisy)', where no restoration occurs.

(3') (Jirka) (potkal.CO) CONJ (pozdravil.CO) (Marii)

Sentence (4) is an example of the addition of a node that stands for a whole structure; in such a case this 'restored' node carries the label ELEX (for an expanded deleted item), see (4'):

(4) Jirka potkal Marii včera   a   já dnes.
     George met   Mary yesterday and I today.

(4') ((Jirka) potkal.CO (Marii) (včera)) CONJ ((já) potkal.ELEX.CO (dnes))

(ii) In Cases of So-Called 'General Participants': Among the items that are often deleted in the surface, there is the case of an Actor or another argument (inner participant) of a verb with the meaning of 'general' (coming close to the English *one* or German *man*, as for the subject). This argument is represented in the TGTSs as a node with the lexical value 'Gen'; cf. the following examples, for which we adduce linearized representations:

(5) Ten dům byl postaven ve dvacátých   letech.
     That house was built   in the-twenties years.

(5') ((ten.Restr) dům.Pat) (Gen.ELID.Act) postavit ((rok.Temp (dvacátý.Restr))

(6) Ta   trouba dobře peče.
     That oven   well bakes.

(6') ((ta.Restr) trouba.Act) (Gen.ELID.Pat) péct (dobře.Mann)

(7) Dědeček   dobře vypravuje pohádky.
     Grandfather well tells   fairy-tales.

(7') (dědeček.Act) (Gen.ELID.Addr) vypravuje (dobře.Mann) (pohádky.Pat)

The General Actor can also be expressed by the so-called reflexive passive; in that case the node corresponding to the particle *se* occurring in ATS gets the lexical label Gen with the functor Act (without ELID).

(8) Domy se   stavějí z   cihel.
     Houses Refl built   from bricks.

     (Houses are built from bricks.)

(8') (dům.Pat) (Gen.Act) stavět (cihla.Orig)

(iii) In Case of Zero Subject with Infinitive: The so-called verbs of control take an infinitive as their Object (Patient) and their Actor or Addressee is referentially identical to the (deleted) 'subject' of the infinitive. Thus, the Actor of the main clause is such a 'controller' in (9), and the Addressee in (10):

(9) Jirka slíbil   matce přijít   domů včas.
     Jirka promised mother to-come home in-time.

(9′) (Jirka.Act) slíbit (matka.Addr) ((Jirka.ELID.Act) přijít.Pat (domů.Dir)
      (včas-Temp)

(10) Rodiče žádali Jirku   nechodit tam.
      Parents asked George not-to-go there.

(10′) (rodiče.Act) žádat (Jirka.Addr) ((tam.Dir) (Jirka.ELID.Act) nechodit.Pat)

A similar structure is present if the infinitive is passivized:

(11) Richard se   bál   být   spatřen.
      Richard Refl. was-afraid to-be seen.

(11′) (Richard.Act) bát-se ((Richard.ELID.Pat) (Gen.Act) spatřit)

(iv) Cases of a Deleted "Non-Omissible" Obligatory participant: With
certain verbs, an argument can only be deleted if it is given in the immediately
preceding co-text, cf. (12):

(12) (Potkal Milan   Jirku?)
      Potkal. (Has-met Milan George?) Met-Masc.

(12′) (Milan.Act.ELID) potkat (Jirka.Pat.ELID)

In cases (i) through (iv), full lexical items can be identified as antecedents by
the annotator, and thus they are placed into the positions of the deleted tokens.
With the exception of (iv), the possibility (or necessity) for the relevant item to
be deleted is determined by the grammatical structure of the sentence. In (iv),
the specific lexical value of the restored item reproduces that of the overt item
present in a structurally corresponding position in the immediately preceding
utterance.

2.2   Restoration of a Pronominal (Anaphoric) Element

A prototypical context in which a pronominal rather than a lexically fully specified element is added to the tree structure, is that of zero subjects with finite
verbs (Czech is a so called pro-drop language):

(13) Přišel   pozdě.
      Came-masc. late

      (He came late.)

(13′) (on.ELID.Masc.Act) přijít (pozdě.Temp)

(14) Přišla   pozdě.
      Came-fem. late

      (She came late.)

(14′) (on.ELID.Fem.Act) přijít (pozdě.Temp)

If we compare example (9) above with (15), the respective TGTSs in (9′) and (15′) reflect the difference between two kinds of coreference: one given grammatically by the properties of Czech verbs of control, and the other determined by the context, which may even go beyond the sentence boundary (*he* is not necessarily coreferential with *Jirka*).

(15)   Jirka slíbil   matce, že   přijde   domů včas.
       Jirka promised mother that he-would-come home in-time

(15′) (Jirka.Act) slíbit (matka.Addr) ((on.ELID.Act) přijít.Pat (domů.Dir) (včas
       Temp))

2.3   Borderline Examples

Cases in which an omissible obligatory complementation is deleted constitute a special group of deletions. These cases differ from (12) quoted in Section 2.1(iv) in that they concern a deletion licensed by the valency frame of the given head word: the frame includes the respective complementation (be it a participant or an adverbial modification) as semantically obligatory, but omissible on the surface. In case of its deletion in the surface shape of the sentence, its lexical value is chosen according to the context: e.g., with the verbs přijít 'to come' or odejít 'to leave' the choice is between sem/odsud 'here/from here' and *tam/odtamtud* 'there/from there'. In the TGTSs, this ambiguity is to be resolved, which is possible on the basis of the context (not grammatically); for a characterization of intersentential coreference see [2].

2.4   Special Cases

Among the special cases of adding some information that is not present (or is only implicitly present) in ATSs, there are two that deserve a special mentioning:

Case of Sentence Negation   In Czech, negation of verbs is expressed by a negative prefix *ne-* attached to the affirmative form of the verb. In ATSs, the negative verb is thus treated as a single node. However, the semantics of negation and its relationship to the topic-focus articulation of the sentence makes it necessary to introduce into the TGTSs a special node for the operator of negation derived from the negative prefix of the verb and having the lexical value Neg. The Neg node depends on the verb; if the verb has the value F (contextually non-bound, in the focus) in its TFA attribute, Neg is placed to the left of the verb and has also the value F in the TFA attribute (this is the interpretation of negation in (16)). If the verb has the value T (contextually bound, in the topic) in its TFA attribute, Neg is placed either to the left of the verb and has also the value T in the TFA attribute (situation exemplified by (17)), or to the right with the value F (exemplified by (18)).

(16) (Co je s Honzou? Proč pláče?) Honza nespí únavou.
(What is the matter with Honza? Why is he crying?) Honza doesn't sleep
due to fatique.

(17) (Proč Honza nespí?) Honza nespí, protože je unaven.
(Why doesn't Honza sleep?) Honza doesn't sleep, because he is tired.

(18) (Myslíš, že Honza spí, protože je unaven?) Honza nespí, protože je unaven, ale protože si vzal silný prášek na spaní.
(Do you think that Honza sleeps because he is tired?) Honza doesn't
sleep, because he is tired, but because he took a strong sleeping pill.

Restoring Grammatical Values Rather than Entire Nodes In some cases
it is necessary to add some values of attributes to existing nodes. This occurs
e.g. when the grammatical information is to be derived from function words or
from morphemic forms; in the automatic module of the procedure translating
ATSs to TGTSs, this grammatical information would only be added to one of
the nodes standing in the coordination relation, see (19).

(19) Vláda musela odložit pravidelnou schůzi a svolat zasedání zvláštní komise
pro bezpečnost.
The government had to adjourn the regular meeting and to convene a
meeting of a special committee for security.

The modality expressed by the (function) modal verb *musela* is attached as a
value of the attribute of modality with the verb *odložit*; it is necessary, however,
to fill in the same attribute with the same value also with the (coordinated) verb
svolat.

3   Summary

We have outlined one aspect of the difference between ATSs and TGTSs, namely
the situation when the ATSs do not contain all the information that belongs to
the tectogrammatical structure of the sentence. The restoration of the syntactic
information absent in the surface (morphemic) shape of the sentence is done
in the manual phase of the transduction procedure; however, the 'user-friendly'
environment developed for transduction of ATSs to TGTSs is designed in such
a way that it will be possible to include there automatic procedures that will
fulfil some of the transduction tasks.

References

1. Hajičová E.: Prague Dependency Treebank: From analytic to tectogrammatical
annotations In: *Text, Speech, Dialogue* (eds. P. Sojka, V. Matoušek, K. Pala and
I. Kopeček), Brno: Masarykova univerzita. (1998) 45–50.
2. Hajičová E.: The Prague Dependency Treebank: Crossing the sentence boundary. In:
Text, Speech, Dialogue (eds. V. Matoušek, P. Mautner, J. Ocelíková and P. Sojka),
Plzeň, Springer-Verlag. (1999) 18–25.

Some Types of Syntactic Ambiguity;
How to Treat Them in an Automatic Procedure

Jarmila Panevová and Markéta Straňáková

Faculty of Mathematics and Physics, Charles University, Prague, Czech Republic
{jarmila.panevova, marketa.stranakova}@ufal.mff.cuni.cz

Abstract. Several basic types of prepositional group (PG) syntactic ambiguity is presented in the paper. Criteria for automatic PG-disambiguation are proposed and illustrated: (i) the possible contribution of valency frames of verbs, nouns and adjectives, (ii) the notion of verbonominal expressions, (iii) the possibility of defining some formal criteria (based on word order) and (iv) the reliability of semantic features is discussed.

1 Description of the Problem

Syntactic ambiguity of prepositional groups (PG) belongs to the most difficult problems of any automatic procedure of the syntactic parsing in any syntactic framework (phrase structure as well as dependency structure oriented). The PG placed immediately after another noun (in any morphemic form) may in principle modify any of the preceding nouns (NG-type) or the verb (or deverbative adjective), VG-type. It is important to distinguish these two structures because in general NG-type structures and VG-type structures do not have the same meaning (they usually have different truth conditions). The chance that their occurrences in specific contexts share the same meaning (have the same truth conditions) cannot be predicted in advance. This fact can be illustrated by example (1):

Example (1): *Japonsko a EU budou spolupředsedat mnohostranným rozhovorům o míru na Blízkém východě.* (PDT, bl103js.fs ♯34, shortened)
[Japan - and - EU - will - co-chair - multilateral - talks - about - peace - on Middle East.]
(Japan and the EU will co-chair multilateral Middle East peace talks.)

The PG *na Blízkém východě* [on Middle East] can be treated
– as an attributive local modifier of the closest noun *mír* [peace] or
– as a local modifier of the noun *rozhovory* [talks] (NG-type in both cases)
– either it can depend on the verb *spolupředsedat* [co-chair] (VG-type structure).
Syntactically we have to deal with three different structures represented by three different dependency trees. Two of them, the structures *spolupředsedat na Blízkém východě* [co-chair on Middle East] and *rozhovory na Blízkém východě*

F. Jelinek, E. Nöth (Eds.): TSD'99, LNCS 1692, pp. 50–55, 1999.

[talks on Middle East], denote the same situation (content) due to their implications; on the other hand, the structures *mír na Blízkém východě* [peace on Middle East] and *rozhovory na Blízkém východě* [talks on Middle East] represent two different cognitive situations.

A syntactic parsing procedure is supposed to provide all "possible" syntactic structures. This requirement (within dependency approach based on the input data from morphology) can be satisfied in two ways:

(a)   The procedure gives tree-structures with one and only one governor determined for any node (with exception of the root of the tree) - with no respect to the specific syntactic and semantic properties of particular lexical items.

(b)   The procedure gives syntactic structures based on language-dependent syntactic rules, where not only POS and morphological information, but also some types of syntactic and semantic information is used. One type of such information is illustrated by example (2):

Example (2): *Trosky 49letého zaměstnance, který tam plnil své úkoly, zcela zavalily.* (MF Dnes, 24.7.94)
[Ruins - 49 years old - employee$_{Gen/Acc}$ - who - there - fulfilled - his - tasks - completely - buried.]
(The ruins buried a 49 years old employee who fulfilled his tasks there.)

The main verb *zavalit* [buried] needs an Object in the Accusative case. This is the reason why the structure is excluded where the noun *zaměstance* with morphological tags Gen/Acc is analysed as an attribute (NG-type) modifier.

In this paper we try to find criteria for the (b)-type analysis and to refine them (as far as possible) using constraints where the word order, semantic features and valency of various parts of speech are considered.

2   Data Collecting

We use the syntactic annotations from the Prague Dependency Tree Bank (PDT) as our basic source of data for Czech. Three basic "suspicious (syntactic) structures" were defined (see Straňáková, [6]):

(a) Syntactic structure where a verb or a noun is modified by another noun which is modified by a PG (i.e. (V/N (N (PG))) in the linearized form).

(b) Syntactic subtree where a verb or a noun is modified by a PG; another noun appears as a brother of this PG (i.e. (V/N (N PG)) in the linearized form).

(c) Syntactic structure where a noun is modified by an adjective which itself is modified by a PG (i.e. (N (Adj (PG))) in the linearized form).

Two macros for Graph editor (i.e. editor used for the purposes of PDT) were designed which enable to search for sentences whose syntactic structure contains the defined subtrees.

At the first stage of our inquiry we focussed on PGs with prepositions *na* [on], *v* [in] (both with Accusative and Locative). A sample of 1000 sentences from PDT was tested and about 150 sentences with "suspicious" structures were received for each preposition.

Word Order Criteria. The word order criteria limit the positions of potentially ambiguous PG (as noun or verb/adjective modification) - PG as a noun modifier stands in principle only after its governor (some deviations are mentioned and analysed in Sect. 3.2). Therefore three basic types of the surface word order position of PG are not considered in the further analysis - the type of ambiguity illustrated in Sect. 1 is not present there:
- V (resp. N at the beginning of a clause) immediately followed by a PG
- PG at the very beginning of a clause
- Adj and PG are separated by a verb (i.e. (PG - V - Adj - N) and (Adj - N - V - PG)).

On the other side, PG in the following positions represents the string where the "suspicious" NG and VG-type ambiguity is present:
- PG between noun and verb (from the left to the right) (i.e. (N - PG - V))
- (V - N - PG)
- (V - PG - Adj - N) and (V - N - Adj - PG).

(The types with adjectives are not discussed in this paper.)
(There can be string of nouns instead of single N in all of these word order types.)

"PP Attachment" in English. Examining the methods proposed for syntactic parsing of English (see Allen, [1]) we found that there are two principles used for "PP attachment":

(a) "Minimal Attachment Principle" – with preference to the lower number of higher branches;

(b) "Last Attachment Principle" - according which a PG is attached to the most recently established node.

None of these principles is valid for Czech. (Example (3) clearly demonstrates wrong results given by the (b) principle.)

Example (3): *Návrh na zachování vlastnického práva obce na pozemek, na němž stojí příslušný činžovní dům, Sněmovna zamítla.* (PDT, bcb01aba.fs ♯15)
[Proposal - on - preserving - (of) proprietary - right$_{Gen}$ - (of) municipality$_{Gen}$ - to - land - on - which - stands - relevant - rent - house - Lower House - rejected.]
(A proposal to preserve the proprietary right of a municipality to land where the relevant rent house stands was rejected by the Lower House.)

3   Proposition of the Criteria

Before introducing some criteria for the reduction of syntactic outputs for the "suspicious" structures the pre-requisites for their application must be presented here.

The system of syntactic rules proposed here supposes the existence of a lexicon with a detailed information: the valency slots of verbs, nouns and adjectives (i.e. the "labels" specifying the type of a modification/participant - as Actor, Patient, Local, Direction, Manner etc. - the lexical item requires, with their respective morphological expressions) are included. Moreover, the theoretically based notion of valency must be extended in favour of the aim of syntactic parsing. Morphemic forms for modifications commonly used with the particular lexical item will be listed in its frames too, being called "second order" valency (e.g. *trvalý pobyt na území ČR* [permanent residence on the territory of the CR]). Semantic features of nouns will be stored in the lexicon (for examples see below, Sect. 3.4) as well as the possible semantic functions of prepositions.

The rules proposed here are supposed to operate on the results of the morphological analysis; the analysis of compound (analytical) verb forms, elementary noun groups and frozen collocations is presupposed, too.

3.1   Valency Frames

Valency frames of verbs and nouns play a crucial role for the PG-disambiguation. We have demonstrated it by Example (3), where the "first order" verbal valency slot (requiring noun in Accusative) was filled by the noun with Gen/Acc ambiguity. Examples (4) and (4') demonstrate the usefulness of the "second order valency" being introduced into the lexicon (see Panevová, [4]):

Example (4): *Zvýšení silniční daně na všechna vozidla - to je otázka, o níž podnikatelé nejčastěji hovoří.* (PDT bcb01aba.fs ♯37, shortened, modified)
[Increasing - (of) road - tax - to - all - vehicles - that - is - question - about - which - entrepreneurs - most often - speak.]
(The increasing of the road tax to all vehicles - that is the question the entrepreneurs speak about most often.)

Example (4'): *Rozšíření silniční daně na všechna vozidla - to je otázka, o níž podnikatelé nejčastěji hovoří.* (PDT bcb01aba.fs ♯37, shortened)
[Extension - (of) road - tax - to - all - vehicles - that - is - question - about - which - entrepreneurs - most often - speak.]
(The extension of the road tax to all vehicles - that is the question the entrepreneurs speak about most often.)

In (4) the PG will be determined as an attribute of the last noun *daň* [tax] (which has the form *na*+Acc in its frame as a "second order" valency). In (4') there are two candidates for the governor of PG - the deverbative noun *rozšíření* [extension] requires *na*+Acc form as its "first order valency" while *daň* [tax] requires one of three alternative forms *na*+Acc, *za*+Acc, *z*+Gen as a "second order" valency. Their competition is solved in favour of the valency of the "higher" priority (the both structures must be preserved, the rule serves to an evaluation of the structures).

3.2   Verbonominal Expressions

The identification of the occurrence of the verbonominal expressions seems to be very effective as a reasonable restriction of the PP attachment ambiguity.

The verb (usually with a very general meaning) and a noun are combined in a (semi)idiomatic unit (e.g. *mít/potřebovat/vynaložit peníze na co* [have/need/ spend money on st.], *mít/poskytnout/získat prostředky na co* [have/provide/gain means for st.], *podat odvolání proti čemu* [submit appeal against st.]). The noun part of such collocation has usually some valency slot filled in by the PG. Analysing this PG as an attribute we often receive a non-projective structure (with a deviation from the usual word order of the noncongruent attribute). Though the verbonominal expression is represented by two nodes the choice of governor between verbal and noun part is irrelevant. See example (5):

Example (5): *Na jejich opravu noví majitelé domů potřebují peníze.* (PDT, bl104js.fs ♯15, modified)
[For - their - reconstruction - new - owners - (of) houses - need - money.]
(New houseowners need money for their reconstruction.)

3.3   Word Order Configurations and Formal Criteria

After having applied valency information of single words of the sentence we must concentrate on word order configurations again. Some formal criteria are valid:
(a) "Separation Principle": If there is a chain consisting of V - N_i/PG_i - N_j/PG_j - N_k/PG_k in the sentence where the N_j/PG_j group has been analysed as a VG-type structure (verbal participant/modifier), the N_k/PG_k cannot be a daughter of N_i/PG_i. See example (6):

Example (6): *Vysoký činitel ... přiměl Palestince k ústupkům na poslední chvíli.* (blc02zu.fs, shortened, modified)
[Top - official - ... - induced - (the) Palestinians - (to) concessions - at - last - moment.]
(The top official ... induced the Palestinians to retreat at the last moment.)

The nouns *Palestinci* [the Palestinians] and *ústupky* [concessions] are analysed as Addressee and Pat, i.e. as verbal modifiers. According to separation principle PG *na poslední chvíli* [at the last moment] cannot be treated as an attribute of the first noun *Palestinci* [the Palestinians]; it can modify either the second noun *ústupky* [concessions] or the verb *přimět* [induce].
(b) "Clitic Position": The position of clitic plays an important role in PG-disambiguation – clitic can separate one noun group from the other (see Uhlířová, [7]), as is illustrated by example (7):

Example (7): *... vláda a následně Parlament ČR se po dlouhých odkladech pokusí stanovit pro podnikání seriozní a legislativně jasná pravidla hry.* (PDT, bce17zua.fs ♯42, shortened)
[... government - and - subsequently - Parliament (of) CR - *clitic* - after - long - delays - will try - (to) lay down - for - enterprising - serious - and - legislatively - clear - rules - (of) game.]
(... the government and subsequently Parliament of CR after long delays will try to lay down serious and legislatively clear rules for enterprising.)

3.4 Semantic Features

If the nouns in the lexicon are provided by the semantic features (criteria for semantic features are given e.g. in Buráňová, [2]) these features can be efficiently used as a supplementary criterion for the solution of the PG ambiguity.

Example (8): *Vypsal výběrové řízení na budovu na Brusnici.* (PDT, bm102zua.fs ♯18, shortened, modified)
[(He) advertised - tender - for - building - at - Brusnice.]
(He opened a tender for a building at Brusnice.)

In (8) the VG analysis is not probable, but there are two candidates for the governor of PG *na Brusnici* [at Brusnice]. If there is a string of nouns in the sentence - the first one, *budova* [building], (provided with the feature "concrete") is followed by the PG *na Brusnici* [at Brusnice] (consisting of a noun with the feature "local+name" and of a preposition with possible local meaning) - the latter will be analysed as a local attribute of the former.

4 Conclusions

In this paper we have tried to find and to illustrate some criteria for PG-disambiguation. They are based on word order constraints, on valency of various parts of speech and on semantic features of single words.

The formal criteria - concerning word order restrictions, the position of the clitics and the separation principle - are of high reliability. Also the application of valency information is relatively very safe though sentences were found where the valency of nouns led to inadequate structures. On the other hand the advantage of rules based on semantic features is disputable, they can be used (if at all) only as supplementary criteria.

Acknowledgement. The research reported on this paper has been partially supported by the grants GAČR 405/96/K214 and VS 96/151.

References

1. Allen, J.: Natural Language Understanding. The Benjamin/Cummings Publishing Company (1987).
2. Buráňová, E.: Ob odnoj vozmožnosti semantičeskoj klassifikacii suščestvitel'nyx. Prague Bulletin of Mathematical Linguistics **34** (1980) 33–43.
3. Panevová, J.: K otázkám homonymie a neutralizace ve stavbě věty. Jazykovedné štúdie **16** (1981) 85–89.
4. Panevová, J.: Nesoglasnoje opredělenije s točki zrenija analiza dlja mašinogo perevoda. Prague Studies in Mathematical Linguistics **1**. Prague: Academia (1966) 219–239.
5. Sgall, P., Hajičová, E., Panevová, J.: The Meaning of the Sentence in its Semantic and Pragmatic Aspects. Dordrecht:Reidel and Prague: Academia (1986).
6. Straňáková, M.: Ambiguity in Czech Sentences, its Classification and Searching for it. In: Proceedings of WDS'98. Prague: Matfyz press (1998) 165–170.
7. Uhlířová, L.: Knížka o slovosledu. Prague: Academia (1987)

Semantic Annotation of (Czech) Corpus Texts*

Karel Pala

Faculty of Informatics, Masaryk University Brno
Botanická 68a, 602 00 Brno, Czech Republic
pala@fi.muni.cz

Abstract. In the presented paper we deal with the issue of semantic tagging of the (Czech) corpus texts. An attempt has been made to take advantage of the grammatical tagging and relabel some of the tags as semantic and pragmatic. Then the notion of the enriched valency frame is introduced – we call it lexical valency frame.

1 Introduction

Several levels of annotation can be found in the present corpora. Annotating consists of adding selected linguistic information to an existing corpus of written or spoken language. Typically, this is done by some kind of coding being attached (semi)automatically or manually to the electronic representation of the text. Here we should mention at least the following types of tagging:

1. Structural, metatext tagging displays how the corpus text is segmented and thus it indicates the titles, headings, paragraphs, sentences in the text.
2. Grammatical or POS tagging represents "canonical" form of annotation and consists in attaching lemma, POS and grammatical information to each word form in corpus.
3. Syntactic tagging assigns a phrase marker or labelled bracketing to each sentence of the corpus usually in the manner of a context-free or dependency grammar parsing (Prague Dependency Corpus, [8]). Thus, syntactically parsed corpora are usually called *treebanks*.
4. Semantic tagging should be an obvious next step: it would substantially help in solving of what is now called *word sense disambiguation* [3]. Basically, it may consist in associating the word forms in corpus with a selected set of semantic labels (features). In this way we can later characterize the sense(s) the word form displays in the corpus text.

2 What Is Needed?

Some of the grammatical tags obtained in the course of morphological analysis can be immediately used as semantic tags. Particularly, this is true for some

---

* This research has been partially supported by the Czech Ministry of Education under the grant VS97028.

POS tags: e.g. tags like k6xM, k6xL and k6xT denoting adverbs of manner, place and time respectively, in Czech tagset used in grammatically annotated corpus DESAM [6]. Also tags signifying the category of animateness and inanimateness, associated in Czech with masculine and femine nouns and person and nonperson with pronouns, can be treated as semantic ones. The full list of these tags can be found below.

3   Types of Semantic Tags (Labels)

3.1   Grammatico-Semantic Tags (Labels)

In this respect we can take advantage of the existing Czech tagset that has been developed for Czech lemmatizer and morphological analyzer [5,9] and used in grammatically annotated corpus DESAM [6]. We should like to point out that our tagset reflects the highly inflected nature of Czech in two important respects:

a) the number of the tags is much higher than in the similar tagsets for English (or similarly for French and German): if for English, typically, that number does not exceed 120 tags, then for Czech we have to work with the tagset containing about 1100 tags (in DESAM) and this is confirmed also by Hajič [2]. The number of all the potential tags is even higher and can reach 2000.

b) since the Czech tagset is rich and detailed the following tags (primarily introduced to capture all the inflectionally expressed grammatical categories) can be used for semantic tagging as well:

nouns:
- gM: animateness (masculines),
- gI: inanimateness (masculines),
- gF: the distinction of animateness and inanimateness can be automatically derived for the feminines from the inflectional patterns allowing to form possessives),

pronouns:
- gP: person (this tag is typically associated with some pronouns), e.g. *kdo (who)*, *někdo (someone)*
- gT: nonperson (the same as above), e.g. *něco (something)*, *nic (nothing)*

adverbs:
- k6xM: adverbs of manner, e.g. *rychle (quickly)*, *silně (strongly)*
- k6xL: adverbs of place, e.g. *tady (here)*, *doma (at home)*
- k6xT: adverbs of time, e.g. *dnes (today)*, *ráno (in the morning)*, *měsíčně (monthly)*

verbs:
- mI: imperfective aspect, impferfective verbs denote an action limited in time, e.g. *skočit (to jump)*
- mP: perfective aspect, perfective verbs denote an action that is understood as not finished, still going on, eg. *skákat (keep jumping)*.
- tP: present tense, gives the information about the moment of speech and reference point
- tF: future tense, as above but with the different arrangement

– tM: past tense

mC: conditional, yields an information about the reality or non-reality of the verb action

– nS: singular, a grammatically expressed quantifier signifying individuals, individual objects and entities

– nP: plural, a grammatically expressed quantifier denoting sets, groups or collections of individuals, objects, entities.

Altogether there are 16 tags belonging presently to this group.

3.2   Pragmatic Tags

If we have a closer look at the grammatical tags it can be seen that some of them cannot be classified as semantic tags since they express grammatical categories that do not refer to the meaning of the word forms they are associated with but to their pragmatic aspects. As such they play the decisive role in the discourse and dialogue analysis – thus it is definitely useful to have them accessible in a transparent and fast way. In the Czech tagset used in DESAM corpus the following five tags of this sort can be found and processed systematically:

verbs:

– p1, p2, p3: first, second and third person, determining who is a speaker or hearer in the discourse or dialogue, or whether a person, thing or anything else is being referred to,

– mI, mR: the indicative and imperative, indicating the attitude of the speaker to the propositinal content of the sentence.

3.3   Conversion Rules for Grammatico-Semantic and Pragmatic Tags

As we have indicated in the preceding section some of the tags produced on the morphological level can be interpreted as semantic or pragmatic ones and relabeled as such. Since they reflect some grammatical categories in Czech and form a part of the grammatical tag, it is not too difficult to do that automatically. The conversion rules allowing for that may take e.g. the following shape:

gTag → $sTag$,

gTag → $pTag$.

They can be simply understood as a sort of reduction rules converting the respective grammatical tag to the semantic or pragmatic one.

In fact, the experimental version of the above mentioned corpus DESAM running under CQP corpus manager (installed in NLP Laboratory at FI MU) already contains the indicated semantic and pragmatic tags obtained in this way and it is possible to exploit them for several separate research purposes, especially with regard to word formation explorations, lexicographical needs and also to discourse analysis.

3.4   Lexical (Sense) Tags

If we are looking for appropriate semantic tags one possible way would probably be to turn to an appropriate machine readable dictionary and attempt to take advantage of it trying to exploit the information about lexical units there.

Unfortunately, since there is no appropriate Czech machine readable dictionary at hand, another source has to be looked for. The only solution that appears to be interesting and feasible is based on the fact that we have the Czech Wordnet at our disposal, which is being built within EuroWordNet-2 project [10]).

This leads us to a proposal of the hierarchical tagset containing semantic features in the WordNet fashion and based on EuroWordNet-1 Top Ontology and the set of Base Concepts [10]. It can be seen that they typically occurr as higher nodes in the hypero/hyponymical trees associated with the particular senses as they can be found within the individual synsets. Our suggestion is to select the following features as the appropriate candidates at the moment:
HUMAN BEING, PERSON, ANIMAL, PLANT (VEGETABLE, TREE, ...), ARTEFACT, PHYSICAL OBJECT, PLACE (noun expressions like *place, side, top, bottom, ...*), TIME (noun expressions for time intervals – *second, minute, hour, day, month, year, ...*), BUILDING, PSYCHOLOGICAL FEATURE, INSTRUMENT, IMPLEMENT, QUALITY, PROPERTY, SOCIAL GROUP, SOCIAL EVENT, COGNITIVE PROCESS, CAUSE, INDIVIDUAL, ACT, ACTIVITY, VEHICLE, INFORMATION, COMMUNICATION, KNOWLEDGE.

Systematic tagging of the word forms in corpus text with these semantic features can obviously start with the corpus word forms which are associated with one sense in Wordnet lexical database. This part of the tagging using the mentioned semantic features can be done automatically. To give an example of the one-sense words that can be tagged immediately: *střechýl (icicle)* or *střídavý proud (alternating current, AC), stroncium, substantivum (noun)* in a grammar.

The next step: tagging of the ambiguous word forms found in corpus texts is the most difficult part of the present task. Typically, the ambiguous word forms will be associated with more than one sense which means that successful tagging amounts to the word sense disambiguation of a given corpus word form.

There are many attempts [3,4] to answer this question positively though we are aware of the fact that no straightforward automatic procedure that would be able to search corpus texts and annotate the ambiguous word forms quite reliably can be easily built.

However, we hope that some small steps that will eventually lead to this goal can be envisaged and we would like to pay attention to some of them.

3.5   What Can and Should be Done?

A brief analysis of the corpus texts, e.g. concordancies containing ambiguous words as *strana:1 (side), strana:2 (page), strana:3 (political party), strana:4 (aspect of sth.), strana:5 (other party, opponent)* shows that it is reasonable to consider the local context within sentence boundaries. An appropriate way how to deal with the local context would be to perform a partial syntactic analysis of NG's or VG's in which the examined ambiguous word occurs. Thus, we may start with:

a) partial parsing yielding the relational information about a noun or prepositional group containing the examined word as a head,

b) partial parsing providing the structure of a verb group containing the ambiguous verb and the respective relational information using valency frames enriched with the lexical information that can be obtained from the corpus text.

It is our strong feeling that the valency frames filled with the lexical information about the particular verb participants provide one of the keys to the solution of semantic tagging and possibly word sense disambiguation as well.

Normally, a typical verb frame includes the verb and its arguments usually described by means of semantic cases which indicate the semantic relations between the verb arguments or, in other words, main sentence constituents.

The same or similar can be said about the valency frames [7] which will be preferred here since we deal with an inflectional language – Czech: in comparison with the Fillmore's verb frames [1] they are suited better to provide the information not only about the semantic relations between the verb arguments (participants with the labels like agent, patient, etc.) but their usefulness lies especially in their ability to capture the inflectional (surface) cases in Czech: Nominative, Genitive, Dative, Accusative, Locative and Instrumental (and Vocative as well).

3.6   L-Valency Frames

The most important improvement can be seen in the fact that in our view the valency frames should be enriched in such a way that apart from the semantic roles of the participants given symbolically they will include also the typical lexical items reflecting the most frequent lexical collocates occurring with the given verb. Thus in our opinion the valency frames have to be devised as richer structures than the case frames in their standard shape. The enriched valency frames will be called l-valency (lexical) frames and they may take the following form:

```
otevřít (open)
=1 kdo/ag/osoba co/obj/(láhev, pivo, krabici)
=2 kdo/ag/osoba co/část-těla/(oči, ústa)
=3 kdo/ag/osoba komu/adr/člověk co/obj/(dveře, )
=4 kdo:co/(osoba, instituce) co/obj/(školu, obchod)
```

It can be seen that the l-valency frames are getting closer to the data structures that typically occurr in the dictionaries. However, it is our experience that almost no lexical resource offers the valency (or case) frames capturing systematicaly the relations between the various combinations of verb arguments and their senses. Even in the WordNet which is regarded to be a well structured lexical resource the description of these relations is definitely neglected. Therefore it is our aim to build the list of l-valencies in such a way that they will systematically capture the relations of verb arguments and its senses.

At the present moment for Czech the situation is quite favourable – we are starting with a reasonably large list of Czech basic verb valency frames containing approximately 15 000 items [7] (the estimated number of verbs in Czech is about 36 000 items). We plan to prepare the tentative list of l-valencies in the close future and include it in Czech WordNet lexical database.

4 Conclusions

We have explored some aspects of the semantic tagging of (Czech) corpus texts and proposed to take advantage of the rich grammatical tagset used in corpus DESAM. Some components of the grammatical tags in DESAM can be automatically relabelled as semantic or even pragmatic tags – there is a version of DESAM corpus containing word forms tagged in this way installed at NLP Lab. at FI MU. In the second part of the paper the lexical valency frames are introduced and a way in which their list can be built is outlined. In our view the l-valency frames seem to represent a tool that can facilitate word sense disambiguation in an reasonable way.

References

1. Charles J. Fillmore. The Case for Case. *Universals in Linguistic Theory*, New York, 1968, pp 1–88.
2. Jan Hajič, Barbara Hladká. Probabilistic and Rule-Based tagging of an Inflective Language – a Comparison. Technical Report No.1, UFAL MFF UK, November 1996, Prague.
3. Nancy Ide, Jean Véronis. Word Sense Disambiguation: The State of Art. Computational Linguistics, Vol.24, No.1, March 1998, pp 1–40.
4. Claudia Leacock, Martin Chodorow, George A.Miller. Using Corpus Statistics and WordNet Relations for Sense Identification. Computational Linguistics, Vol.24, No.1, March 1998, pp 147–167.
5. Klára Osolsobě. Algorithmic Description of Czech Morphology. Dissertation, Brno 1995.
6. Karel Pala, Pavel Rychlý, and Pavel Smrž. DESAM – Annotated Corpus for Czech. In *Proceedings of SOFSEM'97*. Springer-Verlag, 1997.
7. Karel Pala, Pavel Ševeček. Valencies of Czech Verbs. Studia Minora Facultatis Philosophicae Universitatis Brunensis, A45, 1997.
8. Prague Dependency Treebank. Technical Report, UFAL MFF UK, Prague 1998.
9. Pavel Ševeček. *LEMMA — a lemmatizer for Czech*. Brno, 1996. (manuscript, programme in C).
10. Piek Vossen (ed.). *EuroWordNet General Document – Version 2*. Amsterdam, June 1999. (Draft of the resulting CD).

The General Principles of the Diachronic Part of the Czech National Corpus*

Karel Kučera

Institute of the Czech National Corpus Faculty of Philosophy and Arts,
Charles University, Prague
Karel.Kucera@ff.cuni.cz

Abstract. The diachronic part of the Czech National Corpus (CNC) has been organized as a general basis for the study of the entire history of Czech (from the 2nd half of the 13th century to 1990). It has been built around four principles, namely representativeness, authenticity, transcription, and preservation of maximum amount of information contained in the text. The diachronic part of the CNC includes the corpus, a bank of transcribed texts, a bank of transliterated texts, a text archive, a language database, a dictionary database, and a control database storing information about the texts. The diachronic part of CNC now includes about 1.5 million tokens.

The Czech National Corpus, now encompassing more than 100 million running words, has been built since 1994 at the Institute of the Czech National Corpus at Charles University in Prague, Czech Republic. The goal of the project is to create and continually update a representative textual basis of several hundred million running words which would meet both the scientific and general cultural needs of its prospective users (cf. [2,3]). The core of the CNC is, of course, its synchronic part co nsisting of contemporary texts (journalistic and technical texts since 1990, prose and poetry since 1960); a sample of this corpus, about 20 million running words, is accessible at URL http://ucnk.ff.cuni.cz. Since its beginning, however, the CNC project has had a complementary diachronic part, the goal of which has been to build a representative basis for the study of the entire history of the Czech language, from its first preserved textual records (2nd half of the 13th century) up to the time covered by the synchronic part of the CNC. Understandably, the starting points, targets and partly even the methods of building the diachronic part of the CNC differ, to a certain extent, from those of the synchronic par t: there is the linguistic variety given by the time span of seven centuries, the marked changes of the writing system, in stylistic and functional characteristics of Czech texts etc. And, of course, there are also purely practical specifics of handling and representing old texts electronically resulting from the differences in traditional approach of the linguist to the texts of different age (in the old texts the linguist generally expects a much more critical analysis of the written or printed record, more information about their formal

* This research was supported by the GACR, Grant Nr. 405/96/K214.

characteristics, and he focuses on a number of problems different from those he concentrates on when dealing with the contemporary language).

Considering these general facts, the diachronic part of the CNC (DCNC) has been built around four pivotal principles, namely representativeness, authenticity, transcription, and preservation of maximum amount of information contained in the original text or its edition. The essence of the principles is as follows:

1. The crucial concept of representativeness, largely avoided both in corpus and non-corpus linguistics, has been lately defined in general terms for the synchronic part of the CNC and is being gradually elaborated through a number of surveys (according to this definition, a representative corpus is one that "records the language broadly and faithfully as an open, ever-changing phenomenon in its various manifestations and variations"; [3]). Although this is hardly a disputable statement, the goal of a diachronic, or in fact any historical, corpus is somewhat different: in this case, the representativeness of the sum of the texts gathered in the corpus can be measured neither against the totality of the contemporaneous communication, nor against the living linguistic consciousness and feeling of its users; all it can be confronted with is the more or less limited body of preserved texts, which in most historical periods is seriously skewed stylewise, genrewise and dialectwise. The representativeness o f a diachronic or historical corpus can be measured against this body of texts only, and the goals of such corpora are consequently more limited and specific than those of the present-language corpora (historical corpora often concentrate on particular texts, authors, or aim at converting into electronic form all the preserved texts from a given period). The plans of the DCNC for the foreseeable future are to attain representativeness in the following steps:

- Within years, the DCNC should reach full representativeness for the period of the oldest Czech written records, i.e. it should include all authentic texts up to 1400; within decades, rather than years, it should also reach full representativeness for the 15th century.
- The period from the 16th through the 20th centuries will be covered by stratified random sampling of texts. The number of texts should grow in the course of time, but the representation of certain styles and genres will always be unrealistic in the older part of this period (16th through 18th centuries, reflecting the skewed stylistic, genre and dialectal proportions in the preserved body of texts. However, the closer to the 1960s the more similar will the proportions in DCNC be to those in the sync hronic corpus. To gradually link the two parts of the CNC in this way is indispensable since sooner or later parts of the synchronic corpus will be transferred to the DCNC as the synchronic corpus grows older.

2. The principle of authenticity reflects the conviction that reliable study of language development may be based on authentic texts only, i.e. on texts free from elements imported from other language states and other dialects. This means that neither the texts preserved only in copies that were written or printed decades or even centuries later than the originals nor the texts copied by a

speaker of a different dialect become part of the diachronic corpus. However, if such texts are offered to the DCNC in computer readable form, they are kept in a different part of DCNC, in the bank of transcribed texts which encompasses all historical texts possessed by CNC, even those that have not been, for different reasons, accepted to the corpus.

3. Texts written or printed before the last marked changes of the Czech orthography in 1849 enter the DCNC in transcribed form. This principle reflects the major practical problems that would arise if we tried to search a body of texts using all the historical varieties of the Czech orthographic system. There seem to be only two ways of carrying out such a search: one can either keep a database of all the various spellings of all the forms of all the lemmata in the corpus (which is, in fact, an extension of lemmatization), or to use a searching program which would generate – and consequently search for – all the possible spellings of the given word, form, morpheme or whatever one is looking for. Both of these approaches are prohibitively ineffective in Czech, although they have been used successfully in historical corpora of some other languages (cf. [1]). The radical changes the Czech writing system has undergone during the seven centuries, as well as the rich Czech morphology and its many historical changes would – so it seems now, after a few cursory tests – cause a number of problems and considerably slow down the process of building the DCNC. However, to meet the linguists' possible specific needs, DCNC keeps a small specialized bank of parallel text samples both in paleographical transliteration and transcription; the texts in the bank now add up to about 100,000 running words.

4. The transformation of old texts into electronic form has been designed to preserve maximum amount of linguistically relevant information contained in the text or its edition. Although the electronic texts in the DCNC are not intended to replace the originals or their critical editions, they include information that is not a standard part of most contemporary language corpora, like the author's name, the title of the work, the time of its origin, pagination or foliation and other structuring or localizing codes indicating special function and location of some parts of the texts, like marginalia or footnotes. There are also codes distinguishing the parts of the text that are written in Czech from the parts written in other languages (mostly Latin and German), and codes marking editorial notes (mostly short notes indicating that the word was misspelled, partly illegible, that there was a typographical error etc.).

The effort to preserve maximum linguistically relevant information about the texts and their language is also reflected in the structure of DCNC. There is, of course, the core, the corpus, which is a realization of a certain conception of representativeness, but there are other sections containing more information and more texts. There is the above-mentioned small bank of texts in paleographical transliteration, and the bank of transcribed texts, also mentioned above, which includes both the corpus texts an d texts that are not part of the corpus (these are, above all, inauthentic texts and texts belonging to the style, genre, dialect etc. which has been already amply represented in the corpus). Thus, the corpus is in fact only one possible selection from the bank, and in the future other selections should be available: the bank should serve as a basis for virtual corpora tailored to the specific needs of various users of DCNC, including possible different concepts of authenticity. Among other parts of the DC NC is the archive which includes the electronic texts in their original form, without codes and corrections, but sometimes preserving bits and pieces of information that do not enter the corpus and may be interesting from a special linguistic, cultural or other point of view. Part of the design of DCNC is also the language database, intended to keep translations or explanations of uncommon words and phrases, and the dictionary database which will, in time, encompass the material of Old and Middle Czech dict ionaries. The last, but important part of the DCNC is a control database storing information about each text in the DCNC (details about the manuscript, print or edition which served as the basis for the electronic text, stylistic and genre characteristics of the text, file name under which it is stored etc.).

The building of the DCNC is a slow process, since most texts are keyed in manually or scanned. As the corpus extends over its present 1.5 million running words, it may well prove advantageous to add more specialized parts to the DCNC, such as a historical terminological database. The project of DCNC is of course open to such meaningful additions.

References

1. Beltrami, P. G.: Norme per la redazione del Tesoro della Lingua Italiana delle Origini. In: Bollettino dell'Opera del Vocabolario Italiano, 1998, pp. 277–330.
2. Čermák, F.: Jazykový korpus: Prostředek a zdroj poznání. Slovo a slovesnost, 56, 1995, pp. 119–140.
3. Čermák, F., Králík, J., Kučera, K.: Recepce současné češtiny a reprezentativnost korpusu. Slovo a slovesnost, 58, 1997, pp. 117–124.

Performing Adaptive Morphological Analysis Using Internet Resources*

Marek Trabalka and Mária Bieliková

Department of Computer Science and Engineering
Slovak University of Technology
Ilkovičova 3, 812 19 Bratislava, Slovakia
trabalka@decef.elf.stuba.sk, bielik@elf.stuba.sk

Abstract. In this paper, we describe an approach to an adaptive morphological analysis based on lexicon corpus acquired from Internet. We focus on automating categorization words into a morphological paradigm in flexive languages. It is done by inducing possible word forms using morphological knowledge base and by looking for word forms of possible inflections in a morphological lexicon.

We developed a prototype system based on the proposed approach. Our system is general (it respects language but it performs better on a flexive language). We tested the system for the Slovak language. System's lexicon is built by means of browsing Internet pages. Parsed texts, recognized to be written in Slovak, are used to establish database of Slovak words with their frequencies in texts.

1 Introduction

Most of todays research and practice in natural language processing is concerned with the English language. The concentration on English has resulted in advances in solving of its linguistic problems [5]. Morphological analysis (the process of describing a word in terms of the prefixes, suffixes, and root forms that comprise it) is such a phenomenon that in English has a little importance due to morphological simplicity of the English language. Limited amount of morphological information is often used with part-of-speech disambiguation techniques (taggers) [2,6,1]. By associating each word with a unique tag, tagging helps in disambiguating words with multiple parts of speech, hence tagging can provide information on ways in which each word is used.

Many other languages use more of morphology than English (e.g., German language; Slavic languages such as the Slovak, Czech, or Russian languages). Words in these languages have usually several different morphological forms that are created by changing a suffix. Each form represents some of morphological attributes like case, number or gender. Such languages with a rich morphology are called flexive.

---

* The work reported here was partially supported by Slovak Science Grant Agency, grant No. G1/4289/97.

F. Jelinek, E. Nöth (Eds.): TSD'99, LNCS 1692, pp. 66–71, 1999.

In flexive languages morphology strongly influences syntax. In English sentence, meaning is determined by a word order while e.g., in Slovak it depends primarily on a word form and order of words is not so strict. On the other hand syntactical analysis of the sentence in flexive language is after successful morphological analysis simplier task than in English.

Bellow we illustrate richness of flexive languages morphology. In an example sentence below you can see two different forms of Slovak word *ruka* (hand):

In Slovak: *Ruka* drží *ruku.*
In English: *Hand* holds *hand.*

A word form (*ruka* versus *ruku*) determines its role in the sentence. In this case, *ruka* is the subject and *ruku* is the object. Following example illustrates a variability of the order of words:

Alternative 1 in Slovak: *Ruka* drží *nohu.*
Alternative 2 in Slovak: *Nohu* drží *ruka.*
In English: *Hand* holds *foot.*

where both alternatives in Slovak have the same meaning. Again we can see that determination of the word form is necessary for sentence parsing and understanding.

2   Method of Adaptive Morphological Analysis

Usually, dealing with morpohology is to define a set of paradigms expressing type of inflection and to create a morphological lexicon containing as many words as possible with reference to their appropriate paradigms [4].

Creating a morphological lexicon is a big investment in time and money. Moreover, it is an endless task. If we omit a problem with manual setting word-paradigm pairs there still exists problem of special terms and new words that should be added. General purpose applications (e.g., text editors, using the lexicon for spell-checking) usually only allow user to add a word on his own.

To support an analysis of unknown words (morphology learning) we propose a method of adaptive morphological analysis. The idea is as follows:

We give to computer multitude of word forms and paradigm definitions and the program should determine an appropriate paradigm for each word.

The method uses paradigms and heuristics to compute an acceptance probability of a particular word. The acceptance probability expresses a degree of belief in the correct determination of paradigm for the word.

Input to the method is a word, which is subject to the morphological analysis. Output is determination of a form for this word (a morphological tag) with an acceptance probability and acceptance probabilities for derived morphological tag alternatives of the analysed word. These alternatives and their acceptance probabilities are computed by means of linguistic knowledge. Linguistic knowledge is either filled in advance by an expert, or learned by the system during the previous analysis.

The method consists of the following steps:

1. Analysis of possible forms that the analysed word can match. Set of form/ paradigm pairs is composed.
2. Performing inflection for each alternative. Set of groups, one group of words for each form/paradigm pair is composed. Each group contains *all* word forms that can be inflected using the given paradigm.
3. Looking for all word forms created in step 2 in a lexicon. Each word form found in the lexicon is marked-up.
4. Computing the acceptance probability for each form/paradigm alternative. It strongly depends on the number of forms found in the lexicon and on heuristics applicable for a particular word form. All these partial acceptance probabilities of word forms of each group are composed to resulting acceptance probability of form/paradigm pair.

In order to perform an analysis, four different knowledge types are used:

- *Attributes* represent morphological attributes such as gender (musculine, feminine or neuter) or number (singular or plural).
- *Forms* express morphological forms. They are defined by specifying values for some subset of attributes. Comparing with English tag-sets [6] the attribute is similar to a simple tag and a form represents typical group of tags used together.
- *Paradigms* are sets of forms and relations between them. Each relation describes how to create one word form from another (e.g., by adding a specified suffix).
- *Heuristics* are sets of quantitative information that are used to express an acceptance probability of forms and paradigms.

The method of adaptive morphological analysis usually creates several morphologically non-correct forms deduced from selected paradigms alternatives. Computed acceptance probabilities distinguish these non-correct forms from the correct ones. For example in the process of analysing the word *otcom* (form of the word *otec*, which means father) 19 different forms of the analysed word are inflected: *otec, otcovia, otcov, otca, otcom, otcovi, otcoch, otcami, otc, otcu, otcmi, otcomy, otcomov, otcoma, otcomom, otcomu, otcomoch, otcome, otcommi.*

Only first 8 of them are morphologically correct Slovak words (the analysed word *otcom* is among them).

3   Acquiring Vocabulary from Internet

If computation of acceptance probabilities is based solely on knowledge base (attributes, forms, paradigms and heuristics) distance from acceptance probabilities of correct forms to acceptance probabilities of incorrect forms is sometimes very close. This distance can be increased by using a lexicon. The more (correct) forms are found in the lexicon, the bigger is this distance.

Vast repositories of useful textual information in machine-readable form are now commonly available, especially with advantage of Internet and its World Wide Web service (WWW). We have on mind that the Internet contains generally not the quality but quantity of texts. Nevertheless we used it as our primary source for creating a language corpus due to its availability.

Typical problem of electronic documents in the Slovak language (and many other languages including almost all Slavic languages) on the Internet is diacritic. Texts contain letters that standard Latin-1 charset does not contain. We should take into account this fact when design the process of text selection for analysis. Our approach is based on statistical research of the Slovak language [3]. The proportion of letters with diacritical signs is an important criterion for considering text suitable for learning (i.e., it is Slovak text).

Building of a lexicon is automated by a multiagent system that browses WWW pages and extracts words from the texts. System consists of a global synchronization server and one or more local agents. The global server functions as a centralized storage the local agents are synchronized with. The system also offers an interface for maintaining the global lexicon.

Local agents can run on a variable number of machines. They work separately and synchronize their contents with a global server. Each agent maintains its own set of hyperlinks and a private lexicon.

The process of HTML page analysis is as follows. Primary objective is to decide whether the page is written in the Slovak language. At first the HTML page is downloaded and character set is determined, it must be dedicated to Eastern Europe. Contents is re-coded into Win-1250 charset and amount of letters with diacritical signs is computed. A common Slovak text contains specific subset of charset and some minimal level of diacritics. Hence texts written without diacritics are refused. Note that nowadays a lot of Slovak texts are published on Internet without diacritic. Decision about considering text for further processing is based also on proportion of specific Slovak and non-Slovak words. Maintainer of a global lexicon adds information of whether word is specific Slovak or non-Slovak manually.

The selected text is used both to extract URL references and to create list of words that the page contains. All hyperlinks found in HTML document are added to the database of links with computed probability that it contains Slovak text.

Synchronization with a global server is done in both directions. Agent sends words with their occurrences to the server and the server sends back its complete marked-up lexicon. Agent remembers last synchronization of each word so it sends only occurrences added after the last synchronization.

Final result of such a system is a collection of words forms together with number of their occurrences in texts.

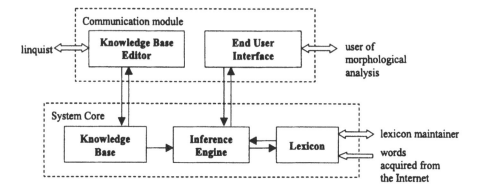

Fig. 1. Architecture of the system for morphological analysis.

4   Experiments

We developed a system that is based on the proposed method of adaptive morphological analysis. The overall structure of the system is depicted on Figure 1. In order to support creation and modification of morphological knowledge we developed also an editor of the morphological knowledge base. It is used especially by language experts. The knowledge base editor allows to insert or modify paradigm and form definitions together with heuristics used by the inference engine to perform morphological analysis.

The system core implements our method of adaptive morphological analysis. It is implemented as a series of programs in interpreted PERL. The system offers functions for analysis and synthesis of morphological forms. The result of analysis is:

- possible forms that match the input word and also
- acceptance probability of each case.

Co-operation of the two developed systems (a system for morphological analysis and system for automation of building a lexicon) is important. Our adaptive morphological analysis obviously needs the lexicon of morphologically correct words. But also the system for acquiring the lexicon can use morphology to improve its decision process by inflection of non-marked words and searching for their other forms.

We developed a useful tool that uses both of these systems. We made a special World Wide Web site containing searching engine interface where user can search for all or selected word forms. The user simply puts a word and selects the searching engine. Our system recognizes word paradigm, creates all word forms and sends them to desired searching engine, e.g. Altavista.

Although system is still under development it already offers some usable results. System for automation of building a lexicon started with the initial lexicon containing about 800 words only. After approximately 100 hours of running, the agents processed more than 25.000 links. Only 510 of them successfully passed tests and were reporting to be regular Slovak texts with diacritics. More than 200.000 of words were found on these pages that finally represent 50.000 unique word forms. We manually checked most frequent words (around 20.000). The analysis showed that none of the words that occurred in more than 5 different pages were incorrect.

After testing output quality of acquired lexicon we began to use it for testing proposed adaptive morphology concept. We selected manually from acquired lexicon words marked as Slovak and used it for testing morphology based on a set of paradigms for Slovak language [4]. We tested for example influence of multiple occurrences word forms on the final result of morphological analysis. The lexicon was modified by removing some word forms to recognize changes it was causing.

By increasing the number of forms in the lexicon we observed an increase of the acceptance probability of the input word form. Correct forms had usually doubled acceptance probability. But if the only few word forms are in the lexicon, result success strongly depends on which word form is in the lexicon.

5   Conclusion

In this paper, we describe a method of adaptive morphological analysis based on usage of real texts as a source of word forms. We developed a software system that implements proposed method and scans Internet for texts used in process of lexicon construction. Searching interface is currently under development.

The advantage of the developed system is that it also works well for unknown words (not included in the lexicon yet). The system recognizes possible paradigms and creates a list containing all possible forms.

The developed system can be primarily used to enhance full-text searching regardless of form of searched word. It also sets up a basis for a syntactic analysis, which we are currently working on.

References

1. Brill, E.: A simple rule-based part of speech tagger. In Proceedings of the Third Annual Conference on Applied Natural Language Processing. 1992.
2. Chandrasekar, R., Srinivas, B.: Using syntactic information in document filtering: A comparative study of part-of-speech tagging and supertagging. In Proceedings of RIAO'97, Montreal, pp. 531–545. 1997.
3. Mistrík, J.: Frequency of forms and constructs in Slovak. Bratislava, Veda. 1985.
4. Páleš, E.: Sapfo: Natural language paraphraser for Slovak language. Bratislava. 1993.
5. Shinghal, R.: Formal concepts in AI: Fundamentals. Chapter 6. Natural Language Processing: A prescriptive grammar. Chapman & Hall, London. 1992.
6. van Guilder, L.: Automated part of speech tagging: a brief overview. Georgetown University. 1995.

Automatic Text-to-Speech Alignment: Aspects of Robustification

R. Schmidt and R. Neumann

Institut für deutsche Sprache, D-68161 Mannheim, Germany

Abstract. This paper outlines the architecture of a text-to-speech aligner based on Hidden Markov Models (HMMs) and describes the problems arising from of applications. For each problem presented in this paper we demonstrate how it can be solved. It was the purpose to develop this text-to-speech aligner to generate automatically links between text and audio files.

1 Introduction

For many years linguists of our institute have gathered a great amount of audio and video recordings to study a number of aspects of spoken German and German dialects. Many of these recordings are transcribed mainly using orthographic transcription. Meanwhile there exists the request to be able the hear the results of a query to a discourse database (see [4]). This can only be done if there are links between text and speech signal. However it is a hard and time consuming procedure to generate and incorporate these links manually. Therefore it is quite natural to claim a tool that could do all this without any efforts.

It is the task of a so-called aligner – a system that automatically time aligns speech signals and the corresponding text – to deliver these links. The task of such an aligner is closely related to the automatic speech recognition (ASR) task. In speech recognition we are interested what words have been spoken and do not care about the start and end points of each segment (generally a word or a symbol for an utterance like pause, music etc.). For alignment however, the start and end points of each segment are of interest while the string of spoken words is supposed to be known. Therefore, it seems quite natural to modify an existing speech recognizer to do the job. Of course, the recognizer must be speaker independent and must handle continuous speech. As time is the only degree of freedom the aligner has to handle, it is not necessary to install a language model like n-gram statistics. Thus a large portion of the time searching the solution can be reduced.

2 Architecture

Our system was developed using version 2.0 of Entropic's Hidden Markov Toolkit and a grapheme-to-phoneme converter. The input consisting of orthographic text

F. Jelinek, E. Nöth (Eds.): TSD'99, LNCS 1692, pp. 72–76, 1999.

and sampled speech data is preprocessed and then given to the HTK Viterbi decoder which does the actual alignment. The grapheme to phoneme converter partly allows to handle morphological pecularities of German which influence the pronunciation and it can easily be adapted to any other natural language. It consists of 740 context-sensitive rules for pronunciation and about 500 rules formulated according to a right linear regular grammar to describe morphology. Further information will be given within the next section.

HTK depends on a regular grammar for the Viterbi decoder to prune the search space of possible word sequences. For alignment, the regular grammar is simply the linear concatenation of phonemes resp. words. The only variability is given by pauses between words. The grammar is generated in such a way that it represents this phenomena adequately. A language model does not have to be considered.

Furthermore, Viterbi decoding needs vectors that describe the sampled speech data in terms of features. Our system uses spectral information represented by 12 mel frequency ceptral coefficients plus overall energie and their first and second derivatives, giving a total of 39 parameters per vector. The vectorframes have a distance of 10 ms and are calculated using a Hamming weighted window of 25.6 ms length.

The HMMs were trained by the ERBA (Erlanger Bahnauskunft) material consisting of 40 speakers both male and female and 100 sentences per speaker. The training material consists of six hours length with a sample rate of 16 kHz and a resolution of 16 bit per sample. The phonemes are modeled by context independent left-to-right HMMs with 3 emitting states, single mixture gaussian output probability density functions and a diagonal covariance matrix. There aren't any skip transitions per model and the model for speech pauses has the same topology.

3 Aspects of Robustification: Problems and Solutions

The discourses we have to align are mainly interviews, conversations and talk-shows. That means, our data is realistic with all the deficiencies occuring in daily life. The problems arising from these tasks are

- dialect, especially when not clearly transcribed
- unclear pronunciation, especially within fast utterances
- noise, music, clapping, laughing, coughing etc.
- acoustic artefacts, e.g. voice in the background
- simultaneous speech

3.1 Dialect

Because of the great variety of dialect forms, it does not make sense to make them part of a pronunciation dictionary. There is no convention how to transcribe them "correctly" and there exists the request to transcribe special kinds of utterances without using the IPA symbols, i.e. a pseudo transcription that does not obey to any orthography. Thus a rule-based phonetization tool has to be part of the aligner to be flexibel enough to handle these requirements. The tool we developed transcribes words according to context-sensitive rules of pronunciation. Most of these rules are formulated according to the standard pronunciation of German and work with both morphological segmented and not segmented words. In general dialect forms are not morphological segmented, but until now the phonetization tool has proved to work good enough to enable the aligner to run on audio files of a length of 20 minutes which are completely uttered and transcribed in dialect. The results of these runs can be integrated into the database without any further correction.

3.2   Colloquial Language

Obviously, utterances made in a quite natural atmosphere present a lot of variants. For instance,

- strongly pronounced vowels are pronounced weakly
- weakly pronounced vowels aren't pronounced at all
- *g* at the end of a segment may be pronounced as [k] or [C]
- *er* at the end of a segment as [6] , [er] or simply [r]
- *o* in words of English origin as [oU], [o] or [o:]
- *afe* in words of English origin as [eIf] or [e:f]
- *are* in words of English origin as [E:6] or [E]
- *anc* in words of French origin as [a s] or [aNs] etc.

where the SAMPA notation has been used to describe the phonemes.

We can handle this aspect by enabling the phonetization tool to produce multiple pronunciation variants for each rule. This makes it possible for the aligner to detect differently pronounced utterances with the same accurancy. We have not measured any significant longer run-time according to the increase of the pronunciation dictionary, thus we suppose that the use of the more precise variants must have a favorable influence in searching to compensate the volume of the dictionary.

3.3   German Peculiarities

Furthermore, German pronunciation strongly depends on morphology. This is because there are many ways to construct words and the rules of pronunciation require that each letter of the string to be transcribed belongs to the same morpheme. Therefore, the string *sch* in the words *Verkehrschaos* and *Unfallschaden* is completely differently pronounced. These words can be segmented as *Verkehr-s-chaos* and *Unfall-schaden*, thus only in the second word *sch* belongs to the same morpheme and is transcribed as [S] while in the first word it is transcribed as [sk]. We tackle the problem by preprocessing the words to be transcribed by a procedure that segments the words into morphemes and then apply the proper phonetization. For the segmentation procedure it proved to be convenient to use a right linear regular grammar to describe morphological rules. A profound description of aspects of rules for German morphology can be found in [2]. An extensive list of rules for German pronunciation was published in [1]. Similar subjects are treated in [3] and [5].

3.4   Nonverbal Signals

When aligning talkshows there often arises the problem of how to handle non-verbal signals. There are segments of clapping, laughing, coughing, or music that should be separated from speech. The only way to handle this problem seems to be to create special HMMs and train them. We only recently began to devote us to this work and thus at the moment don't have experience enough to talk about.

3.5   Simultaneous Speech

The more participants a discourse has the more likely are situations when more than one person speaks. We get the words to be aligned from our discourse database [4]. This database contains partly nonsequential parts of speech which can be displayed in a score style analog to the notes in music. A special converting tool extracts the data from there and eliminates the data of all speakers but one in periods of simultaneous speech. The strategy of the converting tool is to select that speaker who was already speaking. It may happen that the part of speech of the selected speaker is not dominating but nevertheless, it has pointed out to be practical and the results are acceptable in general.

Furthermore, we have marked parts of simultaneous speech as a special phoneme and trained its HMM. Until now the results are not so good as we expected them to be. We suppose the reason might be the great variety of phenomena in comparison to the amount of training material. Further work has to be done to verify this assumption or to find out the real reason.

3.6   Poor Quality

Sometimes the quality of the recordings is so poor that even filtering methods aren't successful. This can happen when the whole spectrum of noise is within the spectra of vowels. In most cases even unpruning the search path of the Viterbi decoder doesn't deliver the wanted results for audio files of the length of 20 minutes. For instance, it might happen on a Sun Ultra 1 with 288 MB memory that after a 30 hours run one gets the message that it couldn't find a solution. The failure can be avoided even with recordings whose quality are so poor that a native speaker has trouble to understand them. This can be achieved by splitting the audio file into a number of small pieces and running the alignment on each of them without (or with a very large) pruning factor.

4 Conclusion and Further Work

We have a text-to-speech aligner that is able to deliver satisfactory results for most applications within twice realtime on a SUN Ultra 1 (143 MHz single processor machine), i.e. a 20 minutes audio file can be aligned in about 40 minutes. For some of the problems mentioned above the solutions still must be improved. Especially, we consider an extensive training of non-verbal signals to be neccessary. There, the automatic fade-out of vocal music seems to be hard to tackle.

References

1. Heinecke, J.; Wothke K.: Letter-to-Phone-Rules for German Taking into Account the Morphological Structure of Words. IBM Germany, Heidelberg Scientific Center Technical Report TR 75.92.03, 1992.
2. Pachunke, T.; Mertineit, O.; Wothke, K.; Schmidt, R.: Linguistic Knowledge in a Morphological Segmentation Procedure for German. Computer Speech and Language, 8, pp. 233–245, 1992.
3. Schmidt, R.: Technical Aspects of a Phonetization System Considering Morphological Peculiarities. IBM Germany, Heidelberg Scientific Center Technical Report TR 75.91.22, 1991.
4. Schmidt, R.; Neumann, R.: The Digital Discourse Processing System DIDA. Proc. of the 2nd SQEL Workshop on Multi-Lingual Information Retrieval Dialogs, pp. 145–148, 1997.
5. Wothke, K.; Schmidt, R.: A Morphological Segmentation Procedure for German. Sprache und Datenverarbeitung, 1, pp. 15–28, 1992.

Czech Translation of G. Orwell's '1984': Morphology and Syntactic Patterns in the Corpus

Vladimír Petkevič

Charles University in Prague, Faculty of Arts,
Celetná 13, 110 00 Prague 1, Czech Republic
Vladimir.Petkevic@ff.cuni.cz
http://utkl.ff.cuni.cz/~petkevic/welcome.html.cz

Abstract. The paper describes the morphologically annotated corpus formed by the Czech translation of George Orwell's novel *Nineteen-Eighty Four*. It also presents frequencies of some morphosyntactic features and focuses on syntactic structure of noun-ended prepositional phrases in the corpus with emphasis laid on grammatical concord in these structures. The study of these structures serves for the development of the formal grammar used for morphosyntactic rule-based tagging of Czech texts.

1 Introduction

One of the main results of the MULTEXT-EAST project (*Multilingual Tools and Resources for Central and East European Languages* – cf. [3] and [1]) which was a spin-off of the MULTEXT (cf. [6]) project was the development of the parallel corpus which consists of morphosyntactically tagged translations of G. Orwell's novel *Nineteen Eighty-Four* to all participating languages, i.e. Bulgarian, Czech, Estonian, Hungarian, Romanian and Slovene, the translations being pairwise sentence-aligned to the corpus of the original English text as the hub. In this paper, the Czech version of this parallel corpus is described, especially its morphosyntactic annotation and some (morpho)syntactic patterns whose study serves for the development of the grammar for the rule-based tagger[1] being currently developed for Czech. The paper consists of the following parts:

1. global description of the corpus and its annotation
2. basic statistical results
3. syntactic structure of prepositional phrases ended by a noun.

2 Global Description of the Corpus and Its Annotation

In the parallel *1984* corpus there are three levels of annotation (cf. [5] and [1]):

---

[1] The development of this tagger and related research is funded by the *Grant Agency of the Czech Republic*, grant No. 405/96/K214.

F. Jelinek, E. Nöth (Eds.): TSD'99, LNCS 1692, pp. 77–82, 1999.
© Springer-Verlag Berlin Heidelberg 1999

1. SGML markup up to the paragraph level including the markup for sentence boundaries (so-called *cesDoc* encoding based on the *TEI Guidelines*);
2. morphosyntactic annotation of individual words with the identification of sentences in which the words appear (so-called *cesAna* encoding);
3. pairwise linking of the corresponding sentences in the original version and in its translation, i.e. a sentence-level alignment.

A short example should show how the morphosyntactic annotation of the Czech version of the parallel corpus is performed:

```
<s from='Ocs.1.1.3.5'>
  ...
    <tok type=WORD>
     <orth>k</orth>
     <disamb><base>k</base><msd>Spsd</msd></disamb>
    </tok>
    <tok type=WORD>
     <orth>oknu</orth>
     <disamb><base>okno</base><msd>Ncnsd</msd></disamb>
     <lex><base>okno</base><msd>Ncnsd</msd></lex>
     <lex><base>okno</base><msd>Ncnsl</msd></lex>
    </tok>
  ...
```

The underlying text ... *k oknu* ... (E. ... *to the window* ...) is part of the sentence 'Ocs.1.1.3.5' (i.e. part 1, chapter 1, paragraph 3, sentence 5). Each word in the text is described by the token section *tok* containing the word form itself (*orth* element), by possible morphosyntactic interpretations in the *lex* elements formed by a *base* (lemma) and a morphosyntactic description (*msd*) – cf. [2],[2] and by the appropriate variant contained in the *disamb* element.

Each *msd* has the form of a string of values of morphosyntactic categories identified with the positions in the string (eg. Ncnsd encodes common (c) noun (N), neuter (n), singular (s), dative (d)).

The annotated version of the whole multilingual parallel corpus was created by the language-independent software tools which were developed within MULTEXT and MULTEXT-EAST and which used language-specific resources as data. The most important language resources were the *word form lexicon* and *morphosyntactic description* developed for each language involved (cf. [2]). By means of this lexicon in which each word form was assigned its lemma and possible morphosyntactic descriptions in the form of the strings exemplified above,

---

[2] The morphosyntactic description of all the languages involved was quite detailed. For Czech, not only basic morphological categories (such as *case, number, gender* for nouns, adjectives, pronouns and numerals; *person, tense, voice, mood, gender, number, negation* etc. for verbs; *degree of comparison* for adjectives and adverbs) were included but also more specific ones (fine-grained distinctions among pronouns such as *nominal* vs. *adjectival type* of pronouns, *clitic* vs. *nonclitic forms* of pronouns, *long* and *short forms* of adjectives etc.).

the lemmatization of the Czech part of the corpus (and other corpora) was performed. Then the corpus was *morphosyntactically disambiguated* by a stochastic tagger developed outside the MULTEXT-EAST project (cf. [4]). As the accuracy of this stochastic tagger was 93.07%, the result of the automatic tagging had to be manually checked and hand-disambiguated so that in its present state it is almost free of errors.

3 Basic Statistical Results

In this section, the frequencies of structural elements and some morphosyntactic features in the annotated corpus are presented. The corpus is hierarchically structured (similarly as the underlying text) as follows (the frequencies are given in parentheses): parts (4) > chapters (28) > paragraphs (1297) > sentences (6751). Basic frequencies follow.

Number of word forms:	79861
Number of punctuation marks:	20497
Number of morph. unambiguous word forms:	33729 (42 %)
Total number of distinct word forms:	17689
Total number of bases (lemmas):	9125
Total number of distinct tags used:	959

Table 1. Part-of-speech distribution of word forms

TOTAL:	79861	100 %			
Nouns:	19354	24.23 %	Prepositions:	7542	9.44 %
Adjectives:	7797	9.73 %	Conjunctions:	7107	8.89 %
Verbs:	16805	21.04 %	Interjections:	23	
Numerals:	1491	1.86 %	Particles:	313	
Pronouns:	11192	14.01 %	Abbreviations:	20	

Table 2. Most frequent occurrences of morphosyntactic tags

TOTAL	79861 (100.00 %) – 959 distinct tags			
Adverb	7664	(9.59 %)	Subord_conjunction	2327 (2.91 %)
Coord_conjunction	4419	(5.53 %)	Verb-infinitive	1851 (2.32 %)
Verb-past_partic.,masc.sg.	3365	(4.21 %)	Preposition (gen. case)	1742 (2.18 %)
Preposition (loc. case)	2753	(3.45 %)	Noun-masc.sg.nom.	1638 (2.05 %)
Refl. pronoun	2475	(3.10 %)	Noun-fem.sg.nom.	1586 (1.98 %)

In Table 1 we see that each fourth word in the corpus is a noun, each fifth word is a verb and each seventh word is a pronoun (the high frequency of pronouns is primarily due to the reflexive pronoun *se*, E. *-self*). Table 2 and Table 3 show the most frequent morphosyntactic tags and pairs of tags, respectively. The leading position of the adverb is given by the fact that the inflectional parts of speech are very fine-grained (eg. nominal descriptions break up into many distinct tags with lower frequencies). Therefore more objective information is provided by Table 4.

Table 3. Most frequent pairs of adjacent morphosyntactic tags

Adverb – Adverb	962
Coord_conjunction – Adverb	795
Adverb – Verb_past-participle_masc_sg	655
Prep_with_locative – Noun_fem_sg_locative	517
Adj_masc_sg_nominative – Noun_masc_sg_nominative	489
Adj_fem_sg_nominative – Noun_fem_sg_nominative	457

Table 4. Most frequent pairs of adjacent parts of speech

Adjective	Noun	5012	6.57 %
Pronoun	Verb	3850	5.04 %
Preposition	Noun	3765	4.93 %
Noun	Verb	2868	3.75 %
Verb	Pronoun	2639	3.45 %
Adverb	Verb	2397	3.14 %
Verb	Noun	2308	3.03 %

According to Table 4 *Adjective–Noun* and *Preposition–Noun* patterns are very frequent; the high frequency of the second pair in the table, *Pronoun–Verb*, is to be ascribed to pronominal clitics often taking up the second syntactic position in the sentence with the finite verb very often immediately following the clitic. In particular, the word-order relationship of adjacent nouns and verbs can be considered interesting from the viewpoint of typological properties of Czech with respect to the SVO, SOV, OVS ... classification.

4 Syntactic Patterns of Prepositional Phrases Ended by a Noun

The annotated corpus can be exploited for studying various syntactic structures, eg. prepositional phrases ended with nouns. Such structures begin with a preposition and end up with a noun; moreover, all case-specified elements in the phrase, namely *adjectives, pronouns* and *numerals* agree *in case* determined by the preposition (except for nouns modified by some cardinal numerals) and all adjuncts of the noun agree with the noun in *gender* and *number*. Thus, if in some non-disambiguated corpus the pattern *Prep Adjunct$_1$ ... Adjunct$_n$ Noun* is detected the morphological disambiguation consisting in the unification of the *case, gender* and *number* values of the elements involved discards a number of wrong interpretations. The adjuncts can also be modified by adverbs and/or they can be coordinated but being unspecified for agreement categories the adverbs, conjunctions and/or commas do not cause unification to fail. One of the adjuncts can, however, be modified by a nominal group or even another prepositional phrase (!) (eg. an adjective is modified by a temporal nominal group) but I do not study this case here. Let us now look how this reasoning reflects the prepositional structures in the real corpus data.

In the Czech *1984* corpus 7542 prepositions have been detected (see Table 1. Table 5 presents a survey of patterns of prepositional phrases ended by a noun with their frequencies. Each prepositional phrase in the survey starts with a preposition and ends with the closest noun (or nominal coordination) in the case required by the preposition. Such structures do not contain intervening verbs, other prepositions, punctuation marks ending a sentence and semicolons. In the patterns below, [ ] stands for a single element, [ ]* denotes an arbitrary number of elements and | is an 'or' symbol. The abbreviations are mostly self-explanatory (eg. *PronDem*, *PronTot* and *NumCard* denote demonstrative, total [type: Cz. *vše*, E. *all*] pronouns and cardinal numerals, repspectively). In boldface there are the basic types followed by the survey of more detailed patterns grouped according to the main types.

Table 5. Most frequent noun-ended prepositional phrase patterns in the corpus

TOTAL: Prep [ ]* N	**5933**	**100 %**
Prep N	3769	63.51 %
Prep Adj [ ]* N	1386	23.36 %
Prep Pron [ ]* N	498	8.39 %
Prep Num [ ]* N	246	4.14 %
Prep Adv [ ]* N	34	0.57 %
Prep Adj N	1236	
Prep Adj Adj N	88	
Prep Adj (COMMA — Conj) [ ]* N	15	
Prep PronDem N	180	
Prep PronPoss N	104	
Prep PronPoss Adj N	30	
Prep PronTot [ ]* N	84	
Prep PronIndef [ ]* N	65	
Prep NumCard [ ] N	218	
Prep NumCard N	195	
Prep NumOrd N	66	
Prep Adv Adj N	32	

The table shows only the most prominent patterns out of the total of 56 distinct patterns identified in the corpus. These patterns form a solid basis for the construction of a pilot formal grammar covering the structures in question.

Of course, such a grammar can hardly be considered appropriate for the noun-ended prepositional phrases in Czech in general because it reflects the structure of a text which is only a translation (from English) and which is relatively short. However, the grammar is planned to be later tested on the texts of the *Czech National Corpus* and appropriately modified.

Let us now show at least two examples of the patterns given above.

Pattern: *Prep Adj Adj N*

```
muže asi pětačtyřicetiletého, <s hustým černým knírem>, drsných ..
of a man of about forty-five, <with a heavy black moustache> and r
```

Pattern: *Prep PronPoss Adj N*

```
k němu přicházela <z jeho raného dětství>, ale s níž se člověk ...
like an emanation <from his early childhood>, but which one did ..
```

5   Conclusion

The paper described the Czech version of the parallel morphologically annotated corpus formed by the G. Orwell's novel *Nineteen-Eighty Four*. It also presented a pilot corpus-based study of noun-ended prepositional phrases in Czech without claiming that the results obtained from the small *1984* corpus formed by a translation can be generalized as to reflect the syntax of Czech prepositional phrases. The study is, however, an initial basis for further research.

References

1. Dimitrova, L., Ide, N., Petkevič, V., Erjavec, T., Kaalep, H.J., Tufis, D.: Multext-East: Parallel and Comparable Corpora and Lexicons for Six Central and Eastern European Languages. Proceedings from COLING-ACL'98. Montreal (1998) 355–359.
2. Erjavec, T., Monachini, M. (eds.): Specifications and Notation for Lexicon Encoding. Deliverable D1.1 F. Multext-East Project COP-106 (1997). http://nl.ijs.si/ME/CD/docs/mte-d11f
3. Erjavec, T., Ide, N., Petkevič, P., Véronis, J.: MULTEXT-EAST: Multilingual Text Tools and Corpora for Central and Eastern European Languages. Proceedings of the Trans European Language Resource Infrastructure First Conference. Tihany (1996) 87–98.
4. Hajič, J., Hladká, B.: Tagging Inflective Languages: Prediction of Morphological Categories for a Rich, Structured Tagset. Proceedings from COLING-ACL'98. Montreal (1998) 483-490.
5. Ide, N.: Corpus Encoding Standard: SGML Guidelines for Encoding Linguistic Corpora. First International Language Resources and Evaluation Conference. Granada, Spain (1998). See also: http://www.cs.vassar.edu/CES/.
6. Ide, N., Véronis, J.: MULTEXT (Multilingual Tools and Corpora). Proceedings of the 14th International Conference on Computational Linguistics. COLING'94. Kyoto, Japan (1994) 90-96.

Handling Word Order in a Multilingual System for Generation of Instructions

Ivana Kruijff-Korbayová and Geert-Jan M. Kruijff

Institute of Formal and Applied Linguistics, Charles University, Czech Republic,
{korbay,gj}@ufal.mff.cuni.cz,
WWW home page: http://ufal.mff.cuni.cz

Abstract. Slavic languages are characteristic by their relatively high degree of word order freedom. In the process of automatic generation from an underlying representation of the content, we have to ensure that a semantically and contextually appropriate word order is chosen. In this paper, we elucidate *information structure* as the main factor determining word order in Slavic languages, and we present an approach to handling word order in text generation in the context of the AGILE project [4].[1]

1 Introduction

In natural language communication, each participant can only process the elements one by one, in a linear order. However, the content that is being communicated is by nature multidimensional. Linguistic structures of all degrees of complexity therefore need to be projected into one dimension, and the individual elements ordered sequentially according to certain rules. The rules and schemes of linear ordering can be studied at every stratum of a given language, including the lowest ones, i.e. graphemes, phonemes, morphemes etc. The rules and schemes pertaining to the linear ordering of elements constituting a clause, i.e. phrases and groups, are usually referred to as *word order* (even though the elements are not only individual words). The linear ordering of clauses within a complex clause (sentence) can be referred to as *clause order*.

Various factors in the language system in general can be discerned that play an important role in expressing a given content in a linear form. The inventory of these factors contains at least the following: *information structure, grammatical structure, intonation, rhythm* and *style*. These factors are very general, and can therefore be considered language universals, at least within the family of indo-european languages. However, the individual factors may have different importance for the linear ordering, i.e. word order and clause order, in a given language.

---

[1] AGILE (Automatic Generation of Instructions in Languages of Eastern Europe) is supported by the European Commission within the COPERNICUS programme, grant No. PL961104. We would like to thank our colleagues from the Charles University, the University of Brighton, the Saarland University, the Bulgarian Academy of Sciences and the Russian Research Institute of Artificial Intelligence.

F. Jelinek, E. Nöth (Eds.): TSD'99, LNCS 1692, pp. 83–88, 1999.

Information structure is considered the main factor determining the linear ordering within a sentence in the Slavonic linguistic tradition. It is considered relevant for both word order and clause order. Hereby, we are using the term information structure as a general term for various notions employed in contemporary theories of the syntax-semantics interface, notions that reflect how the conveyed content is distributed over a sentence, and how it is thereby structured or "perspectivized". Within the Czech linguistic tradition, information structure has been referred to as "aktuální členění", functional sentence perspective and topic-focus articulation. Another terminology was introduced by Chafe in the 70's, where information structure is called *information packaging*. Halliday makes a distinction between the *thematic structure* of the sentence and its *information structure*.

In every approach to information structure, the clause is considered to consist of (at least) two parts. The often used oppositions are, e.g. *Theme-Rheme*, *Topic-Focus*, *Background-Focus*, *Ground-Focus*. Sometimes, the authors introduce further sub-divisions. For a discussion of some of the differences between various approaches see [5].[2]

In the AGILE project, we cannot consider all these different approaches to information structure. In the end, we have to work with just one approach for the sake of our linguistic specifications. Currently we restrict ourselves to two approaches: (i) Halliday's thematic structure [3] as developed in the Systemic Functional Grammar framework (SFG) is chosen because our grammars are based on the SFG framework; (ii) the topic-focus articulation approach developed in Prague within the framework of Functional Generative Description (FGD, [8]) serves us to elaborate the SFG approach towards a more flexible treatment required for languages with a higher degree of free word order than English, especially because Halliday's approach is not sufficiently specific with respect to the ordering of non-thematic constituents.[3]

In this paper, we concentrate on the issue of controlling word order in the course of automatic natural language generation. We begin by an illustration of the differences in semantic and contextual appropriateness of sentences with varying word order. Then we present the main principles which enable us to determine a suitable word order when generating a sentence conveying a particular content in a given language in a particular context. Finally, we present our word ordering algorithm, based on combining the SFG and FGD insights, which applies to the constituents of a clause.

---

[2] For a detailed list of references to various approaches to information structure see [1].

[3] Due to the restricted size of this paper, we cannot overview and compare the SFG and FGD approaches in detail, so the interested reader should consult [1].

2 Word Order Variation

We first present some word order variations in the three languages using an example from the set of texts generated in AGILE, which are adapted from a CAD/CAM system user guide. Let us consider the context given in (1).

(1) Open the Multiline styles dialog box using one of the following methods.

In (2) we show the following sentence from the manual in its original word order. In (3) through (7) are the permutations of its main syntactic groups:

(2) **Cz** Z menu   Data vyberte   *Multiline Style.*
     Ru V menju   Data vyberite punkt *Multiline Style*
     Bu Ot menjuto Data izberete   *Multiline Style*
     En From|In menu Data choose *Style Multiline*

(3) **Cz** Vyberte z menu   Data   *Multiline Style.*
     Ru Vyberite v menju   Data punkt *Multiline Style.*
     Bu Izberete ot menjuto Data   *Multiline Style.*
     En Choose from|in menu Data *Multiline Style*

(4) **Cz** Vyberte   *Multiline Style* z menu   Data.
     Ru Vyberite punkt *Multiline Style* v menju   Data.
     Bu Izberete   *Multiline Style* ot menjuto Data.
     En Choose *Multiline Style* from|in menu Data

(5) **Cz**   *Multiline Style* vyberte z menu   Data.
     Ru Punkt *Multiline Style* vyberite v menju   Data
     Bu   *Multiline Style* izberete ot menjuto Data.
     En *Style Multiline* choose from| menu Data

(6) **Cz** Z menu   Data   *Multiline Style* vyberte.
     Ru ? V menju   Data punkt *Multiline Style* vyberite
     Bu * Ot menjuto Data   *Multiline Style* izberete.
     En From|In menu Data *Multiline Style* choose

(7) **Cz**   *Multiline Style* z menu   Data vyberte.
     Ru ? Punkt *Multiline Style* v menju   Data vyberite.
     Bu *   *Multiline Style* ot menjuto Data izberete.
     En *Multiline Style* from|in menu Data choose

(2) – (7) all constitute grammatically well formed sentences in Czech as well as in Russian. In Bulgarian, (7) and (6) are ungrammatical. It is thus apparent that for sentences with a moderate number of syntactic groups there are usually quite a few grammatically well formed word order variants. However, one should not interpret the high degree of freedom in word order as arbitrariness.

In Czech and Russian, it appears that both versions (3) and (4) could be used instead (2) in the context of (1); however, the remaining versions (5) – (7) could not. (5) does not fit into the context of (1) because it presupposes contextual familiarity of *Multiline Style*, which it is not in this context. (6) and (7) in Czech could only be used felicitously in a rather restricted type of contexts, namely those where the verb or the action referred to is to be interpreted as contrasting some other verb or action with the same participants and circumstances. In Russian, (6) and (7) sound strange, though they are grammatical.

Thus, sentences which differ only in word order (and not in syntactic realizations of constituents) are not freely interchangable in a given context (cf. [1,6] for more discussion). The comparison also illustrates that the degree of word order freedom is not the same in across the three languages under consideration. This means that in the process of automatic generation of continuous texts from an underlying representation of the content, we have to ensure that a semantically and contextually appropriate word order is chosen in every language.

3   Information Structure and Word Order

According to [3], a clause as a message consists of a Theme combined with a Rheme, and in this configuration, the Theme is the ground from which the clause is taking off. As noted earlier, Halliday distinguishes between the thematic structure of a clause and the information structure. The latter is the distinction between Given and New within an information unit: the speaker presents information to the listener as recoverable (Given) or not recoverable (New). The thematic structure and information structure are closely related but not the same. Whereas the Theme is what the experiential items the speaker chooses to take as the point of departure, the Given is what the speaker believes the listener already knows or has accessible.

The notion of Theme tells us a number of things about "the first" position in the clause, but it does not tell us much about the word order of "the rest" of the clause. Presumably, Halliday leaves this to be decided by the grammatical structure. However, in languages with a high degree of free word order the grammar is not very strict about the placement of the groups and phrases within the clause. The examples we discussed above showed that ordering in our languages is to a great extent determined by what is presumed to be salient in the context. This means ordering depends on information structure. These issues have been studied in detail in the Praguian FGD framework [8]. We incorporate the most essential ideas into the AGILE account of word order in Slavonic languages.

FGD works with a notion of information structure consisting of one dichotomy, called Topic-Focus Articulation (TFA). TFA is defined on the basis of a distinction between contextually bound (CB) and non-bound (NB) items in a sentence. The motivation behind this distinction corresponds to that underlying the Given/New dichotomy in SFG. A CB item is assumed to convey some content that bears a contextual relationship to the discourse context. Such an item may refer to an entity already explicitly referred to in the discourse, or an "implicitly evoked" entity (see [2] for a summarizing discussion).

The ordering of NB items in a sentence follows the so-called *systemic ordering* (SO). SO is a language specific ordering of complementations, i.e. "arguments" and "adjuncts", of verbs, nouns, adjectives or adverbs which corresponds to neutral word order. It may differ from one language to another, but is considered universal within a given language. SO in Czech has been studied in detail (see [8]). We expect the SO for the main types of complementations in Russian and Bulgarian is similar to the Czech one, but there can be differences [1].

As the starting point for specifying the principles of word ordering in the context of AGILE, we combine this FGD-based strategy which reflects information structure with the possibility of thematization in the usual SFG spirit. For Czech and Russian, we need to allow for more freedom in word order (i.e., a looser relation between ordering and grammatical structure) than in Bulgarian. Namely, in Bulgarian there seems to be the restriction that only one experiential element may precede the verb. Other restrictions or preferences can be included in the grammars of the specific languages.

A specific point concerns the placement of clitics in Slavic languages: they must appear in what is called the Wackernagel position. In the sentences we are generating in AGILE, it appears useful to use the Theme to identify the position for clitics.

4   Word Ordering Algorithm

The algorithm is presented in abstract form in Figure ??. Using ^ for linear precedence, it can be schematized as follows:

```
Theme^Clitics^Rest-CB^Verb^Rest-NB
```

The Theme is determined by text organization. If no element is explicitly chosen as Theme, the thematic position is filled by the first CB element. Any clitics are placed after the Theme. Their mutual order is determined by the grammar. For ordering of the non-thematic constituents within a clause, for which their order is not determined by the syntactic structure, we use systemic ordering in combination with the CB/NB distinction. The NB elements are ordered by SO. The ordering of the CB elements can be (i) specified on the basis of the context, (ii) restricted by the grammatical structure, (iii) follow SO. The verb is placed between the last CB and the first NB element, unless it is itself the Theme.

The proposed ordering algorithm is the same for all the three languages under consideration. What differs across the languages are constraints on which elements can be ordered rather freely in accordance to information structure, and which ones are subject to ordering requirements posed by the syntactic structure.

The current approach is satisfactory to the extent that is it satisfactory to consider word order as a "second order" phenomenon, i.e. as an ordering that is applied to constituents in a structure that has already been generated. Such approach does not enable information structure to influence the grammatical structure of the constituents that are generated.

```
Given: a list G of ordering constraints imposed by the grammar,
       a list L1 of constituents that need to be ordered,
       a list Delta giving ordering of CB constituents,

       create empty lists LC and LN % LC for CB items, LN for NB items
       repeat for each element E in L1
              if E is CB,
                   then add E into LC,
                   else add E into LN.
       if the verb is CB,
            then    Order the verb at the end of LC
                    Order the remainder according to D
            else    Order all elements in LC according to D
       % thus, if e precedes f in D, then e precedes f in LC except
       % for the verb.
       if G is not empty then
            Order elements in L1 using ordering constraints in G
```

Fig. 1. Abstract algorithm for word order

References

1. Elena Adonova, John Bateman, Nevena Gromova, Anthony Hartley, Geert-Jan Kruijff, Ivana Kruijff-Korbayová, Serge Sharoff, Hana Skoumalová, Lena Sokolova, Kamenka Staykova, and Elke Teich. Formal specification of extended grammar models. AGILE project deliverable, University of Brighton, UK, http://ufal.mff.cuni.cz/~agile/reports.html, 1999.
2. Eva Hajičová. *Issues of sentence structure and discourse patterns*, volume 2 of *Theoretical and computational linguistics*. Charles University, Prague, Czech Republic, 1993.
3. M.A.K. Halliday. *An Introduction to Functional Grammar*. Edward Arnold, London, 1985.
4. Ivana Kruijff-Korbayová. Generation of instructions in a multilingual environment. In *Proceedings of the Conference on Text, Speech and Dialogue (TSD'98), Brno, Czech Republic, September 1998*, pages 67–72, 1998.
5. Ivana Kruijff-Korbayová and Eva Hajičová. Topics and centers: a comparison of the salience-based approach and the centering theory. *Prague Bulletin of Mathematical Linguistics*, (67):25–50, 1997. Charles University, Prague, Czech Republic.
6. Ivana Kruijff-Korbayová and Geert-Jan Kruijff. Contextually appropriate ordering of nominal expressions. In *Proceedings of the ESSLLI99 workshop "Generating Nominals", August 1999*, 1999.
7. Ivana Kruijff-Korbayová and Geert-Jan Kruijff. Text structuring in a multilingual system for generation of instructions. In *Text, Speech, Dialogue* (eds. V. Matoušek, P. Mautner, J. Ocelíková and P. Sojka), Plzeň, Springer–Verlag. (1999) 87–92.
8. Petr Sgall, Eva Hajičová, and Jarmila Panevová. *The meaning of the sentence in its semantic and pragmatic aspects*. Reidel, Dordrecht, The Netherlands, 1986.

Text Structuring in a Multilingual System for Generation of Instructions

Ivana Kruijff-Korbayová and Geert-Jan M. Kruijff

Institute of Formal and Applied Linguistics (ÚFAL)
Faculty of Mathematics and Physics, Charles University
Prague, Czech Republic
{korbay,gj}@ufal.mff.cuni.cz
WWW home page: http://ufal.mff.cuni.cz/

1 Introduction

The approach presented in this paper has been developed in the context of the international project called AGILE (Automatic Generation of Instructions in Languages of Eastern Europe)[1]. The overall aim of the project is to develop a multilingual system for generating continuous instructional texts in Bulgarian, Czech and Russian [2]. The project is concerned mainly with (i) the development and adaptation of linguistic resources for the chosen languages, and (ii) the investigation and specification of text structuring strategies employed in those languages for the given type of texts. In the current paper, we concentrate on the latter issues.

What we are particularly interested in is the following. In the application scenario of the system developed in AGILE, the user provides a definition of a text content in the form of an A(assertion)-box. from this A-box we are then to generate a text, in either of the languages we are concerned with. The problem we address in this paper is how to go about this for relatively simple instructional texts about working with AutoCAD.

To that end, we begin with a brief discussion of the texts (and their language-dependent styles) we would like to be able to generate (Section 2). As we already pointed out, these texts are generated from (or "realize") a text content specified by an A-box. In Section 3 we show how we can use the concepts, employed in an A-box, to set up a structure for a text that would realize the content given by that A-box. Finally, we discuss in Section 4 how the approach outlined in the previous sections is implemented in the KPML multilingual generation environment [1] that we use in the context of AGILE.

---

[1] AGILE (Automatic Generation of Instructions in Languages of Eastern Europe) is an international project supported by the European Commission within the COPERNICUS programme, grant No. PL961104. We would like to thank our colleagues from Charles University, the University of Brighton, the University of the Saarland, the Bulgarian Academy of Sciences and the Russian Research Institute of Artificial Intelligence for their cooperation.

F. Jelinek, E. Nöth (Eds.): TSD'99, LNCS 1692, pp. 89–94, 1999.

2 Texts and Text Styles

The texts considered for generation in the current stage of the AGILE project are simplified versions of routine passages occurring frequently in the user guide for the AutoCAD system, namely descriptions of step-by-step procedures for performing a task. An example of such a description is given in Figure 1.

To create a multiline style
First open the Multiline Styles dialog box using one of these methods:

i. **Windows** From the Object Properties toolbar or the Data menu, choose Multiline Style.

ii. **DOS and UNIX** From the Data menu, choose Multiline style.

(1) Choose Element Properties to add elements to the style.

(2) In the Element Properties dialog box, enter the offset of the first line element.

(3) Select Add to add the element.

(4) Choose Color. The the Select Color dialog box appears. Then select the element's color.

(5) Choose Linetype. Then select the element's linetype from the Select Linetype dialog box.

(6) Repeat these steps to define another element.

(7) Choose OK to save the style of the multiline element and to exit the Element Properties dialog box.

Fig. 1. Example "To create a multiline style".

The corpus investigations we carried out for the languages under study in AGILE provided empirical support for the view that in the instructional texts under investigation we can observe alternations that can be classified into two main groups, as follows:

1. *Choice of grammatical means*: Whether lexico-grammatical choices are pre-determined in certain ways by the style of text.
2. *Distribution of content and lexical realization*: Whether information is realized explicitly in the most straightforward way given the A-box, (so that there is only a minimal inference load required to interpret a piece of text), or whether information is left more or less implicitl; and, what the realization is of the explicitly conveyed information.

The first type of alterations concerns the following kinds of choices: (i) Expressing or hiding the intentional agent of an action (agency, voice), (ii) ways of addressing the readership (person, number, voice), and (iii) modes of realising instructions (mood, voice). The second type of of alterations is reflected in the level of detail in expressing the content explicitly. A nice example is whether or not the text mentions any side-effects that will occur as a result of performing a particular action.

In the next sections we develop a view on how to deal with such alterations in a principled fashion.

3   Text Structuring

In this section we discuss how we can set up a structure for a text that is to realize a given A-box, by looking at how the concepts in the A-box are organized in order to specify a particular content. We begin by making a distinction between *text structure elements* being the elements from which a (task-oriented) text is built up, and *text templates*, which condition the way text structure elements are to be linguistically realized (Section 3.1). Subsequently, we focus on the relation between concepts on the one hand, and text structure elements on the other hand. We are specifically interested in those T(erminology)-box concepts that are used to configure the content specified in an A-box. This is because we want to have a close connection between how the content can be defined in an A-box and how that content is to be spelled out in a text using a particular organization, so as to ensure that the intended content is indeed reflected by the text (Section 3.2).

3.1   Structuring and Styling

Above we already alluded to the idea that texts may vary in their distribution and lexico-grammatical realization of content. Here we propose two notions that should enable us to capture these differences in a principled fashion. These two notions are the notion of **text structure element** and the notion of **text template**.

A text structure element is a predefined component that is to be filled by one or more specific parts of the user's definition of the content to be spelled out by the text. Using the reader-oriented terminology common in technical authoring guides, we will distinguish a small (recursively defined) set of text structure elements, given in Figure 2. Orthogonal to the notion of text structure element is the notion of text template. Whereas text structure capture *what* needs to be realized, the idea of a text template captures *how* that content is to be realized. Thus, a template defines a style in which content is to be realized. As we shall discuss below (§4), we will define text templates in terms of constraints on the realization of specific (individual) text structure elements. For example, whereas in Bulgarian and Czech headings (to which the TASK-TITLE element corresponds) are usually realized as nominal groups, the Russian AutoCAD manual realizes headings as partial purpose clauses.

Task-Document A TASK-DOCUMENT has two slots:

  i. TASK-TITLE (obligatory)

  ii. TASK-INSTRUCTIONS (obligatory), being a list of at least one INSTRUCTION.

Instruction An INSTRUCTION has three slots:

  i. TASKS (obligatory), being a list of at least one TASK.

  ii. CONSTRAINT (optional)

  iii. PRECONDITION (optional)

Task A TASK has two slots:

  i. INSTRUCTIONS (obligatory)

  ii. SIDE-EFFECT (optional)

Fig. 2. Text Structure Elements.

3.2   Relating Domain Model Concepts and Text Structure

Our aim is to create a structure for a text that is to realize the content specified by an A-box. We already have defined the elements that build up such a text structure (cf. Figure 2). In this section we begin with defining the concepts that are used to specify content in an A-box. Thereafter we relate these concepts to the text structure elements so as to enable us to build a text structure by going by the way content has been structured in an A-box.

An A-box, short for "Assertion-box", instantiates concepts defined in a T-box, or "terminology box". In KPML, and AGILE, these concepts are generally referred to as *domain model concepts*. For AGILE we employ four concepts to define the admissible A-box configurations. These configurations provide the structured contexts in which actions and events "take place". The concepts are given in Figure 3 [2].

The following mapping between text structure elements (Figure 2) and concepts (Figure 3) enables us to provide a flexible definition of rules for text structuring of a content specified by an A-box:

TASK-TITLE ↔ GOAL of topmost PROCEDURE
TASK-INSTRUCTIONS ↔ METHODS of topmost PROCEDURE
SIDE-EFFECT ↔ SIDE-EFFECT of a PROCEDURE
TASK ↔ GOAL of a PROCEDURE
CONSTRAINT ↔ CONSTRAINT of a METHOD
PRECONDITION ↔ PRECONDITION of a METHOD
INSTRUCTIONS ↔ METHODS
INSTRUCTION ↔ METHOD

---

[2] These concepts were defined by Richard Powers.

Procedure A procedure has three slots:

   i. GOAL (obligatory, to be filled by a USER-ACTION)

   ii. METHODS (optional, to be filled by a METHOD-LIST)

   iii. SIDE-EFFECT (optional, to be filled by a USER-EVENT)

Method A method has three slots:

   i. CONSTRAINT (optional, to be filled by an OPERATING-SYSTEM)

   ii. PRECONDITION (optional, to be filled by a PROCEDURE)

   iii. SUBSTEPS (obligatory, to be filled by a PROCEDURE-LIST)

Method-List A METHOD-LIST is essentially a list of METHOD*'s:

   i. FIRST (obligatory, of type METHOD*)

   ii. REST (optional, of type METHOD-LIST)

Procedure-List A PROCEDURE-LIST is essentially a list of PROCEDURE's:

   i. FIRST (obligatory, of type PROCEDURE)

   ii. REST (optional, of type PROCEDURE-LIST)

Fig. 3. Domain Model concepts defining admissible A-box configurations.

4 Text Structuring Module

In AGILE we are *ultimately* aiming at so-called *end-to-end generation*. That is to say, a user specifies an A-box using an authoring interface, and obtains as output a text realizing that A-box, in a particular text style. Crucial to this enterprise is the Text Structuring Module that we describe in this section. The Text Structuring Module takes as input a user-specified A-box, and generates a set of specifications of the content of individual sentences (SPLs). These SPLs then serve as input to a language-specific tactical generator that actually generates these sentences. At the basis of the Text Structuring Module (or TSM, hereafter) are the ideas regarding text structure elements, domain model concepts, and text templates that we discussed in the aforegoing sections.

4.1 General Architecture of the TSM

The general architecture of the TSM we conceive of is as follows:

Systemic networks for text structuring. By means of the text structuring networks, an appropriate text structure is formulated in terms of hierarchically related text structure elements, given the A-box.

A program dividing the A-box's content over smaller A-boxes. Given the A-box and the text structure that has been generated, this program divides the A-box into parts that are associated with the text structure elements from which the text structure is composed. In the setting of KPML [1] this can easily be done using so-called ID-inquiries.

A program that generates SPL-code from a given A-box. The aim here is to generate SPLs, using the parts of the A-boxes individuated by the previous program and the realization constraints imposed by text templates.

Once we have individual SPLs, each specifying a sentence, we can use a tactical generator to generate the sentences, which altogether make up the text realizing the content specified in the A-box.

4.2   Systemic Networks for Text Structuring

The major component of the TSM is formed by the systemic networks for text structuring, and therefore we will spend a few more words on describing what these networks look like. In the spirit of [1], we construct a region that defines an additional level of linguistic resources for the level of *genre*. The region enables the composition of text structures in a way that is very similar to the way the lexico-grammar builds up grammatical structures. In fact, by using KPML to implement the means for text structuring, it easily facilitates smooth interaction between global level text generation (strategic generation) and lexico-grammatical expression (tactical generation)[3].

The region, called CADCAM-INSTRUCTIONS, follows out the viewpoint that text templates and text structure elements are essentially orthogonal ideas. Therefore it consists of two parts. One part deals with interpreting the A-box in terms of text structure elements. By traversing the network that the systems of this part make up, we obtain a text structure for the A-box conforms the definitions we provided above. The other part of the region imposes constraints on the realisation of the text structure elements that are being introduced by traversing the other part of the region. Naturally it will depend on our choice of a particular text template (style) which constraints will be imposed. These choices are made through interaction between the user and the system.

References

1. John A. Bateman. Enabling technology for multilingual natural language generation: the KPML environment. *Natural Language Engineering*, 1(1), 1997.
2. Ivana Kruijff-Korbayová. Generation of instructions in a multilingual environment. In Petr Sojka, Václav Matoušek, Karel Pala, and Ivan Kopeček, editors, *Text, Speech and Dialogue (Proceedings TSD'98)*, Brno, Czech Republic, September 1998.

---

[3] Thus, we overcome the notorious problem known as the *generation gap* in which a text planning module lacks control over the fine-grained distinctions that are available in the grammar. In our case, both text planning and sentence planning are integrated into one and the same system (albeith organized in a stratificational manner).

Leveraging Syntactic Information
for Text Normalization

Deborah A. Coughlin

Microsoft Research
1 Microsoft Way
Redmond, Washington 98052 USA
deborahc@microsoft.com
http://research.microsoft.com/nlp

Abstract. Syntactic information provided by a broad-coverage parser aids text normalization. This paper introduces a text normalizer for text-to-speech (TTS) and language modeling that makes use of syntactic information to improve output quality. This normalizer takes in raw text and outputs text that has abbreviations, numerals, and symbols spelled out as words. Part-of-speech ambiguous abbreviations, ambiguous abbreviations in coordinated structures, and quantified measure abbreviations in text input can be correctly rewritten when syntactic information provided by the parser is considered.

1 Introduction

Most text normalization modules in text-to-speech and language modeling components make use of surface-level information to inform their rewrites. Though most text normalization modules do not make use of syntactic information, the need for syntactic analysis to disambiguate some tokens has been discussed [3,1]. This paper discusses the benefits of using syntactic information to aid text normalization when the text input is ambiguous. Three types of ambiguous input, part-of-speech ambiguous abbreviations, ambiguous abbreviations in coordinated structures, and abbreviations in quantified measure noun phrases, will be examined. This discussion will be limited to orthographic expansion; phonological representations will not be discussed.

2 Background

Text normalization, in this context, is the rewriting of raw text input into a form that is accessible to a speech synthesizer. Numerals, abbreviations and other symbols are rewritten in full word form. The following input

1377 Dr. Gardner Dr., Seattle WA 98125

is rewritten as

F. Jelinek, E. Nöth (Eds.): TSD'99, LNCS 1692, pp. 95–100, 1999.
© Springer-Verlag Berlin Heidelberg 1999

*thirteen seventy seven Doctor Gardner Drive, Seattle Washington nine
eight one two five*

and

coughlin_cash@msn.com

is rewritten as

coughlin underscore cash at m s n dot com.

These examples illustrate that numerals are interpreted differently depending on
their context and symbols, when appropriate, are spelled out.

The Natural Language Analysis component of our Natural Language Processing system includes an empirically-based, broad-coverage, rule-based chart
parser. This parser makes use of augmented phrase structure grammar rules, described in Jensen [2]. Lexical records, built from information found in a machine-
readable dictionary[1] and further processed by morphological and factoid rules,
provide the parser input. The parser output is represented by a highly embedded structure of attribute-value pairs that can be viewed as traditional syntactic
tree structures with information-rich nodes. The text normalization module is
positioned after a syntactic parser. Information from each preceding level of
processing is available to the text normalizer.

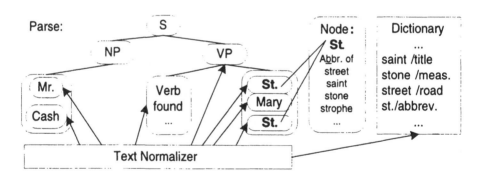

Fig. 1. Diagram of text normalizer working with parse and dictionary information.

Figure 1 shows a simplified diagram of how the text normalizer functions. It
visits nodes on the tree, looking for relevant information stored in each node.
Each node consists of structured attribute-value pairs that are, when appropriate, added to at each level of processing. In this example *St.* is ambiguous. The
St. record contains all possible words associated with that abbreviation. Once

---

[1] An augmented machine readable form of American Heritage Dictionary of the En-
glish Language, Third Edition, copyright 1992 by Houghton Mifflin Company and
Longman's Dictionary of Contemporary English, copyright 1978 by Longman Group
Limited.

the first *St.* is identified as a title and the second *St.* is identified as a road type, finding the appropriate unabbreviated form is done by looking in the dictionary for entries tagged with that role.

3   Part-of-Speech Ambiguous Abbreviations

In a broad coverage system, abbreviations often have multiple possible rewrites. In our dictionary *dist.* has four possible rewrites: *distant, distance, district,* and *distribute.*

1. He came in a dist. second. ⇒ [2] He came in a *distant* second.
2. Dist. by and copyright of Hunt-Wesson, Inc ⇒ *distributed* by and copyright of Hunt-Wesson, incorporated .
3. Hampton Dist. Office is on Butler Farms Rd. ⇒ Hampton *District* Office is on Butler Farms Road .

In the process of constructing a syntactic representation, our parser is able to determine that *dist.* should be an adjective in example 1, a verb in example 2 and a noun in example 3, allowing reasonable rewrites for *dist.*

4   Abbreviations in Coordinated Structures

If coordinated structures are identified, even highly ambiguous abbreviations have some chance of being correctly rewritten. *CA* in isolation could represent *California, calcium, cancer, capital account, cardiac arrest,* etc. Given the highly ambiguous nature of *CA*, rewriting it as *California*, except when clearly in an address context, would often generate incorrect results. Because the parse identifies coordinated structures and, in the example below, all the members of the coordinated structure share one property (US state), rewriting *CA* and the other state abbreviations is reasonable.

4. The appropriate state tax will be added for orders shipped to the following states: AZ, CA, CT, DC, FL, GA, HI, ID, MA, MD, NJ, NM, NV, NY, PA, UT, VA, VT, and WA. ⇒ The appropriate state tax will be added for orders shipped to the following states Arizona, California, Connecticut, District of Columbia, Florida, Georgia, Hawaii, Idaho, Massachusetts, Maryland, New Jersey, New Mexico, Nevada, New York, Pennsylvania, Utah, Virginia, Vermont, and Washington.

Figure 2 provides a window into the richness of the syntactic parse representation. The Coord attribute of the noun phrase node links each coordinated leaf node. Determining whether a leaf node shares a particular feature with its coordinated sister nodes is done by questioning the noun phrase node to determine if all Coords share that feature.

---

[2] ⇒ indicates *rewritten as*

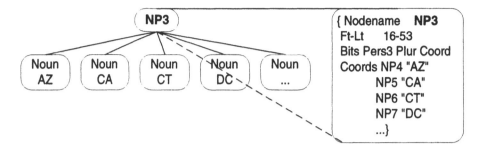

Fig. 2. A noun phrase node for example sentence 4 showing coordination represented in the noun phrase node.

If expanding to state names were not possible for this example, the result would be a sentence that lacks fluidity and would most likely be difficult to understand when spoken by TTS systems. For example sentence 4, without access to parse information, our system outputs a long list of initials.

Rewrite of example 4 when parser information is not utilized:

> The appropriate state tax will be added for orders shipped to the following states A Z, C A, C T, D C, F L, G A, H I, I D, MA, M D, N J, N M, N V, N Y, PA, UT, V A, V T, and W A.

5   Quantified Measure Abbreviations

Identification of noun phrases, including identification of the head of the noun phrase, is vital to good text normalization.

> 5. The pole is 8' long. ⇒ The pole is eight *feet* long.
> 6. Nolan has several 8' long poles. ⇒ Nolan has several eight *foot* long poles.

Without it, determining whether a measure abbreviation should be normalized as a plural or singular becomes tricky at best. Is it *eight foot* or *eight feet*? Our parser is able to determine that in example 5 the head of the noun phrase is *8'*, but in example 6, *pole* is selected as the noun phrase head, making the choice between *feet* and *foot* trivial. Figure 3 illustrates the noun phrase structures constructed by our parser for examples 5 and 6. The * next to *Noun* in the figure indicates that node is the head of the noun phrase.

There are surface clues that can help. A measure abbreviation followed by a noun, for example, can suggest use of the singular rewrite. This is not foolproof however. In a broad coverage system like ours, *long* has adjective, noun, verb and adverb parts of speech. *Pole* is both a noun and a verb. Examples like the following also illustrate why surface clues are insufficient.

> 7. Brianna has a $5 bill framed above her desk. ⇒ Brianna has a five *dollar* bill framed above her desk.

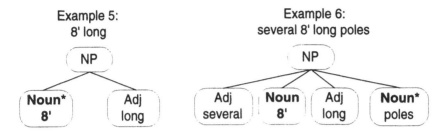

Fig. 3. Noun phrases for relevant parts of examples 5 and 6. (* indicates head of phrasal node.)

8. If it is over $5 bill me. ⇒ If it is over five *dollars* bill me.

In example 7, the ideal interpretation for *$5* is *five dollar*, with *dollar* in the singular. In example 8, the preferred interpretation is *five dollars*, with *dollar* in the plural. Again, the information generated by the parser makes the choice a trivial operation. Our parser correctly identifies *bill* as the head of a noun phrase in example 7 and as a verb in example 8. Figures 4 and 5 illustrate the parses that make these correct spellouts possible.

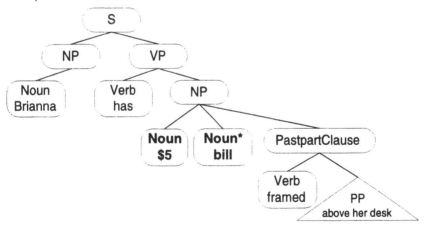

Fig. 4. Simplified parse tree for example 7. Notice that *$5* and *bill* are in the same noun phrase with *bill* as the head of the noun phrase. (* signifies head of phrase.)

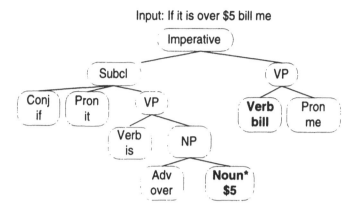

Fig. 5. Simplified parse tree for example 8. Notice that *$5* and *bill* are in separate constituents.

6   Conclusion

The examples discussed in this paper illustrate the contribution of syntactic parse information to text normalization quality. In a broad-coverage system, most abbreviations are ambiguous and surface clues provide limited assistance. Though syntactic information alone is not sufficient to solve all ambiguities faced by text normalization modules, it can contribute significantly to text normalization ambiguity resolution.

References

1. van Holsteijn, Y.: TextScan: A preprocessing module for automatic text-to-speech conversion. In: van Heuven V.J., Pols L.C.W. (eds.): Analysis and Synthesis of Speech: Strategic Research towards High Quality Text-to-Speech Generation. Mouton de Gruyter Berlin (1993) 27–41.
2. Jensen, K.: PEG: the PLNLP English Grammar. In: Jensen K., Heidorn, G., and Richardson, S. (eds): Natural Language Processing: the PLNLP Approach. Kluwer Academic Publishers Boston (1993) 29–45.
3. Sproat, R.: Further Issues in Text Analysis. In: Sproat, R. (ed.): Multilingual Text-to-Speech Synthesis. Kluwer Academic Publishers Boston (1998) 89–114.
4. Sproat, R., Mobius, B., Maeda, K., Tzoukermann, E.: Multilingual Text Analysis. In: Sproat, R. (ed.): Multilingual Text-to-Speech Synthesis. Kluwer Academic Publishers Boston (1998) 31–87.

Automatic Structuring of Written Texts*

Marek Veber, Aleš Horák, Rostislav Julinek, and Pavel Smrž

Faculty of Informatics
Masaryk University
Botanická 68a, 60200 Brno, Czech Republic

Abstract. This paper deals with automatic structuring and sentence boundary labelling in natural language texts. We describe the implemented structure tagging algorithm and heuristic rules that are used for automatic or semiautomatic labelling. Inside the detected sentence the algorithm performs a decomposition to clauses and then marks the parts of text which do not form a sentence, i.e. headings, signatures, tables and other structured data. We also pay attention to the processing of matched symbols in the text, especially to the analysis of direct speech notation.

1 Introduction

In order to reduce the time and memory demands of syntactic analysis, POS tagging, aligning parallel corpora and other NLP tasks, one first needs to divide the analyzed text into parts which are then analysed separately. The first suitable division points are paragraph boundaries. After appropriate pre-analysis it is possible go even deeper and segment the text to sentences and then to particular clauses. The analysis is facilitated by demarcation of those word groups that cannot or should not be divided any further like data, personal names, URL addresses etc.

In sentence boundary labelling we meet the problem of meaning ambiguity of the full-stop mark (a dot). Either it can denote a sentence end or it can be a part of an abbreviation or it can even bear both of these meanings (according to statistical results in English [3]: 90 % — sentence end, 9.5 % abbreviation and 0.5 % both; in the Czech corpus DESAM we have 92 % — sentence end, 5.5 % abbreviation and 2.5 % both meanings). Common approaches to solving the problem of labelling these hierarchical structures use regular expressions or finite automata with look-ahead that bear on several simple clues in text (like capitalisation) with a list of abbreviation and exceptions (see e.g. [1]). Other approaches are based on regressive trees [2] and artificial neuron networks [3] which make use of contextual information about POS tags in the surroundings of the potential structure boundary. However, those approaches cannot be easily applied in the analysis of Czech language because of the extent of the Czech tagset [4,5].

* The research is sponsored by the Czech Ministry of Education under the grant VS 97028.

F. Jelinek, E. Nöth (Eds.): TSD'99, LNCS 1692, pp. 101–104, 1999.
© Springer-Verlag Berlin Heidelberg 1999

2   Structuring Algorithm

Our approach not only uses all the common methods, it also takes advantage of hierarchical processing in several adjoint phases. The algorithm first verticalises the input plain text to elementary lexical symbols (words, numbers and punctuation marks). Then it joins basic groups of elementary symbols into complexes that form further indivisible parts. Besides information contained in the text the analysis exploits the morphological information of particular symbols that is either determined during disambiguation or, when the information is missing, it is acquired by means of the morphological analyser. Above all we are interested in symbols, which are potential abbreviations, coordinate and subordinate conjunctions or which can form a verb group. The selection of the candidates for division runs hierarchically with the use of backtracking.

The heuristic rules generate possible text divisions together with appropriate probabilities, which then enables us to ask for consultation with human expert (linguist) only in cases where the division probability overpasses a given threshold value. The expert can then approve the result or reject it. This possibility can be advantageously used in semiautomatic annotation of training corpora.

Tagging relies on the fact that sentence boundaries do not exceed the paragraph limits and the clause boundaries stay between the sentence tags. The text is processed one paragraph at a time, where we seek for sentence boundaries. Every found sentence is then processed by the clause separation algorithm. Thus at the beginning and the end of paragraph we obtain positions that do certainly form the beginning or the end of a sentence and the same for clause boundaries at the sentence bounds.

Within the scope of the paragraph we mark all the possible candidates for a structural mark. Then we apply a set of partial parsing rules that increase or decrease the values of the positions of selected candidates in the block. Eventually we divide the whole paragraph according to the strongest candidates so that the required conditions would be satisfied. If such division is not possible, we try a different division according to some weaker candidates to the sentence boundaries.

3   Partial Parsing Rules

Sentence boundary The potential candidates for a sentence boundary are the positions consisting of a full-stop ('.'), a question mark ('?'), an exclamation mark ('!'), three dots ('...') followed by a word beginning with an upper case letter or a non-alphanumeric character (an opening quotation mark or a left parenthesis) and a closing quotation mark which is preceded by a full-stop, a question mark or an exclamation mark. The candidates are not sought among positions inside of any matched characters like parentheses, quotation marks or structural marks denoting data or e-mail addresses.

During this run we also seek for the candidates for direct speech, which are found as pairs of quotation marks that contain at least one punctuation character and more than two positions in between. The direct speech candidate is marked by <sx> tag and after this step the block of direct speech text is divided to sentences and clauses.

Signature First, we determine whether the input block of text is formed by a signature. The condition for the block to be a signature is that the whole text must be formed by proper names or abbreviations that denote an academic or a military title. If the whole block satisfies this condition, we mark it with the <sign> tag.

Heading The fact that the block represents a heading is recognised by the condition that the block does not end with a full-stop and contains only one candidate for a sentence end. Such paragraph is marked as <head>.

4   Data Set and Results

Our results are based on the facts from the DESAM corpus (see [4,5]) which is a tagged corpus consisting of more than 1 200 000 positions collected from Czech newspaper articles.

The following two kinds of tags are used in the DESAM corpus: *structural* and *grammatical* tags. The structural tags mark boundaries of documents, paragraphs, sentences, headers and signatures. Each position in the corpus is tagged with the following two grammatical tags: a *lemma* — the basic form of a particular word and a *tag* — representing its grammatical categories. Both the grammatical and structural tags have been manually disambiguated in the DESAM corpus.

The automatic structure tagger uses five tags: <s> for regular sentences (<sx> for direct speech and <c> for clauses — these tags were not included in the manual structure tagging), <head> for headings, <sign> for signatures and <table> otherwise. We have compared an output of the automatic tagging on the DESAM corpus with a manual tagging of the same texts. The results are summarised in the following table:

tag	number of blocks		number of positions		percentage of whole corpus		average length of one block	
	auto	man.	auto	man.	auto	man.	auto	man.
<s>	54030	51092	1054142	996430	84.6 %	80 %	19.5	19.5
<head>	5936	10802	28134	49821	2.2 %	4 %	4.7	4.6
<sign>	1885	1003	5289	2491	0.4 %	0.2 %	2.8	2.7
<table>	6741	8520	155958	196790	12.5 %	15.8 %	23.1	23.0

The next table displays the percentage of errors made either by automatic tagger or by a human labeller:

percentage of corpus	kind of error	manual tagging	automatic tagging
0.3 %	error in data, full-stop is missing	OK	ERR
0.1 %	error in data, extra (useless) full-stop	OK	ERR
0.8 %	error in data, sentence starts with lowercase	OK	ERR
2.2 %	bad tag	OK	ERR
2.7 %	bad tag or consistency err. (missing matched tag)	ERR	OK

Those differences display the necessity to make corrections not only to the tagging algorithm but also to the training data set. We also need to improve the error detection algorithm, which now only finds errors of matched characters.

The most problematic task (70 % of the differences) for the automatic tagger is the decision whether the selected text should be marked as <table> or <head>. We suppose that this task can be solved if the tagger takes the surrounding context of the block into account.

5 Conclusions

The automatic sentence boundary labelling can be demonstrated on a morphologically disambiguated data as well as on a plain corpus text without tags, in which case a morphological analyser Lemma (see [6]) is used. The results of automatic labelling have been compared to a manually labelled corpus.

In the presented approach we have concentrated on accuracy, efficiency and robustness so as the structuring does not slow down the text processing. Based on comparisons that we have made so far, we can say that the automatic algorithm achieves even better (more consistent) results than the human labellers.

References

1. Cutting, D., Kupiec, J., Pedersen, J., Sibun, P.: A practical part-of-speech tagger. In *the 3rd Conference on Applied Natural Language Processing*, Trento, Italy 1991.
2. Riley, M., D.: Some applications of tree-based modeling to speech and language indexing. In *Proceedings of the DARPA Speech and Natural Language Workshop*, pages 339-352, Morgan Kaufmann 1989.
3. Palmer, D., D., Hearst, M., A.: Adaptive Sentence Boundary Disambiguation. In *The Proceedings of the ANLP '1994*, Stuttgart, Germany, October 1994.
4. Pala, K., Rychlý, P., Smrž, P.: DESAM – Approaches to Disambiguation. Technical Report FIMU-RS-97-09, Faculty of Informatics, Masaryk University, Brno, 1997.
5. Pala, K., Rychlý, P., Smrž, P.: DESAM – Annotated Corpus for Czech. In *Proceedings of SOFSEM'97*.
6. Ševeček, P.: *LEMMA* morphological analyzer and lemmatizer for Czech, program in "C", Brno, 1996. (manuscript).
7. Julinek, R.: Automatic Detection of Sentence Boundaries, Master thesis, Masaryk University, Brno, April 1999.

Implementation of Efficient and Portable Parser for Czech

Pavel Smrž and Aleš Horák

Faculty of Informatics, Masaryk University Brno
Botanická 68a, 602 00 Brno, Czech Republic
E-mail: {smrz,hales}@fi.muni.cz

Abstract. This paper presents our work on an implementation of efficient system for syntactic analysis of natural language texts. We show the application of the system to the parsing of Czech by means of a special meta-grammar. Simultaneously, we have prepared a tool that builds the actual parser, which analyses input sentences according to the given grammar.

1 Introduction

Syntactic analysis of running texts plays a key role in natural language processing. Many researches have contributed to the area of text parsing by systems that reach satisfactory or even excellent results for English [1]. Other languages bring many more objections in attempts at creating a systematic description of the language by the help of traditional sort of grammars (e.g. the situation in German is discussed in [2]). Even more problems arise in free word order, respectively free constituent order languages. The sentence structure of such language defies to be described by a maintainable number of rules. The order of sentence constituents is designated as free, but in a matter of fact the order is driven more by human intuition than by firmly given regulators specified by linguists. The word order plays an important role in communicative dynamism, it expresses the sentence focus. This phenomenon is intensively explored by Prague Linguistic School in the context of Functional Generative Description [3].

2 Czech Language Parsing System

The Czech language (together with other Slavonic languages) is a typical example of free constituent order language. One of the first steps to robust parsing of Czech is described in [4]. This system puts to use a certain kind of procedural grammar. It is based on a formalism called RFODG (Robust Free-Order Dependency Grammar). The system encompasses complex rules for sentence syntax specification written in a Pascal-like form.

In contrast to the procedural grammar approach, we constitute a grammar system that retains the simplicity of rules. Herewith the maintenance of the set of grammar rules is kept under an acceptable limit, so that the modifications can

F. Jelinek, E. Nöth (Eds.): TSD'99, LNCS 1692, pp. 105–108, 1999.

be performed even by those users who do not need to have the perfect knowledge of all the internals of the grammar.

Our tool is based on the public domain parser generator BtYacc [5], which is an open source program written in C programming language and is designed and carefully tuned for efficiency and portability.

BtYacc processes a given context free grammar and constructs a C program capable of analysing input text according to the grammar rules. Natural language processing involves manipulation with grammars that allow more than one possible analysis of the input sentence. BtYacc enables the processing of ambiguous grammar that in case of ordinary LR analysis causes shift-reduce or reduce-reduce conflicts, which are in deterministic systems solved by choosing only one variant according to predefined precedences. For the purpose of working with ambiguous grammar we have implemented an intelligent backtracking support for BtYacc that is combined with routines which take care of successive formation of the derivation tree.

3   Meta-grammar

In the previous version of our system [6] we have used the notation of the grammar rules in the expanded form, that served directly as an input to the grammar parser. Such grammar consisted of more than eight hundred CF rules. Maintenance of the system has shown to be a nightmare of linguists working with the tool. Therefore we have decided to use a special kind of meta-grammar designed to discharge some mechanical constructs that are based on rule patterns which repeatedly occur in rule declarations. The number of rules in the meta-grammar has now radically decreased to less than one fifth of the number of generated rules. We believe that further elaboration of the meta-grammar will improve the reduction ratio even more.

The meta-grammar consists of *global order constraints* that safeguard the succession of given terminals, *special flags* that impose particular restrictions to given nonterminals and terminals on the right hand side and of *constructs* used to generate combinations of rule elements. The notation of the flags can be illustrated by the following examples:

```
ss -> conj clause            /* byl bych býval */
/* budu muset číst */        cpredcondgr ==> VBL VBK VBLL
futmod --> VBU VOI VI        /* musím se ptát */
                             clause ===> VO R VRI
```

The single arrow (->) denotes an ordinary CFG transcription. The extended arrow (-->) supplements the right side with possible intersegments in between each couple of listed elements. The extended double arrow (==>) adds (besides filling in the intersegments) the checking of correct enclitics order. This flag is more useful in connection with the order() and rhs() constructs discussed below. The three character double arrow (===>) provides the completion of the right side to form a full clause. It allows to add intersegments in the beginning and the end of the rule, tries to supply the clause with conjunctions etc.

The global order constraints represent universal simple regulators, that are used to inhibit some combinations of terminals in rules.

```
/* jsem, bych, se */          /* byl — četl, ptal, musel */
%enclitic = (VB12, VBK, R)    %order VBL = {VL, VRL, VOL}
                              /* býval — četl, ptal, musel */
                              %order VBLL = {VL, VRL, VOL}
```

In this example the %enclitic specifies which terminals should be regarded as enclitics and determines their order in the sentence. The %order constraints guarantee that the terminals VBL and VBLL always go before any of the terminals VL, VRL and VOL.

The main combinatoric constructs in the meta-grammar are order(), rhs() and first() which are used for generating variants of assortments of given terminals and nonterminals.

```
/* budu se ptát */
clause ===> order(VBU,R,VRI)
/* který ... */
relclause ===> first(relprongr) rhs(clause)
```

The order() construct generates all possible permutations of its components. The first() and rhs() constructs are employed to implant content of all the right sides of specified nonterminal into the rule prefixed with the attribute of first() that is firmly tied to the beginning, it cannot be preceded by an intersegment neither by any other construct.

4   Lexico-semantic Constraints

The analysis is supported by a set of commonly used grammatical tests that have been described in [6]. In addition to these tests we have extended the valency test functions with lexico-semantic constraints. The constraints take advantage of an ontological hierarchy of the same type as in Wordnet [7]. They enable us to impose a special request of compatibility with selected class or classes in the hierarchy to each valency expression.

An example of the constraints in action can be demonstrated by the following phrase:

Leaseholder	*draws*	*beer.*	
Nájemce	čepuje	pivo.	čepovat
k1gMnSc1245,k1gMnPc4		k1gNnSc145	= sb.<HUMAN> & st.<LIQUID>

The lexico-semantic constraints that are found in the valency list of the verb čepovat (draw) make it possible to distinguish the word pivo (beer) as an object and **nájemce** (leaseholder) as the subject. Considering metonymy and other forms of meaning shifts we do not regard this feature so strictly to throw out a particular analysis. We use it rather as a tool for assigning preferences to different analyses.

The part of the system dedicated to exploitation of information obtained from our list of verb valencies is necessary for solving the prepositional attachment problem in particular. During the analysis of noun groups and prepositional noun groups in the role of verb valencies in a given input sentence one needs to be able to distinguish free adjuncts from obligatory valencies. We have implemented a set of heuristic rules that determine whether the found noun group typically serves as a free adjunct. The heuristics are also based on the lexico-semantic constraints described above.

In the forest he walked only in a T-shirt.

V lese   chodil jenom   v tričku.
<PLACE>   <ARTIFACT>

In this example the expression v `lese` (in the forest) is denoted as a free adjunct by the rule specifying that the preposition v (in) in combination with a <PLACE> forms a location expression and by the valency list of the verb `chodit` (walk).

5   Conclusions

The presented system has the potential to become a launch pad to a robust natural language parser augmented by semantical constraints and case frames. Even the contemporary version has proved to be a highly suitable tool for various kinds of Czech text processing. Future research will aim at extensions of the meta-grammar formalism together with enlarging the coverage of Czech grammatical phenomena.

References

1. Sarkar,A.: Incremental Parser Generation for Tree Adjoining Grammar. In Proceedings of the 34th Meeting of the ACL, Student Session. Santa Cruz, June 1996
2. Volk, M., Schneider, G.: Comparing a statistical and a rule-based tagger for German. http://xxx.lanl.gov:80/ps/cs/9811002, September 1998.
3. Hajičová, E., Sgall, P., Skoumalová, H.: An Automatic Procedure for Topic-Focus Identification. Computational Linguistics 21, pp. 81–94, 1994.
4. Kuboň, V.: A Robust Parser for Czech. Technical Report TR-1999-06, UFAL, Charles University, Prague.
5. Dodd, C.: BtYacc — BackTracking Yacc. http://www.siber.com/btyacc/
6. Smrž, P., Horák, A.: Determining Type of TIL Construction with Verb Valency Analyser. In Proceedings of SOFSEM'98, pp. 429-436, Springer-Verlag. 1998.
7. Miller, G.: Five papers on WordNet. Special Issue of International Journal of Lexicography 3(4), 1990.

Word Sense Disambiguation of Czech Texts

Ondřej Cikhart and Jan Hajič

Institute of Formal and Applied Linguistics,
MFF UK, Malostranské nám. 25, Praha, CZ-11800, Czech Republic
cikhart@ufal.mff.cuni.cz, hajic@ufal.mff.cuni.cz

Abstract. This contribution refers to the project of BYLL Software Ltd. that uses human aided WSD for the annotation of a fulltext database of the Czech law system named ASPI. We used about 3 mil. words of annotated texts from the law system of the Czech Republic since the 60's. The annotated law corpus provides certain text regularity, but at the same time it covers wide range of subjects. The goal has been to save as much of the human intervention during text indexing as possible, measured by the number of queries posed to the human annotator, whilst retaining truly minimal error rate ($\sim$0.5 %) in the automatically disambiguated cases. A combination of Naive Bayes, Decision Lists and (minimal number) of manually written rules has been used. The statistical methods showed up to be appropriate for our intention. The results show that we have saved 80 % of queries to the human annotator, which proved to be enough to warrant the inclusion of the software into a production system.

1 Introduction

In the connection with the development of information technologies, the solution of the word sense disambiguation (WSD) problem is considered important. It is obvious that a good solution of the WSD problem would be a precious help also in machine translation, fulltext database annotation, or web pages annotation areas (and others as well).

This contribution refers to the project of BYLL Software Ltd. that uses human aided WSD for the annotation of a fulltext database of the Czech law system named ASPI. We could use large amount of annotated texts from the law system of the Czech Republic since the 60's – it was about 3 million words, and the statistical methods showed up to be appropriate for our intention. The annotated law corpus provides certain text regularity, but at the same time it covers wide range of subjects.

Our approach is based on both Naive Bayes (NB) (Gale et al., 1992) and Decision List (DL) (Yarowsky, 1994) methods. However, we also use morphology (including lemmatization) developed at UFAL MFF UK. Our objectives were strict: we aimed at practical application, where the acceptable error rate is bellow 0.5 %. Moreover, we needed to be able get even better recall (close to 100 %) at an acceptable loss of precision. Of course, with these practical goals in mind, we don't focus at all on words that are rarely (if ever) used for search in fulltext databases like the pronoun se or some prepositions, conjunctions etc.

F. Jelinek, E. Nöth (Eds.): TSD'99, LNCS 1692, pp. 109–114, 1999.
© Springer-Verlag Berlin Heidelberg 1999

2   The Method

We use almost 3 million words corpus that consists of text from the Czech law system for training and the rest of it for testing. This corpus displays certain text regularity and at the same time it covers quite wide range of subject areas. The data in the corpus were manually disambiguated. Thus every word is provided with a disambiguated lemma. The percentage of words that had to be disambiguated is about 15.5 % – the rest is unambiguous with respect to the law domain.

To get objective results, we made three times random data selection into parts D and T (in fact $D0 + T0$, $D1 + T1$ and $D2 + T2$) the proportion being about $2 : 1$. Dx was the part used for training and Tx was the part used for testing.

Each word with several candidates enters the disambiguation process that leads to three possible results - **1**, **?** and **0**. **1** means match of the true lemma with the winning candidate. **?**, it means there is no winning candidate ("I don't know."). Every **?** counts as "good" for recall computation, incurring appropriate penalty for precision. When the result of disambiguation process gets the **0** mark, there is difference between the winning candidate and the true lemma. In fact, due to the way WSD is used in the application, we are not interested in exact precision as much as in the percentage of the **?** answers vs. definite answers.

The decision system uses two or three layers (vide fig. 1.). The first layer always consists of hand written rules that cover words with zero statistics or words that are often erroneously disambiguated and that are from the user's point of view important. This layer is of the highest priority. The second layer consists of two DL models (in a very similar manner as in the Yarowsky's work) in conjunction. This layer is optional. It's task is to cover words that cannot be reliably disambiguated by the NB layer. The third layer consists of a weighted linear combination of three NB models (they differ in size of their window spans).

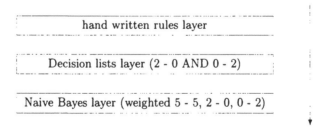

Fig. 1. Three layers of the decision system

To reduce demands on the size of training data set (i.e. to increase the reliability of the estimates), the second and the third layer work on lemmatized context. It means that every word in the context (window span) is replaced with a chain of possible lemma candidates of this word. In this sense the sentence

"Žena je dnem rybníka." is transcribed as *"hnát|žena být|on den|dno rybník."*.
The fact that we use the non-disambiguated context avoids the necessity of any
dynamic programming at runtime.

2.1 Decision Lists (DL) Layer

The DL layer turned out to be a very perspective one. This model seems in
many situations to be superior to NB models. Its application helps to solve
many troubles that are connected with NB models. On the other hand, DL
models tend to increase the error rate and thus we set the DL layer currently
optional: larger training set or more responsible selection of DL items could
help to dispose of this problem. The automatic selection of the lists is very time
demanding, therefore we focus on the words with the worst results in NB layer
only.

The idea of DL is very simple. For each possible candidate lemma we construct such a list of lemmas (non-disambiguated, cf. above), which contains words
the appearance of which in the lemmatized context determines the winning candidate (fig. 2.). If (at runtime) two different candidates are declared to be a
winner, DL model returns ?. We use two DL models with window span (number
of chains on the left – number of words on the right) 2 - 0 and 0 - 2 in the same
way. If these two models declare two different candidates to be winners (or none
is declared), the DL layer as a whole returns – ?. Otherwise the DL layer returns
the winning candidate.

cena-1 (cost)	akcie, dodavatel, kupní, dampingový \ dumpingový darovací
cena-2 (prize)	soutěžní

Fig. 2. Decision lists for candidates cena-1 (cost) and cena-2 (prize)

The DL lists are being selected automatically for each ambiguous set of lemmas. One of the selection criterion is the mutual information between the candidate and the lemma in the window. Another criterion is the entropy of the
possible candidates conditioned on the lemma in the window. If the entropy is
too high, the item is disqualified. This should prevent selection of items that
often co-occur with the candidate, but only randomly.

2.2 Naive Bayes (NB) Layer

Most of the work is done by the NB layer. The essence of the Bayesian approach
(Gale et al., 1992) is to compute the probability of each candidate l of the

word W_i given the context C_i, $P(l|C_i)$, and choose the most probable candidate. $P(L|C_i)$ is computed:

$$P(l|C_i) = \frac{P(C_i|l) \cdot P(l)}{P(C_i)}$$

Thus we aim to maximize the function $F(l|C_i) = P(C_i|l) \cdot P(l)$. We can estimate:

$$P(l) \approx \frac{N(l)}{N_D}, P(C_i|l) \approx \prod_{\omega \in C_i} \frac{N(l, \omega)}{N(l)}, \qquad \text{where}$$

N_D is total number of words in training data set D and ω is lemma chain in context. To get better results, we can use linear interpolation smoothing using one parameter λ. Thus finally

$$F(l|C_i) = \frac{N(l)}{N_D} \cdot \prod_{\omega \in C_i} \left(\lambda \cdot \frac{N(l, \omega)}{N(l)} + (1 - \lambda) \cdot \frac{N(\omega)}{N_D} \right)$$

To improve results of NB models, we use linear weighted combination of three models with different window span. One of them is a "semantic" one; it means wider window (up to 6 - 6). Another two models are "local" ones with a window span up to 2 - 0 and 0 - 2. Let

$$M(l_j|C) = \frac{F(l_j|C)}{\sum_{l_i \in \omega} F(l_i|C)}$$

Then the final probability of every candidate is set

$$p(l|C) = w_1 \cdot M_1(l|C) + w_2 \cdot M_2(l|C) + w_3 \cdot M_3(l|C), \qquad \text{where}$$

$$w_1 + w_2 + w_3 = 1$$

Weights w_1, w_2 and w_3 are currently set manually. If $p(l|C)$ is greater than a reliability threshold, the candidate l is declared to be a winner, otherwise the NB layer returns ?.

The last step is to set the reliability threshold for each possible candidate set. This could be done automatically. We divided the test data set into two parts. One part was used for the reliability threshold setting and the second for final evaluation testing. The strategy was very simple: for every candidate set we started at a reliability threshold 0.5. Then we were cycling through the data set and kept increasing the threshold until the results over the candidate set would be 1 or ?. We did that over all of three data sets Dx/Tx.

3   Results

The WSD algorithm was cross-tested on the three data sets Dx/Tx. The training data set $D0$ contained 2.1 mil. words, $D1$ contained 2 mil. words and $D2$ contained 1.8 mil. words. The results of various combinations of NB window spans are shown in the Table 1 and 2.

The best results were reached by combinations 5 - 2 - 2 and 6 - 2 - 2. The smoothing parameter λ was set to 0.9. The best results were usually reached by weight settings about 0.32 for w_1 and 0.34 for W_2 and w_3. The results of DL layer were tabulated only with the 5-5, 2-0, and 0-2 combination.

Table 1. Results, without an application of the DL layer. (Combination of NB window spans X - X, Y - 0 and 0 - Z is transcribed as X - Y - Z.) Every training data set Tx contained over 35 000 events.

%	T0			T1			T2		
	1	?	0	1	?	0	1	?	0
4 - 1 - 1	78.8	20.9	0.3	79.5	20.2	0.3	80.1	19.6	0.3
5 - 1 - 1	79.2	20.5	0.3	79.6	20.1	0.3	80.4	19.3	0.3
6 - 1 - 1	79.4	20.3	0.3	79.7	20.0	0.3	80.3	19.4	0.3
4 - 2 - 2	79.4	20.3	0.3	79.8	19.9	0.3	80.1	19.6	0.3
5 - 2 - 2	79.5	20.2	0.3	80.0	19.6	0.4	80.4	19.3	0.3
6 - 2 - 2	79.7	20.0	0.3	80.1	19.5	0.4	80.4	19.3	0.3

Table 2. Results of system with application of DL layer.

%	T0			T1			T2		
	1	?	0	1	?	0	1	?	0
5 - 2 - 2	87.5	11.7	0.8	85.9	13.4	0.7	86.0	13.3	0.7

4   Conclusions

The project reached its goals: got 80 % of definite responses at the error rate 0.4 % (= 100 % - recall). It means saving 80 % inquiries to the operator, which is sufficient for our application. The WSD problem of the Czech texts was deeply explored by statistical methods. We suppose that in our project, the Bayesian approach reached it's limits. Further experiments with other various window spans and weights combinations could return only very limited improvement. The distinct improvement can be, in our opinion, reached only with new and substantially different layers consisting of e.g. rule based models, grammar models, parsers etc.

References

1. Cikhart, O. Lexikální disambiguace českých textů. Master thesis, MFF UK Praha, 1998.
2. Fujii, Atsushi. Corpus-Based Word Sense Disambiguation. PhD thesis, Report No. TR98-0003, University of Library and Information Science, Tokyo Institute of Technology, Japan, 1998.

3. Gale, William A., Kenneth W. Church, and David Yarowsky. A method for disambiguating word senses in a large corpus. Computers and Humanities, 26:415-439, 1992.
4. Laciga, Z. Praktická aplikace lingvistické analýzy při vyhledávání v česky psaných textech. Sbornik konference EurOpen CZ' 99, 1999.
5. Yarowsky, D. Word-sense disambiguation using statistical models of Roget's categories trained on large corpora. In Proceedings of Coling-92, 1992.
6. Yarowsky, D. Decision lists for lexical ambiguity resolution: Application to accent restoration in Spanish and French. In Proceedings of 32nd meeting of the ACL, Las Cruces NM, 1994.

The Acquisition of Some Lexical Constraints from Corpora

Goran Nenadić[1] and Irena Spasić[2]

[1] Faculty of Mathematics, University of Belgrade, Yugoslavia
goran@matf.bg.ac.yu
[2] Faculty of Economics, University of Belgrade, Yugoslavia
irenas@ekof.bg.ac.yu

Abstract. This paper presents an approach to acquisition of some lexical and grammatical constraints from large corpora. Constraints that are discussed are related to grammatical features of a preposition and the corresponding noun phrase that constitute a prepositional phrase. The approach is based on the extraction of a textual environment of a preposition from a corpus, which is then tagged using the system of electronic dictionaries. An algorithm for computation of some kind of the minimal representation of grammatical features associated with the corresponding noun phrases is suggested. The resulting set of features describes the constraints that a noun phrase has to fulfil in order to form a correct prepositional phrase with a given preposition. This set can be checked against other corpora.

1 Introduction

Every natural language processing system needs some linguistic knowledge that is obtained either from grammar books or from corpora. The most frequent case is that the knowledge is obtained jointly from "books" and corpora. In this paper we will present a method of automatic acquisition of some lexical and grammatical constraints for a highly inflective language such as Serbo-Croatian. These constraints are acquired from a large corpus. The corpus we worked with is a newspaper corpus containing text taken from a few Yugoslav daily newspapers presented on the Internet,[1] which is therefore electronically available. This way we explore up-to-date language and collect its corresponding corpus as a complement to existing corpora for Serbo-Croatian, which usually contain texts from literature, poetry, and law. The corpus is automatically collected from the URLs mentioned and then converted from typical Windows 1250 encoding format to ASCII for the purposes of employing the system of electronic dictionaries. Also, as there are other intentions for the corpus, we convert it to SGML/TEI Lite scheme [1] and store it in a database.

Since Serbo-Croatian has a rich morphological system, its processing is based on the initial tagging performed using the system of electronic dictionaries (see

---

[1] These newspapers include "Blic" (http://www.blic.co.yu) and "Politika" (http://www.politika.co.yu).

F. Jelinek, E. Nöth (Eds.): TSD'99, LNCS 1692, pp. 115–120, 1999.

[4], [7]). It provides grammatical information for simple and compound words occurring in an electronic text. When applied on an e-text, this system generates the corresponding initially tagged text, which is, generally, ambiguous. Namely, every word in the text is associated with its lexical interpretation, which represents all possible part-of-speech (POS) tags; some of POS tags are not applicable in the context given. Additional processing can be performed on the initially tagged text in order to lemmatize, disambiguate and/or recognize its syntactical constituents (e.g. noun phrases, verb phrases, or prepositional phrases).

2   Extraction of Lexical Constraints Related to Prepositional Phrases

One of the constraints we studied is the relationship between the morpho-syntactic features of a preposition (PREP) and a corresponding noun (N) or noun phrase (NP) that constitute a prepositional phrase (PP). The idea is to extract a textual environment of a PREP [2] from the corpus, and then to analyze it. The extraction of the environment for a PREP is performed using the INTEX corpus processing system (see [4], [5]). Since Serbo-Croatian is morphologically reach language, a lot of lexical ambiguities arise, and we get lexically highly ambiguous sequence of words (see [2], [3]). Here is an example of initial tagging of a textual environment of the preposition na (Engl. on):

{na.PREP} {apsurd,.Nmsa;Nmsn-;}
{na.PREP} {bijektivnost/preslikavanja,.Nfsa-;Nfsn;}

According to linguistic knowledge, we can define a local grammar (see [5]) which describes the following grammatical constraint: the preposition na has to be followed by a NP either in accusative or locative case (see [6]). This way, other tags associated with NP can be discarded:[3]

{na.PREP} {apsurd,.Nmsa-;}
{na.PREP} {bijektivnost/preslikavanja,.Nfsa-;}

In this paper, we approached this problem in the opposite direction, that is – we wanted to automatically generate, by analyzing a large, initially tagged corpus, a local grammar that is to describe a PP structure. The idea was to compute some kind of the "intersection" of morpho-syntactic features of NPs contained in PPs extracted from the corpus. As the initially tagged text is ambiguous, we want to define a method of extraction of a "minimal" set of morpho-syntactic features that are inherent for every NP that is a part of the PP. Therefore, other tags can be left out so that every NP keeps at least one feature from the "minimal" set.

---

[2] Prepositional phrases have the following structure: PREP NP, and thus we need to extract only the right context of a preposition in question.

[3] Note that the above tags are completely disambiguated. It is not always the case that we get fully disambiguated sequence – in general, local grammars are used only to reduce ambiguities.

We give the formal definition of this notion. Firstly, we briefly overview the necessary definitions introduced in [2] and [3].

A regular morpho-syntactic expression (RME) is a regular expression with *extended morpho-syntactic description* (**EMD**) as a basic component. An EMD is set of *morpho-syntactic descriptions* (**MDs**). A MD is a string that denotes grammatical categories for noun(s) and/or adjective(s). The formal definitions for MD, EMD and RME can be found in [3]. Here we give only a few examples:

N	the MD denoting a noun
Nsg;	the MD denoting a noun (*N*) in singular (*s*), genitive case (*g*)
baza.Nf;	the MD denoting all forms of noun *baza* (f)
baza.Nfsg-;	the MD denoting noun *baza* in genitive form of singular
Nfsg-;Nfpn-;	the EMD denoting a noun in singular, genitive case, feminine OR a noun in plural, nominative case, feminine
baza.Nfsg-;Nfpn-;	the EMD denoting appropriate forms of noun *baza*
N Ag Ng	the RME denoting the sequence of three words such that the first is a noun, the second is adjective in genitive case and the third is noun in genitive case

As a result of the initial tagging of an e-text performed using the system of electronic dictionaries (see [3,7]) one gets the sequence of EMDs. Whether a sequence of initially tagged words satisfies an RME can be concluded by calculating the **match**-relation. In short, two MDs satisfy the **match**-relation *iff* the first includes all grammatical features that are present in the second (in that case, we also say that the first MD is *more restrictive* than the second). The relation **match** can be extented so that the first argument can be an EMD: an EMD matches a morpho-syntactic description D iff there exists morpho-syntactic description $D1$ in the EMD such that **match**($D1, D$) is true, or formally

$$\text{match}(E, D) = \bigvee_{D_1 \in E} \text{match}(D_1, D)$$

Also, we can allow both arguments to be EMDs:

$$\text{match}(E_1, E_2) = \bigwedge_{D \in E_2} \text{match}(E_1, D)$$

For example,

$$\text{match}(\texttt{Nfsg-;Nfpn-;Nfpa-;Nfpv-;}, \ \texttt{Na;}) = \top,$$

since

$$\text{match}(\texttt{Nfpa-;}, \ \texttt{Na;}) = \top.$$

Also,

$$\text{match}(\texttt{Nfsg-;Nfpn-;Nfpa-;Nfpv-;}, \ \texttt{Na;Nn;}) = \top,$$

since

$$\text{match}(\texttt{Nfsg-;Nfpn-;Nfpa-;Nfpv-;, Na;}) = \top$$

and

$$\text{match}(\texttt{Nfsg-;Nfpn-;Nfpa-;Nfpv-;, Ng;}) = \top.$$

Let us, now, define a **represent**-relation (A and B are *sets* of EMDs):

$$\text{represent}(A, B) = \bigwedge_{E_1 \in A} \bigvee_{E_2 \in B} \text{match}(E_1, E_2)$$

If $\text{represent}(A, B) = \text{true}$ and $B \subset A$, we refer to B as a **representation** of A. Note that a representation is by no means unique and that there always exists at least one (e.g. the whole set). Here we are interested in a **minimal representation** (MR) in the sense of its cardinality. Consider an example: let A be $\{\texttt{Na;, Nd;Nl;, Na;Nn;, Nd;Nl:Ni;}\}$. A representation of A can be $\{\texttt{Na;, Nd;Nl;, Na;Nn;}\}$, but the minimal is $\{\texttt{Na;, Nd;Nl;}\}$.

The calculation of the MR for a given set A of EMDs can be done iteratively as follows. Firstly, we extract all EMDs from A that are of minimal "length" (i.e. that consists of minimal number of MDs). These EMDs are a part of MR. Then, we discard all EMDs from the set A that match at least one EMD from the current MR. We apply the same procedure until A is empty.

Consider, now, the following example. The minimal representation of the set $B = \{\texttt{Na;Ng;, Na;Nl;}\}$, according to our definition, is the whole set B. Note that the set $A = \{\texttt{Na;}\}$ matches B, but we do not consider it as a (minimal) representation, although it is minimal both in the sense of its cardinality and the length of EMDs it contains. The reason is the fact that we do not want to prefer tag $\texttt{Na;}$ to $\texttt{Ng;}$ or $\texttt{Nl;}$, because it could be the case that these cases are correct tags (in the context given). That is why we restricted a MR to the subsets of the given set.

The primary motivation for such an approach (and the definition of MR) could be described as follows. We can say that we prefer shorter EMDs to the longer ones. But, at the same time, we do not want to remove any correct morpho-syntactic descriptions. More precisely, we filter out all the EMDs that can be matched by the shorter ones, but the latter must be "witnessed" by the corpus, that is – they have the exact matches in the set of EMDs taken from the corpus. This means that resulting MR depends on those morpho-syntactic tags (from the corpus) that are maximally disambiguated. The more representative the corpus is, the more restrictive the corresponding MR is.

Let us discuss the acquisition of constraints for a PP from a corpus. If we calculate a MR for a set of right contexts for a PREP, we get a set of features that describes the context allowed by the PREP. Consider, for example, the excerption of the list containing corresponding NPs of the preposition na[4]:

---

[4] Note that we have mapped complete MDs to MDs that contain only marks for cases as we are interested in them. Similarly, if we want to discuss some other constraints we can make the corresponding mapping.

```
            . . .
na Nd;Nl;              na Nd;Nl;
na Na;                 na Na;Nn;
na Na;Nv;              na Na;
na Nd;Nl;Ni;           na Na;Ng;Nv;Nn;
                                . . .
```

The MR computed for the set extracted from the corpus is {Na;, Nd;Nl;} which means that a NP that follows the preposition *na* is either in accusative case, or in a form that corresponds to dative and locative at the same time! Although this result seems not applicable, it is understandable if we bear in mind the fact that these two cases (locative and dative) are homographs. In addition, if we compare this corpus-based result to the local grammar mentioned above, we see that our approach did not lose any possible grammatical constraint concerning cases. On the other hand, the "overgeneration" is due to morphological characteristics of Serbo-Croatian, which can not be resolved without additional linguistic knowledge.

The implementation of the algorithm presented is done in the programming language C. Here we present results of processing some prepositions:

PREP	Sample EMDs	MR	Theoretical local grammar
prema	Nd;Nl; Nd;Nl;Ni; Nd;Nl;Ng;	Nd;Nl;	dative or locative
od	Ng; Ng;Na; Ng; Ng;Nn;Na;Nv; Nn;Ng; Ng;Nn;Nv;	Ng;	genitive
na	Nd;Nl; Na; Na;Nn Nd;Nl;Ni Na;Ng;Nv;Nn Na;Nv Nd;Nl;Na	Na;, Nd;Nl;	accusative or locative

The resulting set (MR) can automatically be converted into an INTEX graph and stored for later uses. These include checking the results against other corpora (e.g. corpus on study-books, literature, etc.), as well as automatic disambiguation of the corpus using the INTEX system (see [5]). However, this method of a local grammar generation was applied on one part of the corpus, and then checked against the rest of the same corpus. The resulting MRs did not contradict the rest of the corpus.

We have to stress out the fact that after the extraction of an environment for a PREP, we have applied some local grammars (using the INTEX system, see [5]) which discarded some lexicalized phrases that contained the PREP. These phrases contain either "frozen" parts (e.g. *(cene) na malo, od tada, prema tome,...*) or non-inflective constituents (e.g. numbers, dates,...). Also, the IN-TEX system (for the purpose of this research) removed all instances of PPs that could not be tagged (either because they contained words not stored in the dictionary, or because the available local grammars could not recognize a corresponding NP).

3   Conclusion and Further Research

In this paper we have presented an approach for the acquisition of some grammatical constraints related to prepositional phrases. The similar procedure can be implemented for consideration of constraints related to some classes of adverbs, particularly adverbs that denote some abstract and non-countable quantity (e.g. *nesto* (Engl. some), *malo* (a little), *mnogo* (a lot of), etc.). Also, verb valency acquisition can be discussed in the same manner.

The similar approach can be applied for the acquisition of lexical constituents of some classes of NPs, particularly the structure of companies' names and names of the countries. By analyzing the corpus, we plan to automatically extract possible lexical constituents that can form a name of a factory or of a faculty. For example, one of possible structures of a name of a factory is: *Fabrika za proizvodnju* <NP:g;> where NP is from the set {*odlivak, vijak, automobil, usisivač, bela tehnika, sportska oprema,...*}. The main goal is to automatically generate a graph that can describe name structures found in a corpus.

References

1. Burnard, L. et al: TEI Lite: An Introduction to Text Encoding for Interchange, doc. No: TEI U 5, June 1995.
2. Nenadić, G., Vitas, D.: Using Local Grammars for Agreement Modeling in Highly Inflective Languages, in Proc. of First Workshop on Text, Speech, Dialogue - TSD 98, Brno, 1998.
3. Nenadić, G., Vitas, D.: Formal Model of Noun Phrases in Serbo-Croatian, BULAG 23, Universite Franche-Compte, 1998.
4. Silberztein, M.: Dictionnaires électroniques et analyse automatique de textes: le systéme INTEX, Masson, Paris, 1993.
5. Silberztein, M.: INTEX 3.4: Reference manual, LADL, Universite Paris 7, 1996.
6. Stanojčić, Ž., Popović, Lj.: Gramatika srpskoga jezika, Zavod za udžbenike i nastavna sredstva, Beograd, 1994. (in Serbo-Croatian).
7. Vitas, D.: Mathematical Model of Serbo-Croatian Morphology (Nominal Inflection), PhD thesis, Faculty of Mathematics, University of Belgrade, 1993. (in Serbo-Croatian).

Run-Time Extensible (Semi-)Top-Down Parser*

Michal Žemlička and Jaroslav Král

Department of Software Engineering, Faculty of Mathematics and Physics,
Charles University, Prague, Czech Republic
{zemlicka,kral}@ksi.mff.cuni.cz

Abstract. When reading a text or listening to a speech the words are processed by humans in the order they come. Intuitively there are some mental actions just after morfologic analysis of any newly recognized word. This mental action helps understanding of the given word (or positioning the word within the frame of the — still not complete — sentence). Within parsing of formal languages the closest to this idea is the top-down parsing that is usually used only together with LL grammars. Top-down parsing of programming languages has the advantage that it is possible to implement it by recursive descent parser — i.e. by a system of procedures that may recursively call each other. Such a system may be "tuned" by hand made changes. The usage of LL grammars is not always possible, because of the grammars of programming languages may have left recursive symbols. Programming language grammars are intuitively "close" to LL grammars. A good model for such grammars are kind grammars studied in this contribution. Kind grammars preserve all the important features of LL grammars advantageous for parsing.

1 Introduction

Parsing may be seen from different points of view: As a task to determine the structure of the text, as a task to recognize given texts for long period, or as an engineering task to create and support tools for sooner mentioned tasks.

Advantages of top-down parsing are well known to everybody who tried to realize production quality compiler. The main advantage is that it is easy to generate parser as a program formed by a set of procedures recursive calling each other. There is a one-to-one correspondence between nonterminals of the given grammar and the procedures. Similar correspondence is between parts of computation and right parts of productions. The constructor generating a parser from the grammar is able to insert semantic actions into generated program. Resulting product may be tuned by hand — it is (usually) needed. It concerns semantic actions and also the error recovery.

It is known for a long time that programming languages (especially the modern ones) are almost LL. Unfortunately their grammars often contain left recursive productions that cannot be productions of LL grammars. This observation

---

* Supported by Grant Agency of Czech Republic, Grant-No. 201/99/0236.

F. Jelinek, E. Nöth (Eds.): TSD'99, LNCS 1692, pp. 121–126, 1999.

led to attempts to develop constructor that to every LR(k) grammar generates recursive descent parser (i.e. it generates sytem of recursive procedures — see [4,3]). Such a constructor generates cumbersome programs if the grammars are not almost LL. The produced parsers are not easy to enhance by new definitions in the style discussed in [8].

The role of the left recursion is very restricted in the grammars of programming languages — usually only to the syntax of the lists and of the expressions. Kind grammars discussed below were designed as a reasonable extension of LL grammars containing left recursive symbols. Roughly speaking kind grammars differ from LL grammars by admitting direct left recursive productions (if $A \to^* Ay$ then all memebers of the derivation have form Aw_i), and productions that may have common beginning part (e.g. $A \to BcD$ and $A \to BeF$, where B can be rewritten to string of arbitrary length).

Kind grammars allow to develop constructors generating recursive descent dynamically extensible parsers.

We use notions and notations from the theory of formal languages and compilers — see e.g. [1].

2   Motivation

Examples are the best way how to look into the principles of kind parsing. First it will be shown how to order the production and then how to extend the the parser incrementally.

Example 1 Let be following grammar for integer arithmetical expression written in Backus-Naur form:

$$S \to E \dashv$$
$$E \to E + T \mid E - T \mid T \mid + T \mid - T$$
$$T \to T * F \mid T \text{ div } F \mid T \text{ mod } F \mid F$$
$$F \to (E) \mid id \mid num$$

Note 1. Right sentinel is introduced just to make the explanation more simple. It is possible to cancel it without any change of generality.

The productions are divided into groups. Each group contains productions with the same left hand side. For the given left part there is a group containing all non left recursive productions with the given left-hand side and possibly yet another group containing just all the left recursive productions with the given left-hand side. This way we get classification dividing productions into no more than $2 \cdot |N|$ groups (some of them may be empty).

Every group is transformed into a "production tree" (its edges are labeled by symbols and reading it from root to leaves gives right parts of productions with the same nonterminal in the left part; for the left recursive productions the recurring symbol is omitted).

Doing a few little changes we get structures for the nonterminals similar to the one shown below. It is possible to convert the structure into a program in programming language.

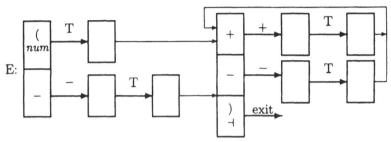

Edge labeled by a nonterminal denotes invocation of its parsing, edge labeled by a terminal denotes its reading. Terminals in frames denote lookahead. Empty frame denote that no lookahead testing is necessary. Parser generated following this scheme may look as follows:

```
PROCEDURE Parse_E;
  BEGIN
    CASE LookAhead OF
      term_num, term_left: Parse_T;
      term_minus: BEGIN ReadTerm(term_minus); Parse_T END;
      END;
    WHILE LookAhead IN [term_plus,term_minus] DO
      CASE LookAhead OF
        term_minus: BEGIN ReadTerm(term_minus); Parse_T END;
        term_plus: BEGIN ReadTerm(term_plus); Parse_T END;
        END;
  END;
```

If semantic actions are put before and after any symbol then their program equivalents may be inserted before and after corresponding command.

Let us label the places for semantic actions within productions by $A_{i,j}$ where A is a nonterminal, i number of production for given nonterminal, and j place in the right part of the production. A_{10} denotes "place" before the first symbol of the right part of the first A-production. Then the parser can produce the following "parse" of the string $(3 + 2) * 7 \dashv$:

S_{10}						S_{11}	S_{12}
E_{30}						E_{31}	
T_{40}					$T_{41}\ T_{11}$	T_{12}	T_{13}
F_{10}	F_{11}			F_{12}	F_{13}	F_{30}	F_{31}
	E_{30}	$E_{31}\ E_{11}$	E_{12}	E_{13}			
	T_{40}	T_{41}	T_{40}	T_{41}			
	F_{30}	F_{31}	F_{30}	F_{31}			
(	3		+	2	)	$*$ 7	$\dashv$

The columns contain sequences of semantic symbols generated at the place indicated in the last row. The parser therefore generates the following string

$$X = S_{10}E_{30}T_{40}F_{10}"("F_{11}E_{30}T_{40}F_{30}"3"F_{31}T_{41}E_{31}E_{11}" + "\ldots \quad (1)$$

Indexed symbols in (1) are *formal semantic symbols* (FSS). Word (1) includes in some sense the richest FSS structure that may be generated from syntactic structure of the arithmetic expression. It is an encoding of the syntactic tree. We call it *complete syntactic description* (CSD).

During the translation of arithmetic expressions nonempty semantic actions are executed in the positions of some FSS. Formaly it can be expressed so that the symbols in (1) denoting empty actions will be replaced by empty symbols and other FSS will be replaced by the symbols denoting the corresponding actions. I.e. a homomorphism h, $h(a) = a$ for $a \in T$ will be performed. $h(X(v))$ will be called *translation* of v. Translation is computable from the left (with lookahead 1) if $h(X(v)) = \tau_0 v_1 \tau_2 v_2 \ldots \tau_{i-1} v_i \tau_i ax$ and $h(X(u)) = \tau_0' v_1 \tau_2' v_2 \ldots \tau_{i-1}' v_i \tau_i' ay$, then $\tau_j = \tau_j'$ for $j = 1, 2, \ldots, i$.

By the kind parsing it is possible to reconstruct all FSS except that ones that are just before left recursive phrases. This restriction is derived from the principles of left recursion and from the parsing from left to right. It is the maximal possible information which can be generated at given point knowing lookahead and left context only.

3   A Bit More Formally

3.1   Production Sets

The productions of the form $A \rightarrow A\alpha$ are A-productions with direct left recursion. Let us write $\text{NLRP}_G(A) = \{A \rightarrow \alpha \mid A \rightarrow \alpha \in P_G \wedge \neg(\alpha \Rightarrow^* A\beta)\}$ and $\text{DLRP}_G(A) = \{A \rightarrow A\alpha \mid A \in N\}$.

3.2   Lookahead Functions

Set of terminal strings of length k that may follow given nonterminal A except its left recursive phrases will be called *non left recursive follow of A (NLRF(A))*. Set of terminal strings of length k that may follow given nonterminal (A) directly after its left recursive phrase will be named *direct left recursive follow (DLRF(A))*.

3.3   Kind Grammars

Definition 1 (k-kind grammar). *A context-free grammar G having only productions without left recursion and productions with direct left recursion is called kind if*

1. *for for every two productions $A \rightarrow \alpha X\beta$, $A \rightarrow \alpha Y\gamma \in P_G$ and $X \neq Y$ then $Pref(k, X\beta) \cap Pref(k, Y\gamma) = \emptyset$, and*
2. *for every nonterminal A $DLRF_G^k(A) \cap NLRF_G^k(A) = \emptyset$.*

Theorem 1. *Any LL(k) grammar is a k-kind grammar. It is possible to create a LL(k) grammar to every k-kind grammar generating the same language. The class of k-kind grammars is a strict superclass of the class of LL(k) grammars and a strict subclass of the class of LR(k) grammars.*

Proof is shown in [10].

4   Kind Parsing Schema

Having lack of space the *kind parsing schema* (see [6]) is introduced just by presenting *kind parsing system* for every context-free grammar. Restricting grammars to the kind ones we get from extended kind parsing to kind parsing.

$$\mathcal{I} \quad = \{[A \to \alpha., i, j] \mid \exists \beta : A \to \alpha\beta \in P, \ 0 \le i \le j\}$$
$$\cup \{[a, i, i+1] \mid a \in T, \ 0 \le i\}$$
$$\mathcal{D}^{Init} \quad = \{\vdash [S, 0, 0]\}$$
$$\mathcal{D}^{Expand} \quad = \{[A \to \alpha., i, j], [a^1, j, j+1], \dots, [a^k, j+k-1, j+k] \vdash$$
$$\exists \beta, \gamma : A \to \alpha B\gamma, \ B \to \beta \in P, \ 0 \le i \le j\}$$
$$\mathcal{D}^{Scan} \quad = \{[A \to \alpha., i, j], [a^1, j, j+1], \dots, [a^k, j+k-1, j+k] \vdash$$
$$[A \to \alpha a^1., i, j+1] \mid \exists \beta : A \to \alpha a^1\beta \in P\}$$
$$\mathcal{D}^{Recurse} \quad = \{[A \to \alpha., i, j], [a^1, j, j+1], \dots, [a^k, j+k-1, j+k] \vdash [A \to A., i, j] \mid$$
$$\exists \beta : A \to \alpha, \ A \to A\beta \in P, \ a^1 \dots a^k \in LRF(A)\}$$
$$\mathcal{D}^{Complete} = \{[B \to \beta., h, i], [A \to \alpha., i, j], [a^1, j, j+1], \dots, [a^k, j+k-1, j+k] \vdash$$
$$[B \to \beta A., h, j] \mid \exists \gamma : B \to \beta A\gamma, \ A \to \alpha \in P, \ a^1 \dots a^k \in NLRF(A)\}$$
$$\mathcal{D} \quad = \mathcal{D}^{Init} \cup \mathcal{D}^{Expand} \cup \mathcal{D}^{Scan} \cup \mathcal{D}^{Recurse} \cup \mathcal{D}^{Complete}$$

5   Extending

Adding one more production (E → +T) to the existing grammar the existing part of parsing diagram change only a bit:

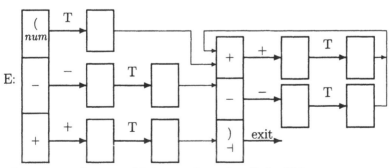

Procedure shown above would change also only a little. This opens a way to a tool allowing extension of programming language syntax by insertion of definitions of new constructions into the program text, see [8].

6   Conclusion

It was shown how to use grammars with direct left recursion for top-down parsing. One of the main disadvantages of top-down parsing was succesfully removed: It is not necessary to redefine some constructs (e.g. arithmetical expression) to allow top-down parsing.

The code generated by our parsers are similar to the code that would be written manually if no constructor were used. The generated code is therefore well understandable. It can be therefore "hand tuned" easily. It is an important software engineering advantage. It was also proven that, under some circumstances, kind parsers are incrementally extensible.

It is possible to derive context-free parsers from kind parsers similarly to how it is with the extended LR parsing (see [7]) and original LR parsing. It is possible to use a parsing schema as a model for their construction.

References

1. Alfred V. Aho, Ravi Sethi, Jeffrey D. Ullmann: *Compilers Principles, Techniques, and Tools*, Addison Wesley, 1986
2. Januš Drózd: *Syntaktická analýza téměř shora dolů.* (in Czech: Semi-Top-Down Parsing.) [Master Theses], Praha, MFF UK 1985.
3. Januš Drózd: *Syntaktická analýza rekurzivním sestupem pro LR(k) gramatiky.* (in Czech: Recursive Descent Parsing for LR(k) Grammars.) [PhD Theses], Praha, MFF UK 1990.
4. Jaroslav Král: *Syntaktická analýza a syntaxí řízený překlad.* (In Czech: Parsing and Syntax Directed Translation. [Tech. report ÚVT ČVUT], Praha, ÚVT ČVUT, 1982.
5. Martin Plátek: *Syntactic Error Recovery with Formal Guarantees I.* [Tech. report MFF UK], Praha, MFF UK, April 1992
6. Nicolaas Sikkel: *Parsing Schemata.* Proefschrift Enschede, 1993, ISBN 90-9006688-8
7. Masaru Tomita, Se-Kiong Ng: *The generalized LR Parsing Algorithm.* (In Masaru Tomita (ed.): Generalized LR Parsing), Kluwer Academic Publishers, 1991, ISBN 0-7923-9201-9.
8. Michal Žemlička: *Překladač rozšiřitelného jazyka.* (In Czech: Extensible Language Compiler.) [Master Theses], Praha, MFF UK, 1994.
9. Michal Žemlička: *Extensible LL(1) parser.* [Poster], SOFSEM'95, Milovy, 1995
10. Michal Žemlička: *Syntaktická analýza rozšiřitelných jazyků.* (In Czech: Extensible Language Parsing.) [Tech. report MFF UK], Praha, MFF UK, December 1996.

Enhancing Readability of Automatic Summaries by Using Schemas

Mariem Ellouze and Abdelmajid Ben Hamadou

Laboratoire LARIS - Faculté des Sciences Economiques et de Gestion de Sfax
B.P. 1088 - 3018 - Sfax - Tunisie
Tel: ++216 4 278 777 Fax: ++216 4 279 139
Mariem.Ellouze@planet.tn
Abdelmajid.Benhamadou@fsegs.rnu.tn

Abstract. The automatic summaries produced by extraction processes have been subject to several critics because of the quality of their textual substance (i.e, lack of cohesion and coherence). In this paper, we propose the use of summary-schemas conceived on the base of an empirical study of the argumentative structure of scientific articles and their author abstracts. These schemas enclose guidelines which help to achieve four properties to the summary textual material: completeness, non-redundancy, canonical organisation, and connectivity. These properties are achieved by means of three operations (addition, suppression and permutation) applied to the initial extract by agents that use various resources and interact with source text.

1 Introduction

The rapid and exponential growth of on-line textual material have provided an important opportunity for broad design of automatic text summarisation systems.

Numerous techniques for building automatic summaries have been proposed. These techniques fall within tow categories: those that rely on text generation and those that rely on passage (sentence) extraction.

The formers have been proposed in [1,2,3], ... They are based on template instanciation and produce a coherent textual unit which represent the abstract. These methods require deep text analysis in order to instanciate the templates. They are domain-dependant.

The second category of techniques, which have been used by [4,5,6], ..., escapes this constraint by identifying "important" or "informative" passages (usually sentences) by the means of some metrics. The collection of these passages (sentences) are deemed to be representative of the key ideas of the source document.

The approach based on extraction seems to be more attractive for automatic summarisation. However, the collection of passages (or sentences) can not be assumed to be a coherent textual unit, a property which is inherent to human abstracts. In fact, they usually present discontinuities. The presence of dangling

F. Jelinek, E. Nöth (Eds.): TSD'99, LNCS 1692, pp. 127–132, 1999.

anaphors, for example, creates discontinuities in the sequence of textual material of the extract. The extracts can also suffer from the lack of some informatinal and semantic material that has not been selected from the document source.

In order to resolve these problems of incoherence in automatic summaries (or extracts) and to attempt to achieve connectivity of their textual material, we propose an approach based on the use of *summary-schemas*. These schemas are different from templates used by the natural language generation processes. They are domain-independent and they are used to provide guidelines to the process of search of coherence in extracts (within the global level) and cohesion in textual segments of the extracts (within the local level). We experiment this approach on technical and scientific texts written in french.

2 Kinds of Discontinuities and Incoherences in Automatic Summaries

We have collected a set of automatic summaries produced by an extraction process and have studied the quality of their textual substance. The shortcomings identified are due to the presence of discontinuities and incoherences that can be caused by:

- the presence of pronominal and lexical references without their antecedents,
- the lack of some informational and semantic material,
- the incompatibility of verb tenses.

2.1 Discontinuities in the Textual Sequence due to the Presence of Pronominal and Lexical References

The presence of discontinuities can be caused by the extraction of sentences that have co-reference relations with others in the original text. Co-reference relations are expressed by means of pronominal and lexical references. Indeed, an extracted sentence can enclose among its constituents a personal pronoun (e.g. "le"), a possesive adjective (e.g. "leur"), a demonstrative pronoun (e.g. "ceci"), a demonstrative adjective associated with a noun (e.g. "cette baisse"). The presence of such sentences in an automatic summary creates some ambiguities and decreases its readability.

2.2 Discontinuities in the Textual Sequence due to Lack of some Informational and Semantic Material

The automatic summaries can also suffer from informational non-fulfilment and semantic inadequacy. Indeed, an extracted sentence can refer to some elements which are not included in any of the other extracted sentences. For example, the extraction system can select a sentence which contains the following prepositional group "dans la figue 1" without including, in the summary, at the same time, the aforementioned figure. A second scenario of informational non-fulfilment can be produced when the extraction system makes a truncature of an enumerated list because of non-saliency of the other elements of the enumerated list. On the other hand, the extracted sentences can enclose frequently additive and conclusive connectors. The presence of these connectors causes of the break of the train of ideas in textual substance of the automatic summary. The textual discontinuities can also be due to the following scenario: for some reasons, the system can select sentences from some sections and jumps others. This phenomenon sometimes causes lacks of some salient informational material.

2.3 Incompatibility of Verb Tenses

Sometimes, a succession of two or several sentences in the automatic summary presents incompatibility of verb tenses. This situation is produced, for example, when a sentence extracted from the introduction section of the original text and having a verb in the present tense, is followed in the automatic summary, for reasons of thematic progression, by a sentence extracted from the conclusion section which has verb in the past tense.

3   Summary-Schemas

To produce effective summaries, K. Spark Jones had proposed in [7] to develop intermediate techniques that include passage extraction and linking. Such techniques require shallow text analysis and can exploit statistical data.

Following the same idea, we propose a method that looks for coherence in extracts at short-dated and for generation of abstracts at long-dated. This method is based on the use of *summary-schemas*.

The schemas are used to instanciate linguistic material stating the main ideas of the full text. We have inspired this process of instanciation, based on predefined frames, from a study on the human behaviour undertaking when understanding texts [8]. Indeed, a human stores in his mind a list of rubrics (items) defining the different themes discussed in each category of texts. Understanding a text means associating to each of these items the suitable main unit.

On the other hand, our idea of instanciating linguistic material of the summary by using schemas joins others present in the literature. Indeed, B. Endres-Niggemeyer noticed in [9] that the expert abstractor uses, to generate an abstract, three views of the document localised at surface, structure and theme levels. The structure and the different aspects of the abstract are enclosed within a schema deduced from these three views. C. D. Paice reports in [10] the necessity of using abstract-frames in order to solve problems of incoherence. S. Teufel argues in [11] that the abstract-worthy sentences are extracted with respect to the rhetorical structure of the source text and she aims at performing classification of these sentences into a set of predefined generic rhetorical roles.

To conceive our summary-schemas, we have examined in a previous work [5] a set of authors' abstracts. We have focused on identifying their structure. We have also looked for identifying the eventual logic links between the abstract and the full document. The results we obtained are the following:

- The structure of the abstracts follows the structure of the full document. This structure encloses mainly a first part (an introductive part) which introduces the general theme of the document and a second part which makes an overview of the document content.
- Generally, the sentences of the abstract have simple syntax structures and the lexical items which make up these sentences are frequently extracted from the source text and, in particular, from titles and sub-titles (headings).

In a follwing step [12], we have emphasised the study of the corpus of the autor abstracts and we have focused on the following aspects: the detailed structure, the semantics, the statement, the syntax and the lexicon. Then, we have defined a summary-schema as a representation of the structural aspects, the semantic roles, the rhetorical relations and the textual types and organisations involved in the generation process of the summary.

The structural aspects of schemas are expressed by the presence of a hierarchy of slots and sub-slots. The semantic roles of these slots can be the following: "Presentation of the purpose of the study", "Method", "Experimental results" and "Evaluation". The organisation of these roles obeys constraints of coherence, which should be respected when producing a summary. The rhetorical relations hold between slots and can be "Elaboration", "Sequence",... Finally, the textual types and organisations involve argumentation, exposition,...

The design of the summary-schemas induces four basic properties: completeness, non-redundancy, canonical organisation and connectivity. The collection of the sentences that form the automatic summary must prove these properties to be viewed as a coherent textual unit.

The completeness must be checked at the global level of the summary. We propose that the completeness is achieved only when all basic slots in the summary-schema have received their informational content from the sentences of the extract. The non-redundancy sees to minimise redundancy of informational material. In fact, an idea mustn't be repeated redundantly in the summary. The canonical organisation must verify if the chronological order of sentences in the extract, which corresponds to their chronological order in the source document, obeys the thematic sequence predefined in the summary-schema. The connectivity is achieved by the presence of links inter and intra slots. Theses links can be implicit or explicit. Thus, in order to enhance readability of the automatic summaries, these four properties must be validated.

4   Devices and Operations to Solve Problems of Cohesion and Coherence

To perform validation of completeness, non-redundancy, canonical organisation and connectivity, we propose three operations: addition, suppression and permutation. An operation must be executed if a relative decision to it is made after an interaction between agents that support guidelines prescribed in the summary-schema and agents that support general knowledge about the scientific texts.

This knowledge is contained in a base and is stored in various forms. We find, for example, a dictionary for common terms, another for functional terms, some basic and inference rules to solve problems of cohesion, ... The combination of some among these devices helps to make decisions to solve ambiguities when looking for readability of extracts (automatic summaries).

The three operations are applied recursively at slot level and at the schema level. So, the local and global incoherences detected in the extract will be eliminated. The process of enhancing readability of automatic summaries is performed in five steps:

Step 1: The agent responsible for completeness of the informational material in the summary verifies if all basic slots will be fulfilled by linguistic material enclosed within the sentences of the initial extract. If one among these slots remains empty, then the agent sends the semantic role of this slot to agents that support general knowledge extracted from scientific texts. These agents look for indicators that correspond to the semantic of the empty slot and use them to find in the source text the suitable textual segment. Once found, the operation of addition is executed.

Step 2: The agent responsible for non-redundancy of informational material measures the semantic distance between sentences that will be enclosed within a slot. Then, if tow sentences are formulated in tow manners but have the same semantic content, one of them will be removed by means of the suppression operation.

Step 3: The canonical organisation of sentences enclosed within slots is tested by a third agent. Initially, the sentences of the extract are listed with respect to their chronological order in the source text. Then, if the chronological order of sentences in the extract does not match the thematic progression described by the summary-schema, the permutation operation will be applied in order to reorganise sentences. The new organisation of sentences will match the semantic configuration of the summary-schema. This scenario takes place when, for example, a sentence that deals with "Purpose study" is extracted from the conclusion section.

Step 4: The fourth agent is responsible for resolving problems of discontinuities and restoring the missing informational units in order to establish connectivity within textual segments of the summary. This agent applies the addition operation to add some informational units to an extracted sentence or to add other sentences to the summary. The addition operation is applied within a sentence when it has for example a truncated enumerated list. In this case, the missing elements of the enumerated list will be added. The addition of new sentences is applied when dangling anaphors are detected in the surface of the textual substance of the summary. By another way, to achieve connectivity of the textual material, the suppression operation can also be used. Indeed, the suppression of a connector enclosed within a sentence can be more suitable than the addition of one or several sentence(s).

Step 5: In this step, the compatibility between verb tenses is checked. If any incompatibility is revealed, then the tense of the verb that had caused the incoherence will be aligned with the other verbs.

5 Conclusion

We have been interested in this paper in the problems of incoherence frequently detected in automatic summaries produced by extraction processes. To solve these problems, we have proposed the use of summary-schemas to produce well-formed summaries. These schemas enclose guidelines that help to achieve four properties: completeness, non-redundancy, canonical organisation, and connectivity of the summary textual material.

The primary tests are done on a small set of extracts. The results are encouraging and point to the general feasibility of the method. However, these results are not reliable enough. In future work, we seek to extend our scale experiment to have more reliable results and to extend our knowledge resources (meta-discourse indicators, rules, . . .) if necessary.

References

1. Dejong, G.: An Overview on the FRUMP System. In: W.G. Lehnert and M. H. Ringle, eds. Strategies for Natural Language Processing. London: Lawrence Erlbaum (1982).
2. Hahn, U.: Concept-Oriented Summarising in the Text Condensation System TOPIC. In Summarizing Text for Intelligent Communications, Hannover, Germany (December 1993).
3. Mckeown, K., Radev, D.: Generating Summaries of Multiple News Articles. In SIGIR'95 Seattle WA, USA (1995).
4. Salton, G., Allan, J., Buckley, C.: Approaches to Passage Retrieval in Full Text Information Systems. In ACM Press SIGIR'93, Pittsbrugh, PA, USA (June 1993).
5. Ben Hamadou, A., Ben Mefteh, E., Jaoua, M.: Une Méthode d'Extraction des Idées Clés d'un Document en Vue de le Résumer. In AI'95-Fifteenth International Conference, Montpellier, France (June 1995).
6. Berri, J., Cartier, E., Desclés, J. P., Jackiewicz, A., Minel, J. L.: SAPHIR, système automatique de filtrage des textes. In TALN'96, Marseille, France (Mai 1996).
7. Spark Jones, K.: Summarizing, Where Are We Now ? Where Should We Go ?. In Intelligent Sacalable Summarization, Madrid, Spain (July 1997).
8. Rossi, J. P., Bert-Erboul, A.: Sélection des informations importantes et compréhension de textes. In Psychologie francaise, N. 36-2 (1991).
9. Endres-Niggemeyer, B., Maier, E., Sigel, A.: How to Implement a Naturalistic Model of Abstracting: Four Core Working Steps of an Expert Abstractor. In Information Processing and Management, Vol. 31, N.5 (1995).
10. Paice, C.-D.: Constructing Literature Abstracts by Computers: Techniques and Prospects. In Information Processing and Management, Vol. 26, N.1 (1990).
11. Teufel, S., Moens, M.: Sentence Extraction and Rhetorical Classification for Flexible Abstracts. In AAAI Spring Symposium on Intelligent Text Summmarization, Stanford (March 1998).
12. Ellouze, M., Ben Hamadou, A.: Utilisation de Schémas de Résumés en Vue d'Améliorer la Qualité des Extraits et des Résumés Automatiques. In RIFRA'98, Sfax, Tunisia (November 1998).

Use of a Weighted Topic Hierarchy for Document Classification*

Alexander Gelbukh, Grigori Sidorov, and Adolfo Guzman-Arénas

Natural Language Laboratory,
Center for Computing Research (CIC), National Polytechnic Institute (IPN),
Av. Juan de Dios Bátiz, CP 07738, Zacatenco, Mexico City, Mexico
{gelbukh, sidorov, aguzman}@pollux.cic.ipn.mx

Abstract. A statistical method of document classification driven by a hierarchical topic dictionary is proposed. The method uses a dictionary with a simple structure and is insensible to inaccuracies in the dictionary. Two kinds of weights of dictionary entries, namely, relevance and discrimination weights are discussed. The first type of weights is associated with the links between words and topics and between the nodes in the tree, while the weights of the second type depend on user database. A common sense-complaint way of assignment of these weights to the topics is presented. A system for text classification *Classifier* based on the discussed method is described.

1 Introduction

We consider the task of classification by their topics: for example, some documents are about *animals*, and some about *industry*. This task is important in information retrieval, classification of document flows, such as incoming documents in a large government office, filtration of document flows, such as Internet news, and in many other applications. In recent years appeared many articles on the theme, see, for example, [1-3], [7-10].

In this paper we consider the list of topics to be large but fixed. Our algorithm does not obtain the topics from the document body, instead, it relates the document with one of the topics listed in the system dictionary. The result is, thus, the measure (say, in percents) of the corresponding of the document to each of the available topics.

A problem arises of the optimal, or reasonable, degree of detail for such classification. For example, when classifying the Internet news for an "average" reader, the categories like *animals* or *industry* are quite appropriate, while for classification of articles on zoology such a dictionary would give a trivial answer that all documents are about *animals*. On the other hand, for "average" reader of Internet news it would not be appropriate to classify the documents by the topics *mammals*, *herptiles*, *crustaceans*, etc.

---

* The work partially supported by DEPI-IPN, CONACyT (26424-A), and REDII, Mexico.

F. Jelinek, E. Nöth (Eds.): TSD'99, LNCS 1692, pp. 133–138, 1999.

2   Topic Hierarchy

In [5] and [6], it was proposed to use a hierarchical dictionary for determining the main themes of a document. Technically, the dictionary consists of two parts: *keyword groups* and a *hierarchy* of such topics.

A keyword group is a list of words or expressions related to the situation described by the name of the topic. For example, the topic *religion* could list the words like *church, priest, candle, Bible, pray, pilgrim*, etc. Technically, our *Classifier* program manages word combinations in the same way as single words.

Note that these words are connected neither with the headword *religion* nor with each other by any "standard" semantic relation, such as subtype, part, actant, etc. This makes compilation of such a dictionary much easier than of a real semantic network dictionary. However, such a dictionary is not a "plain" variant of a semantic network such as WordNet, since some words are grouped together that have no immediate semantic relationship. Thus, such a dictionary cannot be obtained from a semantic network by a trivial transformation.

The other part of the dictionary is the topic tree, which organizes the topics, as integral units, into a hierarchy or, more generally, a lattice (since some topics can belong to several nodes of the hierarchy).

3   Basic Classification Algorithm

The algorithm of application of the dictionary to the task of topic detection also consists of two parts: individual (leaf) topic detection and propagation of the topics up the tree. *The first part* of the algorithm is responsible for detection individual (leaf) topics, i.e., for answering, topic by topic, the question: to what degree this document corresponds to the given topic? Such a question is answered for each topic individually. We call the element that answers such a question for a fixed topic a voter[1]. In our current implementation, a voter is based on a plain list of words corresponding to the topic; however, in general a voter can be associated with a procedure: for example, to detect that a document is an application form relevant to some department of a government office. Then it may be necessary to analyze the format of the document.

In our current system, for each keyword group the number of occurrences of the words corresponding to each (leaf) topic is determined. These numbers are normalized within the document, i.e., divided by the number of words in the document. The accumulated number of occurrences is considered to be the measure of correspondence of the document to the topic. Note that the values for this measure of relevance are not normalized since the topics are not mutually exclusive.

The second part of the algorithm is responsible for propagation of the found frequencies up the tree. With this, we can determine that a document mentioning the leaf topics *mammals, herptiles, crustaceans*, is relevant for the non-leaf topic *animals*, and also *living things* and *nature*.

---

[1] The terms *tester* or *topic agent* could be also appropriate.

Instead of simple lists of words, some numeric weights can be used by the algorithm to define the quantitative measures of relevance of the words for topics and the measure of importance of the nodes of the hierarchy. Thus, there are two kind of such weights: the weights of links in the hierarchy and the weights associated with the individual nodes.

The classification algorithm is then modified to take into account these weights. Namely, for the accumulated relevance of the topics, it multiplies the number of occurrences of a word (or subtopic) by the weight w_k^j of the link between the word and the topic, and then multiplies the result by the weight w^j of the topic itself.

4   Relevance Weights

The first type of weights is associated with the links between words and topics and between the nodes in the tree (actually, the former type is a kind of the latter since the individual words can be considered as terminal tree nodes related to the corresponding topic). For example, if the document mentions the word *carburetor*, is it about *cars*? And the word *wheel?* Intuitively, the contribution of the word *carburetor* into the topic *cars* is more than that of the word *wheel*; thus, the link between *wheel* and *cars* is assigned a less weight. The algorithm of classification takes into account these weights when compiling the accumulated relevance of the topics.

It can be shown that the weight w_k^j of such a link (between a word k and a topic j or between a topic k and its parent topic j in the tree) can defined as the mean relevance of the documents containing this word for the given topic:

$$w_k^j = \frac{\sum\limits_{i \in D} r_i^j n_i^k}{\sum\limits_{i \in D} n_i^k} \qquad (1)$$

by all the available documents D, where r_i^j is the measure of relevance of the document i to the topic j, and n_i^k is the number of occurrences of the word or topic k in the document i.

Unfortunately, we are not aware of any reliable algorithm of automatic detection of the measure of the relevance of r_i^j in an independent way. Thus, such a measure is estimated manually by the expert, and then the system is trained on the set of documents.

As a practical alternative, it is often possible to estimate the weights w_k^j intuitively at the stage of preparation of the dictionary. The choice of the weight is based on the frequency of appearance of the word in "general" documents from the control corpus of the texts on "any" topic; in our case such texts were the newspaper issues.

As another practical approximation, for narrow enough themes we can take the hypothesis that the texts on this topic never occur in the control corpus (newspaper mixture). Then, given the fact that we have included the word in the dictionary and thus there is at least one document relevant for the given topic, we can simplify the expression for the weights as follows:

$$w_k^j = \frac{1}{\sum\limits_{i \in D} n_i^k} \tag{2}$$

since the numerator of the quotient in (1) in case of narrow topics can be considered to be 1. Not surprisingly, this gives the weight of the word "voting" for a specific topic to be the less the more its frequency; for example, the articles *a* and *the* have a (nearly) zero weight for any topic, while the word *carburetor* has a high weight in any topic in which it is included.

Sometimes a rare enough word, say, a noun *bill*, in its different senses is related to different topics (*money, law, birds, geography, tools*). For a more accurate analysis, some kind of competition between senses of the word for a specific occurrence in the document is introduced. For this, the senses of the word are marked in the topic dictionaries (as $bill_1$, $bill_2$, etc.), and the weights of occurrences of such a word are to be normalized by its different senses (though the occurrences of the same sense are independent in different topics), with the weigh of an individual sense in each document being proportional to the relevance of the document for the given topic:

$$w_k \sim \sum_j r_i^j w_k^j$$

$$\sum_k w_k = 1 \tag{3}$$

where w_k is the weight of the k-th sense of the given occurrence of the word in the given document i, w_k^j is the weight of the link between this sense of the word and the topic j, the summation in the first equation is made by all the topics, and in the second by the senses of the given word. Since r_i^j in its turn depends on w_k, to avoid iterative procedure, in practice we calculate r_i^j based on equal weights w_k.

However, the latter technique is not very important for most cases, since usually it does not change the order of the topics for a document, but only makes the difference between different topics more significant.

5   Discrimination Weights

The classification algorithm described above is good for answering the question "is this document about *animals*?" but not the question "what about is this document?". Really, with such an approach taken literally, the answer will be "all the documents are about *objects* and *actions*", the top nodes of the hierarchy. However, a "reasonable" answer is usually that a document is about *crustaceans*, or *animals*, or *living things*, or *nature*, depending on the situation. For a biologist, the answer *crustaceans* would be the best, and for an average newspaper reader the answer *nature*.

Our hypothesis is that the "universe" of the reader is the base of the documents to which he or she applies the search or classification, i.e., that the reader is a specialist in the contents of the current database. Thus, the topic relevance weights in our system depend on the database.

The main requirement to these weights is their *discrimination power*: a topic should correspond to a (considerable) *subset* of documents, while the topics that correspond to nearly all the documents in the data base are probably useless. Thus, the weight w^j of a tree node j can be estimated as the variation of the relevance r_i^j the topic over the documents of the database:

$$w^j = \sum_{i \in D} \left(r_i^j - M \right)^2$$

$$M = \frac{\sum_{i \in D} r_i^j}{|D|}$$

(4)

here M is the average value of r_i^j over the current database D, and r_i^j is determined by the former algorithm, without taking into account the value of w^j.

With this approach, for, say, a biological database, the weight of the topics like *animals, living things, nature* is low because all the documents equally mention these topics. On the other hand, for newspaper mixture their weight is high.

6   Applications

With the approach described above, we have implemented in the system *Classifier* several useful functions.

The system can determine what are the principle topics of the document. This corresponds to the task of classification. Also the system allows viewing the documents by topics, answering the question: for a selected topic, what documents are the most relevant? This roughly corresponds to the task of information retrieval.

An interesting application of the method is classification of the documents by similarity with respect to a given topic. Clearly, a document mentioning the use of animals for military purposes and the document mentioning feeding of animals are similar (both mention *animals*) from the point of view of a biologist, but not from the point of view of a military man they are very different. The comparison is made on the basis of the weights of the topics for the two documents.

7   Discussion and Future Work

Generally, the results obtained in our experiments show very good accordance with the classification made by human experts. However, we encountered some problems with using our method. Most of them are related with ambiguity.

Sometimes, a frequent keyword (taken out of context) proves to be important for a specific topic: the noun *well* is an important term in *petroleum extraction*, the noun *do* is a term in *hair styles*, the noun *in* in *politics*, etc. However, the expression (1) assigns too little weight to such keywords. To solve this problem, we plan to add a part of speech tagger to our system. For a more detailed analysis, we might have to add our syntactic parser to the program; however, this would greatly slow down the system.

Obviously, this does not solve all the problems of ambiguity. As we have discussed, for the words like *bill* a sophisticated and not always reliable algorithm is used; we plan to resolve the ambiguity of this type with more intelligent methods described in [4].

Though there are some problems with the accuracy of the algorithm, the results of experiments show good accordance with the opinion of human experts. The method is practical in the sense of insensibility to inaccuracies in the dictionary and in the sense of using a dictionary with very simple structure, easily trainable on manually classified collections.

References

1. Anderson, J. D., Rowley, F. A.: Building End-user Thesauri from Full Text. In: Kwasnik, B. H., Fidel, R. (eds.): Advances in Classification Research. Proceedings of the 2nd ASIS SIG/CR Classification Research Workshop, Vol. 2. Learned Information, Medford, NJ. (1992) 1–13.
2. Cohen, W. W.: Learning Trees and Rules with Setvalued Features. In: Proceedings of the Thirteenth National Conference on Artificial Intelligence (1996).
3. Cohen, W., Singer, Y.: Context-sensitive Learning Methods for Text Categorization. In: SIGIR'96 (1996).
4. Gelbukh, A.: Using a Semantic Network for Lexical and Syntactic Disambiguation. In: Proceedings of Simposium Internacional de Computación: Nuevas Aplicaciones e Innovaciones Tecnológicas en Computación. Mexico (1997) 352–366.
5. Guzmán-Arenas, A.: Finding the Main Themes in a Spanish Document. Journal Expert Systems with Applications 14 (1, 2) (1998) 139–148.
6. Guzmán-Arenas, A.: Hallando los Temas Principales en un Artículo en Español. Soluciones Avanzadas 5 (45) (1997) 58, **5 (49)** (1997). 66
7. Jacob, E. K.: Cognition and Classification: A Crossdisciplinary Approach to a Philosophy of Classification. (Abstract.) In: Maxian, B. (ed.): ASIS '94: Proceedings of the 57th ASIS Annual Meeting. Medford, NJ: Learned Information (1994) 82.
8. Krowetz, B.: Homonymy and Polysemy in Information Retrieval. In: Proceedings of the 35th Annual Meeting of the Association for Computational Linguistics (1997) 72–79.
9. Lewis, D. D., Ringuette, M.: A Comparison of Two Learning Algorithms for Text Categorization. In: Third Annual Symposium on Document Analysis and Information Retrieval (1994) 81–93.
10. Riloff, E., Shepherd, J.: A Corpus Based Approach for Building Semantic Lexicons. In: Proceedings of the Second Conference on Empirical Methods in Natural Language Processing (EMNLP-2) (1997).

Remarks on Sentence Prosody and Topic-Focus Articulation

Petr Sgall

Faculty of Mathematics and Physics, Charles University, Prague, Czech Republic
E-mail: sgall@ufal.mff.cuni.cz

Abstract. The topic-focus articulation can be understood as one of the aspects of the underlying structure of the sentence and as expressed by sentence prosody and word order. While the intonation center (the final falling or rising-falling cadence of a declarative sentence) is carried by the focus, the topic often has a contrastive value, which is expressed by a high or raising phrasal stress. A focus sensitive particle is associated with the focus of the sentence in the prototypical case, but in secondary cases it occurs within topic and its focus is then marked by a phrasal stress. The recent computer assisted acoustic analyses allow for studying even subtle phonetic differences of pitch or stress, so that the possibility to investigate the functional roles of these differences has become possible. Such investigations are of crucial importance for achieving an integrated description of language.

1 Introductory Remarks

The research in the information structure, or topic-focus articulation (TFA) of the sentence in Czech linguistics has been based on a long-term comparative analysis of typologically different languages such as Czech and English and on discussions concerning several other languages (German, Russian, French, Spanish, Japanese, etc.)[1]. This research, reassuming previous results achieved in France, Germany, Russia, and Poland, and linked up to more recent developments in British and American linguistics, has brought about the conviction that TFA can be understood as one of the aspects of the underlying structure of the sentence and that the oppositions inherent in TFA are expressed in the outer shape of sentences by certain oppositions present in their prosody and word order.

It is a task of fundamental significance for linguistic research to investigate how these relationships and the character of the mentioned means of expression can be specified in an integrated description of language. Especially after the recent results of the computer assisted acoustic analyses, which make it possible to identify even subtle phonetic differences of pitch or stress, it has become important to study the functional roles of these differences and to establish their links with TFA.

---

[1] See esp. [6] and [2].

F. Jelinek, E. Nöth (Eds.): TSD'99, LNCS 1692, pp. 139–145, 1999.
© Springer-Verlag Berlin Heidelberg 1999

2   Topic-Focus Articulation in Underlying Structure

In the Praguian approach to TFA (more details and the motivation of this approach can be found in [6]; [2]), we distinguish (a) the sentence topic (T) as typically denoting "given" information, which is "spoken about" in a declarative sentence, and is supposed by the speaker to be readily accessible in the hearer's memory), and (b) the sentence focus (F), denoting "new" information, i.e. what is being asserted about T in the given sentence. A basic hypothesis usually accepted by the linguists studying TFA assumes that the focus (or its prominent part, the focus proper) is expressed by the bearer of the intonation center (IC) of the sentence, which, in the prototypical case, consists in the final falling (or rising-falling) cadence of a declarative sentence.

A few examples can illustrate the phenomena under discussion and, perhaps, be useful for continuing discussions among those interested in TFA and in sentence prosody, especially for the question whether the hypotheses briefly formulated below can be viewed as plausible (the capitals denote the bearer of IC):

(1)   (Paul's family certainly is not really poor.) His son has bought a new CAR.

Ex. (1) is a prototypical simple sentence whose subject is in T, the object is in F and the position of the verb is ambiguous: it belongs to F on some readings (such as that present in the context given here) and to T on others. Also the word order has its prototypical (primary) shape: T precedes F. In secondary cases F (the bearer of IC) is placed at the beginning of the sentence;

(2)   (Paul still uses the public transport.) His SON has bought a new car. [or: It's his SON who has bought a new car.]

Since both F and T are not necessarily single words (and not even single constituents in a phrase-structure based syntax), it is useful to classify individual words as for being either contextually bound (CB) or not (non-bound, NB); while CB items primarily belong to T, NB items are included in F (exceptions may concern more deeply embedded items, such as *his* in (2), which is CB although it belongs to F together with its head noun). It should be noted that in the typical cases only full lexical (autosemantic) items are understood as 'words' here, whereas function words are handled, in the underlying structure, as belonging to the layer of function morphemes (thus, e.g. both *will work* and *work-ed* are treated as tense forms of *work*).

Another classification of word occurrences in a sentence consists in the degrees of communicative dynamism (CD): it is supposed that items univocally belonging to T display lower degrees of CD than those univocally belonging to F and that the degrees of ambiguous items (such as the verb in (1)) are in between. T proper can then be specified as the least dynamic item and F proper as the most dynamic one. T proper often has a contrastive value (i.e. refers to an entity chosen from a set of alternatives determined by the context), which, according to another hypothesis, is expressed by a certain kind of phrasal stress, denoted here by italics:

(3) (Paul's daughter still uses the public transport.) His *son* has bought a new CAR.

(4) Paul's *son* has bought a new CAR and his *daughter* has got an APPART-MENT of her own.

(5) Paul's *son*, who has *married* recently, has bought a new CAR.

According to the hypotheses linguists usually work with, in (3) a rising contrastive stress on the subject is present, which is similar in the two clauses of (4) and in (5). However, in (5) the stress on *married* probably can be characterized phonetically (and thus also phonologically) as being of a different kind; it functions as a non-contrastive phrasal stress. Ex. (4) also illustrates the fact that compound sentences prototypically include spearate TFA patterns in their individual clauses. Another question concerns the pauses – e.g. those marked by the commas in (5) or that before the conjunction in (4), as well as the possibility of a pause after the subject group in (1): are all of these pauses obligatory, or may they be absent in allegro speech, or are they just optional? Do they differ phonetically (and phonologically) from each other and/or from non-systematic pauses occurring in cases of hesitation or similarly? How can the phonetic difference between (5) and its counterpart with *"Paul's* son" be specified? Questions of this kind should be systematically checked on the basis of acoustico-phonetic analyses.

As mentioned in the comments on (2) above, all complementations following IC belong to T (are CB); as the known example (6) shows, this is not necessarily valid of the verb itself, which may belong to the focus, i.e. may be contextually non-bound):

(6) (What is the news?) KENNNEDY has been killed.

A focus sensitive particle (*only, also*, etc.) is associated with the focus of the sentence in the prototypical case, but in secondary cases it occurs within topic and its focus is then (according to [1] and others) marked by a specific kind of stress, or at least by some other kind of rhythmic prominence ([4]), cf. also the ambiguous position of *only* in his example reproduced here (with a modification of the phrasal stress, discussed in Hajičová et al. 1998, p. 157) as (7a), which has a reading in common with (7b):

(7) a. People who *grow* rice often only EAT rice.

b. People who *grow* rice often EAT only rice.

A general framework for the semantic interpretation of focus can be looked for in B. H. Partee's 'tripartite structures' (with Operator, Restrictor and Nuclear Scope).

3   Prosodic Means of Expression

The intonation center (IC, sentence stress) of a declarative sentence, is understood prototypically to express the focus, or focus proper (i.e. the most dynamic item in the sentence structure). This stress probably can be determined as falling or rising-falling (cf. J. Pierrehumbert's H* L accent; see also the systematic analyses by M. Steedman). There are several other kinds of stress, which are less prominent than the IC and more prominent than the lexical stress of individual words (cf. e.g. E. Selkirk's 'phrasal stress'). It is an open question how their classification and specification as for their positions in one or more hierarchies can be correlated with the individual functions mentioned above (focus proper, contrastive topic, complex phrases[2]) and with other roles. Some of these kinds of stress (not to speak of the rising contour of interrogative sentences) are directly relevant for sentence structure, whereas others express different ('pragmatic', attitudinal) values, which do not directly belong to the system of language (various kinds of emphasis, wonder, aversion, doubts, and so on).

Typical examples of declarative sentences with the primary means of expression of TFA can be seen in the following examples presented by [3] (in the sequel we write just PH), the original number of which we add after their wording:

(8)   The train leaves at seven. (5)
      H*   H*   H* L L%

(9)   George likes pie. (9)
      H*   H* L L%

In (8) and (9), it might be supposed that the H* accent followed by the phrase accent L corresponds to the intonation center of the sentence (carried by F proper), L% is the low boundary tone and the difference between the presence and the absence of the high tone accent at the preceding words should be checked in more detail (with more material) as perhaps corresponding to their further properties:

(a) H* may express the appurtenance of its bearer to the contrastive part of T, or to T proper (with *train* and *George*), and also

(b) to a part of F other than F proper (*leaves*);

(c) the absence of stress on ate could be interpreted as expressing a noncontrastive part of T (other than T proper).

Function words (*the, train*) are excluded from this prototypical layer of prosodic means of TFA, which is in accordance with the assumption that they do

---

[2] Following [5] and analysing in more detail issues of the intonational phrasing in English, [9] shows how this phrasing is restricted by syntactic constraints that exclude e.g. the phrasing present in *On [["Monday] % [morning]] they "left %* or that in *[[John] [Mary] % [and Peter]] %*, where the brackets indicate the syntactic relationships, pitch accents are marked by double primes and intonational phrase boundaries by the sign %. Complex intonational phrases such as e.g. *the passengers were unhurt* in Taglicht's ex. (58), p. 159, may be supposed to be characterized by a specific stress pattern.

not have any lexical counterparts in the (underlying) sentence structure, in which their correlates have the form of indices of basic (autosemantic) lexical items.

In more complex examples the high tone is even more clearly connected with a non-final position, i.e. it perhaps signals that a structure that has the form of a completed sentence is to be followed by another part of the message, e.g. with coordination, as in (10):

(10)   The train leaves at seven or nine twenty-five. (6)
     H*   H*   H* H H*   H* L L%

The boundary tone H% may indicate that the following sentence would "be interpreted as a unit" with the preceding one (PH, p. 287), as in (11); we render this close relationship of the two utterances by a semicolon:

(11)   The train leaves at seven; it'll be on track four. (7)
     H*   H*   H* L H% H*   H* L L%

We certainly cannot claim that H* (along with its function in the question contour) consistently corresponds to NB or contrasted items, H* L to IC and the absence of stress to CB. There are also cases in which PH point out other roles of these features, as in the following examples:

(12)   My name is Mark Liberman. (12)
          H*   H* H H%

(13)   A: It's awfully warm for January.
     B: It's even warm for December. (31)
               L+H* L H%

The end contour in (12) is supposed to "question" the relevance of the message, an implicit continuation being "are you expecting me", or "am I in the right place?" (PH 290), and that in (13) is assumed "to mark a correction or contrast" (PH 296). The often discussed example (14), introduced by R. Jackendoff, is paraphrased by PH (296) by "As for the beans, Fred ate them. As for the other food, other people may have eaten it." This makes it possible to understand this sentence in our terms as containing a NB subject as F, preceding T; it is an open question if *the beans* can be understood here as a contrastive part of T, and if then a notation corresponding to that used in Section 2 above might be that of (15).

(14)   A: What about the beans? Who ate them?
     B: Fred ate the beans
         H* L   L+H* L H

(15)   FRED ate the *beans*.

Discussing an example of E. Selkirk's, Steedman (ms.) shows how the subtle differences in what we call TFA can be handled, cf. the following examples (we again attach his original numbers; he indicates the words bearing nuclear pitch accents by capitals and uses parentheses to mark prosodic phrase boundaries):

(16) Q:I know who prefers bombazine. But who prefers CORDUROY?
      A: (MARY) (prefers CORDUROY). (3)
          H* L   L+H* LH%

(17) Q: I know which material ALICE prefers. But which material does MARY
      prefer?
      A: (MARY prefers) (CORDUROY) (4)
          L+H* LH%   H*LL%

(18) (MARY) (prefers CORDUROY) (5)
       L+H* LH%   H*LL%

(19) MARY prefers *corduroy*.

While in (16) the situation is similar to (14) and our simplified notation from
Section 2 would be that of (19), *in* (17) the subject and the verb constitute the
topic and the object bears the IC, i.e. is in F; in (18) the verb belongs to F,
being NB, so that we are not certain that it can answer the question from (17),
in which the verb was present (an important point may consist in the fact that
in the question the declarative modality of the verb is not present, which may
allow for the NB character of the verb in the positive answer). In any case, it
appears that the rising pattern of L+ H ˘ * ˘ corresponds in these examples to T
proper or contrastive T, and the falling H ˘ * ˘ L is a counterpart of F proper. Not
being able to go here into the details of the relationships between T proper and
contrastive T (which often are assigned to the same item, although this probably
is not always the case), or of the function of the sign + in PH's notation, we
can only offer these examples as interesting starting points for further, more
systematic analysis.

4   Conclusion

As we have mentioned in Section 1, the approach to the topic-focus articulation
(information structure) of the sentence presented here has been found appropriate for the description of typologically different languages. English and Czech
are often supposed to belong to subclasses of languages that require different
theoretical approaches (while a phrase structure based description is understood
as appropriate for 'configurational' languages such as English, a 'flat structure'
or a description based on syntactic dependency is viewed as required by the high
degree of free word order in languages with inflection, such as Czech). However,
with the descriptive framework used here (the motivation and detailed characterization of which can be found in the writings quoted in Note 1) it is possible to
capture both the syntactic relations (subject, object, etc.) and the information
structure in languages of the two types. It seems that also languages such as
Japanese (with an overt topic marking particle) and many other can be handled
by this approach.

It is then important to pay attention to the relationships between the topic-focus articulation and the sentence prosody. A detailed study may help to find out whether the hypotheses presented in the present paper can be seen as plausible or how they should be made more precise.

Discussions on these and further questions have already their tradition, but with the recently gained means of the investigations of sentence prosody they may obtain a new range of explicitness and they can be of high relevance for throwing more light on basic issues of linguistic descriptions. To describe the functions of the relevant features of sentence prosody would mean a significant step on the way towards a unified description of the system of language as a whole, from the phonetic form of sentences to their underlying structure (disambiguated, directly relevant for a semantico-pragmatic interpretation).

Notes * This paper has been formulated in connection with research on the project VS 96/151.

References

1. Bartels C.: Acoustic correlates of 'second occurrence' focus: toward an experimental investigation. In: Context dependence in the analysis of linguistic meaning 2. Proceedings of the workshops in Prague and Bad Teinach, 1995. University of Stuttgart Working Papers. Ed. by H. Kamp and B. Partee (1997) 11–30.
2. Hajičová E., Partee B. H., Sgall P.: Topic-focus articulation, tripartite structures, and semantic content. Studies in Linguistics and Philosophy 71. Dordrecht/Boston/London: Kluwer (1998).
3. Pierrehumbert J., Hirschberg J.: The meaning of intonational contours in the interpretation of discourse. In: Intentions in communication. Ed. by P. Cohen, J. Morgan and M. Pollock. Cambridge, MA: MIT Press (1990) 271–312.
4. Rooth, M.: A Theory of Focus Interpretation. Natural Language Semantics 1 (1992) 75–116.
5. Selkirk E.: Sentence prosody: Intonation, stress, and phrasing. In: Handbook of phonological theory. Ed. by J. A. Goldsmith, Oxford: Basil Blackwell (1995) 550–569.
6. Sgall Petr, Hajičová Eva and Jarmila Panevová: The meaning of the sentence in its semantic and pragmatic aspects. Ed. by J. Mey. Dordrecht: Reidel - Prague: Academia (1986).
7. Steedman M.: Surface structure and interpretation. Cambridge, MA: MIT Press (1996).
8. Steedman M. (ms.): Information structure and the syntax-phonology interface (unpublished manuscript).
9. Taglicht J.: Constraints on intonational phrasing in English. Journal of Linguistics 34 (1998) 181–211.

Speech Recognition Using Elman Neural Networks

L.J.M. Rothkrantz and D. Nollen

Delft University of Technology, Knowledge Based Systems
Zuidplantsoen 4, 2628 BZ Delft, The Netherlands
Phone +312787504 Fax +312787141
L.J.M.Rothkrantz@cs.tudelft.nl

Abstract. The main topic of this paper was to research the use of recurrent neural networks in the process of automatic speech processing. The starting point was a modified version of RECNET as developed by Robinson. This phoneme recognizer is based on a RNN but post-processing is based on Hidden Markov Models. A parallel version of RECNET was implemented on a parallel computer (nCUBE2) and Elman RNN were used as postprocessor. Word segmentation is also realized using an Elman RNN. The network models and results of testing are reported in this paper.

1 Introduction

Since the late 1970's speech technology has been dominated by the technique of Hidden Markov Models (HMM). In the late 1980's Neural Networks (NNs) became very popular. Recurrent NN are used to model temporal variation in speech [1,2]. A better modeling of time variation was achieved by Time Delay NNs. But it turned out that the neural approach could not compete with HMM. From the early 1990's on hybrid speech recognition systems have been developed which combine the advantages of HMMs and NNs. A successful example is the RECNET system [3]. However today's speech recognition technology is still far from being competitive with human skills. A Markov process does not accurately model speech. A model of information processing in the human brain differs completely from HMM models and is more similar to NN models. Inspired by the working of the human brain, we developed an automatic speech recognizer, which only uses NNs.

2 Phoneme Recognition

The system RECNET is a phoneme recognizer, developed at Cambridge University by T. Robinson [3]. In the preprocessor a 32 ms Hamming window is used, with frame spacing of 16 ms. From every frame a 23 valued feature vector has to be extracted. The components are the cube root powers in the first 20 channels derived from the Fast Fourier Transformation, the zero crossing rate, the fundamental frequency and the power of the speech signal.

F. Jelinek, E. Nöth (Eds.): TSD'99, LNCS 1692, pp. 146–151, 1999.

The external input is fed with the 23 channels from the preprocessor. The external output contains the probability distribution of the 61 phonemes available in the phoneme alphabet. The internal part of the network contains 192 nodes, both at the input layer and at the output layer (see Fig. 1).

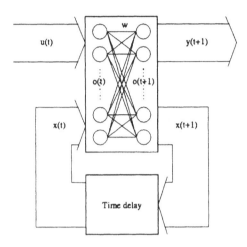

Fig. 1. Architecture of RECNET

The system was trained, using a variant of the backpropagation through time algorithm [3]. To speed up the process of training we implemented the RECNET system on a parallel computer (nCUBE2) with 32 nodes. The BTT algorithm has been made parallel by dividing the complete number of frames between two weight adjustments over the 32 nodes. Because the weights are fixed during the forward pass, the nodes don't have to communicate with each other during the local forward and backward pass. After calculating the average partial derivative it is collected from all nodes by the master node. The master node averages the partial derivative of all nodes and adjusts the weights. After adjusting the weights a new cycle has been started.

The system is trained and tested with the TIMIT database. This set contains 5966 utterances in the training set. After preprocessing the complete training set contains 700 130 frames. The preprocessed frames are labeled with the correct phonemes. Per training cycle over 78 billion floating point calculations have to be realized. The training time on a SUN 14 MHz was 700 hours and on the nCUBE2 70 hours. After approximately 70 training cycles with decreasing stepsize we found that 64.7 % of the frames were labeled correctly. Similar results were reported by Robinson [3].

3 Phoneme Post Processing

The output of the RECNET phoneme recognizer is a vector with the probability distribution of phonemes of every speech sample. One or more probability vectors represent a phoneme, depending of the duration of a phoneme. Successive samples show a lot of variation due to errors in speech recognition and variation in speech. To convert this stream of probability vectors into a stream of single phonemes without repetition different post-processing techniques have been used. To test the performance of these techn iques the complete test set of the TIMIT database was used. This set contains 321 680 frames with 51 681 phonemes.

After applying a postprocessor, a stream of phonemes or words is obtained. To score the recognizer one has to compare the desired outcome with the real outcome. There are three different kinds of errors a recognizer can make:

- Substitution, a phoneme is recognized as a different phoneme
- Deletion, a phoneme is not recognized at all
- Insertion, a phoneme is recognized which wasn't spoken

To match the reference sentence to the hypothesis sentence the errors can be determined. This matching is known as symbol string alignment. The technique used for determining the best alignment is based on Dynamic Time Warping. First we scored the output of the RECNET recognizer directly. Without any post-processing. The repetitions in the phoneme stream are deleted and this stream of single phonemes is optimally aligned with the stream of correct phonemes. Next the numbers of errors were computed (Table 1). From the table we can see that the number of insertion errors is very high. Every error made by the RECNET recognizer disturbed the repetition of phoneme labels during a phoneme and will insert one or two incorrect p honemes.

Table 1. Scoring without post-processing

Number of phonemes	51.681	
Correctly labeled	41.290	79.9%
Insertion errors	23.839	46.1%
Substitution errors	9.072	17.6%
Deletion errors	1.319	2.6%
Total number of errors	34.230	66.2%

In [3] the post-processing is realized by using Hidden Markov Models. The total error rate decreases significantly (Table 2). As expected the improvement is made on the insertion errors. The small irregularities are smoothened out by the HMM.

A different approach in correcting the irregularities in the output of RECNET recognizer is the use of recurrent NN. The architecture of the NN is displayed

Table 2. Scoring of the HMM model

Number of phonemes	51.681	
Correctly labeled	37.644	72.8%
Insertion errors	1.097	3.7%
Substitution errors	10.974	21.2%
Deletion errors	3.063	5.9%
Total number of errors	15.934	30.8%

in Fig. 2, 61 input and output nodes and a context layer of 150 nodes. The goal of the Elman RNN is to transform its input (the output of RECNET) in the correct phoneme, in which the correct phoneme has the highest probability. The recurrent connection and the context layer enable the network to include information of previous frames.

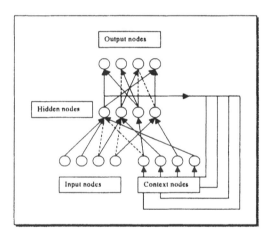

Fig. 2. Architecture of the Elman RNN

In order to train the Elman RNN we had to create a training set. As noted the output of RECNET contains many errors. So if we take the phoneme with the highest probability as the assumed correct output, we make a lot of errors. By definition we assume that the correct output vectors of the RECNET phoneme recognizer contain a one for the correct phoneme and zero for all others. We designed our training set as a linear combination of the "RECNET output" and the "correct output".

$$\text{Mixed output} = \alpha \text{ Normalized RECNET output} + (1 - \alpha)\text{output} \qquad (1)$$

If we feed the output of the RECNET phoneme recognizer to the network and the correct output is feedback ($\alpha{=}0$) the performance was bad. The error rate was reduced to 60.8%. The input has too many errors (35%). So the network wasn't trained to include the repetitive character of the phoneme labels. On the other hand if we simply feedback the correct output of RECNET and use the same vector as the desired output, the network will train itself to exclude the context. If we feedback a combination of th e "RECNET output" and the "correct output", we found better results. In Table 3 we display the results (α=0.5).

Table 3. Scoring of the Phoneme Post-processing RNN

Number of phonemes	51.681	
Correctly labeled	38.812	75.1%
Insertion errors	11.318	21.9%
Substitution errors	10.905	21.1%
Deletion errors	1.964	3.8%
Total number of errors	24.187	46.8%

4   Phoneme Prediction

In [1,2], J. Elman used RNN in a study on character prediction. He used a RNN as displayed in Fig. 2. We made a similar experiment for phoneme prediction. The idea is to feed a RNN with a sequence of phonemes one by one. The output of the network predicts its next input, the next phoneme. The context layer provides the network with some kind of memory. Using the context generated by the previous phonemes, the network will try to predict the next phoneme. The network has an inpu t and an output layer of 6 nodes. The hidden layer has 100 nodes. The input and the output of the network are the current phoneme and the next phoneme. The 63 possible phonemes are coded in a 6 bit binary code.

To create a vocabulary 10 sentences are chosen randomly from the TIMIT database. The 10 sentences form a vocabulary of 105 words. The TIMIT database contains for every sentence the correct phonetic transcription. The network has to be trained on prediction within a word, not between two words. To avoid these word relationship, the word sequences are chosen randomly. The training set is created by sampling 2000 words randomly from the vocabulary of 105 words. In this way we had a continuous stream of 9361 phonemes. The squared distance between the desired phoneme code and the network output was used as the error function. After 60 training cycles the average squared distance limits to 0.3. After training the network was tested on the 10 sentences.

From Fig. 3. can be seen that most of the initial phonemes show a local maximum in the error. So we can use this phoneme prediction network to segment a stream of phonemes. It proved that in 80 % of the situations the parser indicates the right segmentation point.

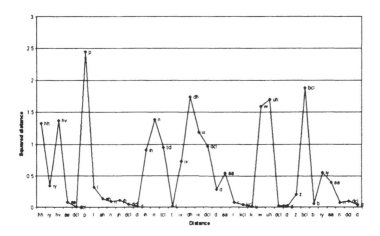

Fig. 3. Output of the phoneme predicting RNN

5   Conclusions

The main goal of our research was to design and implement an Automatic Speech Recognizer based on NN. A prototype is successfully implemented. We implemented a parallel version of the RECNET phoneme recognizer. It proved that 65 % of the phonemes were correctly classified. A RNN phoneme postprocessor was implemented to smooth the RECNET output. The error rate was reduced with 20 %. A HMM postprocessor reduced the error with 36 %. So our RNN is still not optimal. But an advantage is that it performs real tim e. We used RNN to parse the stream of phonemes. On a vocabulary of 105 words the 80 % of a phoneme stream was segmented correctly.

References

1. Elman, J.L.: Finding structure in time, Cognitive Science, 14, (1990), 179–211.
2. Elman, J.L., Bates, E.A., Jonhson, M.H. (eds.): Rethinking Innateness, The MIT Press, Cambridge, Massachusetts, (1996).
3. Robinson, A.J., Fallside F.: A recurrent error propagation network speech recognition system, Computer Speech and Language, Volume 5, Number 3, (1991).

Use of Hidden Markov Models for Evaluation of Russian Digits Pronunciation by the Foreigners

Alexei Machovikov and Iliana Machovikova

Telum Russia, Inc., 77, 44, 11th line, St.-Petersburg, 199178, Russia
abm@unitel.spb.ru

Abstract. In this paper the use of Hidden Markov Models for the evaluation of the Russian words pronunciation by the foreigners is considered. The average distortion of the codebook and the likelihood of the discrete variable duration Hidden Markov Model are used. Such model allow to take into account the temporary characteristics of a speech signal, which are very important for the evaluation of pronunciation. The results for ten Russian digits are reported.

1 Introduction

The rapid progress in the speech processing and creation of systems for non-standard speech recognition (with dialects and accents) allow to apply modern speech technologies to ≪Computer-Aided Language Learning≫ (CALL). Several research groups have started to develop language teaching systems based on such technologies. These works are represented two directions: the pronunciation training and the teaching in dialogue. The first type of systems (for example [1,2]) allow to automatically estimate the deviations of pronunciation and help user to correct them. The second type of systems (for example [3,4]) are intended for the reception of primary colloquial skills.

Earlier, the system of language learning in dialogue has been created by us [5]. A new system for words pronunciation training are described in this paper.

2 HMM-Based Evaluation of Pronunciation

Nowadays, the basic technology for acoustic modeling of speech is the Hidden Markov Models (HMMs). This technology is widely applied to the speech recognition, the speaker verification and identification, also can be used for the estimation of pronunciation quality. Our method of the evaluation of pronunciation is based on the discrete variable duration Hidden Markov Models [6]. Such models allow to take into account the temporary characteristics of a speech signal, which are very important for the evaluation of pronunciation.

Figure 1 show the structure of speech processing system based on the discrete HMMs.

F. Jelinek, E. Nöth (Eds.): TSD'99, LNCS 1692, pp. 152–155, 1999.

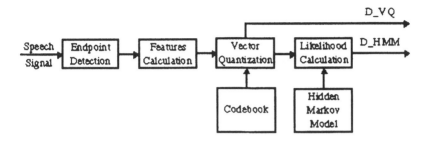

Fig. 1. The structure of speech processing system based on the discrete HMMs.

The speech signal of pronounced word after endpoint detection is coded in some features system. We use an original on-line endpoint detector and LPCC features. The sequence of features vectors enters the block of vector quantization. Thus the vector sequence transformed into symbolical one with the numerical estimation of the average distortion of the codebook (D_VQ). In the likelihood calculation block the numerical estimation (D_HMM) of hypothesis that the symbolical sequence derived from HMM is formed. The codebook and the parameters of HMM are determined by training the system on database represented by native speakers. The codebook is determined by a combined algorithm: the dichotomy + k-means. For HMM training we use the Viterbi algorithm. As appeared, the parameters D_VQ and D_HMM allow to estimate an accent degree in pronouncing a word.

3   System for an Evaluation of Russian Digits Pronunciation

As the base dictionary we have chosen ten Russian digits (from 0 up to 9). This dictionary was chosen because the pronunciation of digits matures at the first stages of the language study and the foreigners pronounce them rather confidently. Besides we have the large database earlier collected for the creation of speaker-independent isolated word recognition system [7].

For training and testing the system we had three databases of ten Russiad digits:

1. Training database – 170 native speakers.
2. Russian testing database – 33 native speakers.
3. Foreign testing database – 33 non-native speakers.

The students of the St.-Petersburg Mining Institute (50% males and 50% females) from China, Southern Korea, Mongolia, Israel, Morocco, Guinea, Lebanon, Mozambique, Nigeria and Somalia participated in creation of the Foreign testing database.

As a result of processing the Training database the codebook and the HMM for each digit were defined. Thus, for each realization of the word two parameters can be calculated:

- the average distortion of the codebook (D_VQ);
- the HMM likelihood (D_HMM).

Hence, each realization of the word from testing databases defines a point on a plane (D_VQ, D_HMM). For example, the results for word «6» are represented at the figure 2. Here, the beginning of coordinates is transferred in the center of the Training database.

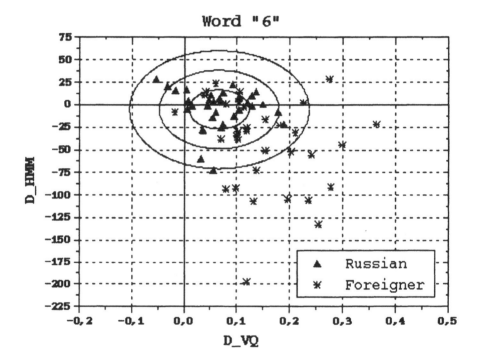

Fig. 2. Representation of testing databases of the word «6» in a plane (D_VQ, D_HMM).

In this figure, the ellipses appropriate single, double and treble standard deviations for the Russian testing database are constructed. It is visible, that the realizations from the Foreign testing database basically are located on the right and below from a beginning of coordinates. The part of them gets in a zone of three standard deviations and the part lays behind its limits. The listening of realizations from the Foreign testing database testifies that the arrangement of a point on a plane directly corresponds with a felt degree of accent.

The summary results for the Foreign testing database are presented in Table 1.

Table 1. Experiment results for the Foreign testing database

Digit	Amount of hits in area			Outside
	1	2	3	
≪ 0 ≫	1	8	10	14
≪ 1 ≫	13	7	6	7
≪ 2 ≫	8	8	8	9
≪ 3 ≫	4	18	4	7
≪ 4 ≫	10	9	9	5
≪ 5 ≫	2	17	7	7
≪ 6 ≫	4	11	4	14
≪ 7 ≫	5	13	8	7
≪ 8 ≫	5	12	8	8
≪ 9 ≫	4	10	5	14

4   Conclusions

The results of this experiments indicate that Hidden Markov Models can be used with some success to distinguish between native and non-native pronunciations of a word. The system functions in real time and allows to receive the adequate estimation of pronunciation with graphically display of results.

References

1. Hiller, S., Rooney, E., Laver, J., Jack, M.: SPELL: SPELL: An Automated System for Computer- Aided Pronunciation Teaching. Speech Communication, 13 (1993).
2. Kawai, G., Hirose, K.: A CALL System Using Speech Recognition to Train the Pronunciation of Japanese Long Vowels, the Mora Nasal and Mora Obstruents. In Proceedings of EUROSPEECH'97, Greece (1997).
3. Najmi, A., Bernstein, J.: Speech Recognition in a System for Teaching Japanese. Journal of the Acoustical Society of America, Vol. 100, No. 4 (1996).
4. Ehsani, F., Bernstein, J., Najmi, A., Todic, O.: SUBARASHI: Japanese Interactive Spoken Language Education. In Proceedings of EUROSPEECH'97, Greece (1997).
5. Dikareva, S., Kosarev, Y., Machovikov, A., Machovikova, I.: Speech Dialogue in Learning Systems for Foreign Language. In Proceedings of SPECOM'98, St.-Petersburg, Russia (1998).
6. Ferguson, J.: Variable Duration Models for Speech. In Proceedings of Symposium on Application of Hidden Markov Models to Text and Speech, Princeton (1980).
7. Kozlov, Yu., Machovikov, A., Maleev, O.: On the Improvements of Speaker-Independent Isolated Word Recognition. In Proceedings of SPECOM'96, St.-Petersburg, Russia (1996).

Allophone-Based Concatenative Speech Synthesis System for Russian

Pavel A. Skrelin

Department of Phonetics, St.Petersburg State University,
Universitetskaya emb. 11, Saint-Petersburg, Russia
paul@phonet.lang.pu.ru

Abstract. Existing concatenative speech synthesis systems developed
for various languages use diphones as their main construction units. The
report describes the Russian synthesis system based on the use of allophones, that is realizations of Russian phonemes in definite phonetic
contexts. In speech synthesis an allophone is a concrete sound which is
tolerant to a number of its right and left contexts. The use of allophones
gives a number of advantages compared to the use of diphones, but at the
same time it sets a number of certain limitations. The report describes
the principles of the allophone extraction from natural speech flow, ways
of forming synthesized speech, modification of the acoustic parameters.

1 Introduction

The experimental allophone-based speech synthesis system for Russian was developed at the Department of Phonetics of Saint-Petersburg State University in
1996–1997 [1]. In 1997 a set of phonetic experiments with the synthesized speech
were carried out using this system as a research tool.

In 1998 as a result of these experiments the system was considerably improved
and adapted for work under Windows'95. The lexicon used for text processing
(stress marking, the calculation of the fundamental frequency contour, pauses)
was extended up to 110 000 units which opened new facilities for the context
analysis, new rules for the automatic transcription of the text were formulated,
the number of sound units was reduced by 10 %; new means for the modification
of physical parameters of the sound stream were developed.

2 Sound Database

The comparison of the allophone- and diphone-based concatenative speech synthesis technologies and principles of the Sound Database formation for the
allophone-based speech synthesis system were described in detail in [2]. At
present the database of the working allophone based system numbers 3200 sound
units. The reduction of the inventory of allophones was achieved by introducing
changes into the segmentation and description principles of the consonant allophones: in the new system the left context of the allophone is not regarded, thus

F. Jelinek, E. Nöth (Eds.): TSD'99, LNCS 1692, pp. 156–159, 1999.
© Springer-Verlag Berlin Heidelberg 1999

the number of the sound units has been reduced by 30 % without affecting the quality of the synthesized speech.

At the same time experiments showed that the acoustic and perceptual characteristics of the Russian allophone /r/ are almost exclusively determined by its right and left vowels. To improve the quality of the synthesized speech special r–forming sound units will have to be formed.

At present the sound units from natural speech flow recorded from normative speakers, are extracted manually, as the quality of the resulting synthesized speech is to a great extent determined by the accuracy of the segmentation. In simple cases like, for example, in defining a boundary between a vowel and a stop or a fricative consonant, an automatic segmentation procedure, based of the spectral analysis of the sound stream, can be used. In more complicated situations, like extracting diphone-like or triphone-like sound units (for example, vowels preceded and followed by sonants) the use of the automatic segmentation should necessarily be accompanied by the audio control. This reduces the efficiency of the automatic segmentation procedure. At the same time automatic segmentation can be more efficient and precise in cases where manual segmentation fails to yield satisfactory results as, for example, in segmenting vowel sequences. Automatic extraction of the formant transitions will considerably facilitate and make more precise the definition of an acceptable boundary between each vowel pair. In developing means for such segmentation procedure we use the LPC analysis, but at present the accuracy of the automatic segmentation is far from satisfactory.

3   Sound Stream Formation

In our system database sound units used for the formation of the synthetic sound stream are structurally different. They include
- sounds representing physical realizations of the Russian allophones;
- sound sequences, representing a physical realization of an allophone and a fragment of the preceding sound;
- sound sequences, representing a physical realization of an allophone and a fragment of the following sound;
- sound sequences, representing a physical realization of an allophone and fragments of the preceding and following sounds;
- fragments of sound realizations of allophones the initial part of which are contained in other sound units;
- fragments of sound realizations of allophones the final part of which are contained in other sound units;
- fragments of sound realizations of allophones both the final and initial parts of which are contained in other sound units;
- symbolic sound units having no physical duration, but necessary for operating the speech synthesis system.

Due to differences in the inner structure of the sound units, in the process of the synthetic sound stream formation real boundaries of the allophones are restored to make further modifications of duration and fundamental frequency possible.

To define real allophone boundaries in the describer of the sound unit information on the boundaries of one or two allophones (in this particular sound unit) is stored. The Sound Database contains information about those sound units the initial or final parts of which are included in other sound units.

Such method of the sound stream formation allows a high precision of regulating of the physical parameters of the allophones and ensures the same high degree of quality of concatenation as the diphone-based method does.

One of the drawbacks of the allophone-based model of concatenative speech synthesis is the necessity to regulate the amplitude parameters of the sound units.

In this model a real phonetic context is described in a very general way; in natural speech situation there may appear a combination of two vowels of the same quality having different degrees of reduction . For example, a stressed and an unstressed vowel may follow a post-stressed /a/; the intensity/amplitude characteristics of these vowels may be drastically different. In these cases the physical transition from one vowel to the other should be accompanied by the regulation of the amplitude parameters of two neighboring vowels. Though the procedure of changing the amplitude parameters is very simple, it requires a number of preliminary operations to be fulfilled, namely, the comparison of amplitude characteristics of concatenated sound units, the definition of the boundaries of the sound string where the amplitude of the fundamental frequency periods is to be leveled, the calculation of the amplitude value for each fundamental frequency period in this sound unit. The fulfillment of these operations may hamper the synthesis process.

The experiments with the synthesized speech show that such leveling needs to be done pretty rarely. The comparison of the amplitude values of the concatenated sound units is not always necessary and it can be accomplished in the process of the automatic transcription of the text or in the process of the formation of the synthetic sound stream, when the sound units with different degrees of quantitative reduction are retrieved from the Database.

4   Tools for Modification of Physical Parameters of the Synthesized Speech

In the allophone-based Russian speech synthesis system tools for modifying the fundamental frequency and duration work independently. First, the duration of each allophone is calculated, then the fundamental frequency is modified, if necessary.

4.1   Modification of Duration

The duration of voiceless plosives is regulated by the changing the duration of the closure; in fricatives the duration of the noise period is regulated; in affricates the duration of the closure and narrowing are proportionally regulated. In voiced plosives modification of duration is achieved by deleting or multiplying certain fundamental frequency periods . Vowel duration is modified by deleting /multiplying fundamental frequency periods of the stationary part of the vowel. Duration of the sound unit is modified with regard to its proposed fundamental frequency values. To simplify the procedure, first the proposed mean Fo of the sound unit is defined, then the number of the fundamental frequency periods required for its duration is calculated. After this the modification of this sound unit's duration is accomplished. As a result, the duration of sounds in synthesized speech deviates from calculated values by no more than 0.5 of the fundamental frequency period (5–6 ms for our speech synthesis system).

4.2 Fundamental Frequency Modification

To modify the fundamental frequency we use the PSOLA- like method. Our method is different from the original PSOLA technique [3] in a slightly different geometry of the window, which is dictated by the boundaries of the fundamental frequency periods: for the beginning of the period we take the first positive Fo value rather than the amplitude maximum value. To modify voiced fricatives and affricates a more comlicated method is used: the sounds are first split into their low frequency (up to 400 Hz) and high frequency components; then the low frequency component is modified and the original high frequency component is added to it.

5 Conclusion

The quality and naturalness of the allophone-based Russian speech synthesis system is very high. New directions in the development of this technology can be connected with changing of the inner structure of the sound units: instead of the whole allophones as in diphone-based synthesis the allophone halves can be used.

This method will make it possible to considerably reduce the number of the basic units and, consequently, the amount of work required for their segmentation. At present work is progress in this direction and the first results look very promising.

References

1. Skrelin P.: The Linguistic Software of the Concatenation Synthesis Systems for Russian Speech. In: Proc. of the SPECOM'96, St.-Petersburg (1996) 165–166.
2. Skrelin P.: Concatenative Russian Speech Synthesis: Sound Database Formation Principles. In: Proc. of the SPECOM'97, Cluj-Napoka (1997) 157–160.
3. Moulines E., Charpentier F.: Pitch-Synchronous Waveform Processing Techniques for Text-to-Speech Synthesis Using Diphones. Speech Communication. 9 (1990) 453–467.

Intonation Questions in English and Armenian: Results of the Perceptual Study

Nina B. Volskaya and Anna S. Grigoryan

Department of Phonetics, St.Petersburg State University,
Universitetskaya emb. 11, Saint-Petersburg, Russia
paul@phonet.lang.pu.ru

Abstract. Melodic patterns of the so-called 'intonation questions' reveal a certain degree of similarity in many languages. A rise in pitch or a higher tone level signal another phonological category, since there are no other indications of the interrogativity in this type of utterances. Phonetic differences in the realization of IQs which pass unnoticed for speakers of one language may lead to a perception of a different sentence type by speakers of other languages. This seems to be the situation with intonation patters of Armenian perceived by native English speakers.

1   Introduction

Melodic patterns of the so-called intonation questions (IQ) reveal certain similarities in a number of languages: a rise in pitch on the tonic syllable or a high tone level of the whole utterance. Other signs of interrogativity, the inverse word order or the use of interrogative words and particles, are generally absent, handing over this function to the intonation.

In our study we investigated the pitch patterns of the IQ in Armenian in mono-, two- and three- syllable words with a different position of the tonic syllable.

The research included listening experiments which were performed for the perceptual evaluation of the Armenian melodic patterns of IQs by the English subjects and for investigation of the effect that production of Armenian question intonation patters bear on native English speakers. .

Perceptual experiments in which Armenian interrogatory pitch patterns were perceived by the English subjects revealed, that neutral from the point of view of Armenian speakers questions are perceived by the English listeners as emotionally colored questions or exclamations.

2   Description of the Experiment

2.1   Experiment 1: Patterns of Intonation Questions in Armenian

The material on the Armenian language was recorded from 4 native (2 male and 2 female) speakers. The age of the subjects ranged from 18 to 21. Recordings

F. Jelinek, E. Nöth (Eds.): TSD'99, LNCS 1692, pp. 160–164, 1999.

were made onto DAT in a sound-proofed studio. The material consisted of very
short 6–8 sentences printed everyday conversations. Subjects were asked to speak
the dialogues, each containing no less than three intonation questions, in a most
natural way. 148 intonation questions were excerpted from the corpus of material
pronounced by the Armenian speakers, transcribed and input into the computer
for further analysis.

As a result of the analysis of the recorded material most frequently used patterns of IQs in Armenian were transcribed by experts in intonation transcription,
schematized and described. Table 1 presents the results of the auditory analysis.

Table 1. Intonation patterns of IQs with frequency of occurrence (%)

Number of syllables	Intonation patterns and frequency of their occurrence (%)					
	Tonic	Post-tonic	Tonic	Post-tonic	Tonic	Post-tonic
	Mid-Rise		High Rise		Fall-Rise	
Monosyllabic words	33 %		58 %		8 %	
Bisyllabic words with a final tonic syllable	75 %		14 %		11 %	
Bisyllabic words with the first tonic syllable	50 % Low		50 % Low			
Polysyllabic words with unstressed pre-tonic syllables	062 %		38 % Low			
Polysyllabic words with unstressed post-tonic syllables	82 % Low		18 % Low			

There are two distinct patterns of IQ that were used by the speakers, namely,
a continuous rise within the tonic syllable observed in stimuli where the accented
syllable is in the final position, and a rise on the tonic syllable, followed by a fall
on the post-tonic unstressed syllables. On-sets and off-sets of the rising movement
on the tonic syllable differ: from low to upper-mid and from medium to the high
level. It is not surprising that final rises are the most frequently observed patterns
in Armenian IQ, as they are traditionally the canonical question markers used
in many languages [1–3]. At the same time, bisyllabic and polysyllabic words
displayed a slightly different pattern in which a rise on the tonic syllable was
followed by a fall in the post-tonic syllable(s).The choice of the pitch pattern
here seems to be bound up with differences in the lexical stress pattern.

2.2 Experiment 2: Perceptual Identification of IQ Patterns by English Subjects

The material was presented over headphones to 7 native English listeners. They
were instructed to decide what sentence type they heard with forced choice from
four alternatives: statement (.), non-finality (,), question (?), or exclamation (!),

by putting a corresponding punctuation mark in a graphed chart. Sentences were presented in random order. Utterances with a fall-rise pattern were excluded from the experiment. Since a terminal rise is an almost sure sign of interrogativity, we expected useful information in the subtleties of phonetic realizations of Armenian IQs which should be reflected in their evaluation by English listeners. Therefore, we introduced a graph in the chart where the subjects stated the neutral/emotional meaning of the utterance they heard. None of the English subjects spoke Armenian or was familiar with its intonation system. Listeners responses are presented in Tables 2 and 3.

2.3   Results

The analysis of English subjects responses to different pitch patterns of Armenian IQs shows that the results differ from what we expected in the way that a great number of intended neutral questions were perceived as emotionally colored questions or exclamations. This type of responses appeared for patterns with a fall in pitch in the post-tonic syllables and in patterns with low pitched pre-tonic syllables. We consider them separately in Table 3. Note also that for patterns presented in Table 2 the subjects actually chose from three of the five offered alternatives, though Table 3 presents a somewhat richer choice.

We observed that the percentage of perceived neutral questions decreased with the increase of the number of low-pitched pre-tonic syllables: for bisyllabic words with a low pre-tonic syllable the average percent of perceived questions was 61% versus 28% of emotionally colored questions; for three-syllable words 50% versus 37%, and for polysyllabic words 40% and 60% respectively.

Table 2. Percent responses to the suggested alternatives for mono- and bisyllabic stimuli with a post-tonic syllable.

N		Intonation patterns					
	Sentence type	Tonic	Post-tonic	Tonic	Post-tonic	Tonic	Post-tonic
		Mid-Rise		*High Rise*		*Rise Low*	
1	Question	75%		40%		25%	
2	Emotional question	25%		40%		25%	
3	Exclamation			20%		50%	

Table 3. Percent responses to each of five alternatives for stimuli with pre-tonic syllables.

	Sentence type	Pre-tonic	Tonic	Post-tonic	Pre-tonic	Tonic	Post-tonic
		low	*rising*	*upper mid*	*low*	*upper mid*	*low*
1	Question		30.2%			57%	
2	Emotional question		52.8%			43%	
4	Non-finality		5%				
5	Statement		12%				

2.4   Experiment 3: Stimulus Analysis

The analysis of the waveforms (duration of the tonic syllable, degree of the F0 excursion and the rate of the F0 change within the syllable) was performed with the help of the computer program developed at the Department of Phonetics of Saint-Petersburg University. 21 stimuli perceived as neutral questions were input in the computer and digitized at 16 kHz. The following parameters were extracted, either automatically or by visual inspection of the signal:
- F0 excursion size (in Hz) as the distance between the F0 peak on the tonic syllable and the preceding F0 minimum.
- Duration of the accented vowel (in ms).
- Mean F0 of the pre-tonic syllables for bisyllabic words with a final tonic syllable.
- F0 excursion size as the distance between the F0 peak on the tonic syllable and the average F0 on the pre-tonic or post-tonic part in bisyllabic and polysyllabic words.

The acoustic data on the utterances perceived as neutral questions are presented in Table 4. Mean excursion size in the tonic syllable of monosyllabic words is 226 Hz for female speakers and 102 Hz for male speakers. In bisyllabic words with an unstressed pre-tonic syllable(s) : 185 Hz for female speakers and 72 Hz for males. In polysyllabic words mean F0 excursion is 182 Hz for female and 72 Hz for males. In post-tonic syllable(s) mean F0 value is 154 Hz and 110 Hz respectively. The obvious tendency here is a larger F0 excursion in females realizations of the IQs.

Table 4. Mean F0 and duration values of the tonic syllable.

Mean values		Female speakers		Male speakers	
F0 [Hz]	*monosyllables*	226		102	
T [ms]		189		174	
F0 [Hz]	*bisyllables*	interval	range	interval	range
T [ms]		185	200	70	110
		189		201	
F0 [Hz]	*polysyllables*	interval	F0 min	interval	F0 min
		182	154	72	88
T [ms]		157	161	136	218

3   Conclusion

From our experiment on the intonation of IQs in Armenian it appears that there is a certain degree of similarity between pitch patterns of this sentence type in English and Armenian. Three types of pitch patterns were most frequently used by our subjects, namely, The Mid-Rise, The High Rise and The Rise + Fall on the post-tonic syllables. For Armenian speakers they seem to be variants of the same interrogatory pitch pattern, their difference is due mainly to place of lexical stress.

The results of the perceptual study show that for some languages, including English, the relevant features of the melodic patterns used in intonation questions are not only the degree of the F0 excursion, but the position of the pre-tonic and post-tonic syllables relative to the F0 level of the tonic syllable. Phonetic differences in the realization of question intonation in Armenian are relevant for the speakers of English and induce listeners to perceive emotion in questions which are emotionally neutral to Armenian speakers.

References

1. J.Haan, V.J. van Heuven. Intonational characteristics of declarativity and interrogativity in Dutch: a comparison, in: Proc. ESCA Workshop on Intonation, Athens, 1997.
2. J.D. O Connor, G.F. Arnold. Intonation of colloquial English. London, Longman, 1973.
3. E.A. Bryzgunova. Zvuki i intonacija Russkoy Rechi. Moskva, 1977.

Methods of Sentences Selection for Read-Speech Corpus Design

Vlasta Radová and Petr Vopálka

University of West Bohemia, Department of Cybernetics,
Univerzitní 22, 306 14 Plzeň, Czech Republic
radova@kky.zcu.cz

Abstract. In this paper methods are proposed which can be used to select a set of phonetically balanced sentences. The principle of the methods is presented and some experimental results are given. In the end of the paper the use of the proposed methods for the Czech read-speech corpus design is described in detail and the structure of the corpus is explained.

1 Introduction

One of the crucial problems that have to be solved when a speech recognition or a speech synthesis system is developed is the availability of a proper speech corpus for the system training and testing. The problem is usually solved in the following way: first a set of suitable sentences is selected from a database of phonetically transcribed sentences, next the set of the selected sentences is read by a group of speakers and, as the last step, the utterances are used to form a training or a test database [2], [3], [6].

The methods which are used to select sentences from the phonetically transcribed database can be divided into 2 groups. One of them consists of methods that enable to select sentences containing all phonetic events with approximately uniform frequency distributions. Such sentences are usually called *phonetically rich sentences* [4]. The other group includes methods that can be used to select "naturally" balanced sentences, i.e. sentences containing phonetic events according to their frequency of occurrence in natural speech. Such sentences are called *phonetically balanced sentences* [4].

Some ideas how to select a set of phonetically rich sentences were presented in our previous papers [7] and [8]. This paper deals with methods of phonetically balanced sentences selection. The principle of the methods together with some experimental results is presented in Section 2. Next, in Section 3, the use of the procedures in the course of the Czech speech corpus creation is described.

2 Methods of Phonetically Balanced Sentences Selection

The only way how to select the best set of sentences from a database of phonetically transcribed sentences is to form all possible sets and then to select the

F. Jelinek, E. Nöth (Eds.): TSD'99, LNCS 1692, pp. 165–170, 1999.

best one. However, such a way may be too time consuming, especially when the number of sentences in the phonetically transcribed database and the number of sentences being selected are high. For that reason various procedures were proposed which are not so time consuming, however, they allow to select only a suboptimal set of sentences. The most known and used procedure is the *add-on procedure* [1], [4]. The procedure works with a phonetically transcribed text database and has the following 3 steps:

1. For each sentence of the phonetically transcribed text database a score S is computed that reflects how well the phonetic events contained in the sentence are represented in the up to now selected sentences.
2. The sentence with the best score is selected and moved to the list of the up to now selected sentences.
3. The steps 1 and 2 are repeated until the desired number of sentences is selected.

It is obvious that the most important problem in the procedure is the score S computation. To select a set of phonetically balanced sentences we propose the score

$$S = -\sum_{i=1}^{I} \left| \frac{m_i}{m} - \frac{n_i + n_i'}{n} \right| \tag{1}$$

where

$$m = \sum_{i=1}^{I} m_i \,, \tag{2}$$

$$n = \sum_{i=1}^{I} (n_i + n_i') \,, \tag{3}$$

I is the number of distinct phonetic events that we wish to have in the selected sentences, m_i is the number of occurrences of the i-th phonetic event in the phonetically transcribed database, n_i is the number of occurrences of the i-th phonetic event in the up to now selected sentences and n_i' is the number of occurrences of the i-th phonetic event in the inspected sentence. Using this score, the sentence with the minimum score has to be selected in the step 2 of the add-on procedure.

The score (1) and the add-on procedure don't assure, however, that all phonemes will occur in the selected sentences. To overcame this disadvantage a *preselective procedure* has to be used before the add-on procedure. We propose the preselective procedure with the following 3 steps:

1. The sentence with the highest number of the distinct phonetic events which don't occur in the up to now selected sentences is selected from the phonetically transcribed text database and moved to the list of the up to now selected sentences. If two or more sentences can be selected in a moment the sentence which contributes mostly to the phonetical balance of the selected sentences is selected.

2. If some sentences can be excluded from the set of up to now selected sentences without decreasing the number of distinct phonetic events in the up to now selected sentences they are excluded and moved back to the phonetically transcribed text database.
3. The steps 1 and 2 are repeated until all phonetic events are present in the up to now selected sentences.

To test the work of the score (1) and the add-on procedure both without and with the preselection we tried to select a set of 40 phonetically balanced sentences from a set of 24 442 phonetically transcribed sentences. The sentences were selected with respect to the coverage of Czech phonemes [5], i.e. the Czech phonemes were regarded as the phonetic events in this experiment. Achieved results are presented in Table 1.

Table 1. The relative number of occurrences of particular phonemes [%]

Phoneme	in the primary set	in the set of 40 selected sentences without pre-selection	with pre-selection	Phoneme	in the primary set	in the set of 40 selected sentences without pre-selection	with pre-selection
e	9.2161	9.2322	9.2369	h	1.2451	1.0969	1.2048
o	7.9041	7.9525	7.9518	ee	1.1815	1.1883	1.1245
a	6.1693	6.2157	6.1847	sh	1.1033	1.0969	1.1245
i	6.1644	6.3071	6.1044	ch	1.0667	1.0055	0.9639
t	5.0174	5.0274	4.9799	x	0.9444	0.9141	0.9639
n	4.6839	4.5704	4.6586	uu	0.9185	0.9141	0.8835
ii	4.5716	4.5704	4.5783	f	0.8954	0.9141	0.8835
s	4.3688	4.3876	4.3373	tj	0.7744	0.8227	0.8032
v	3.9368	3.9305	3.9357	zh	0.7570	0.7313	0.7229
p	3.8056	3.7477	3.7751	rsh	0.7466	0.7313	0.7229
l	3.7974	3.7477	3.7751	ow	0.6589	0.6399	0.6426
r	3.7267	3.7477	3.7751	dj	0.4542	0.5484	0.4016
k	3.6420	3.5649	3.6948	rzh	0.4053	0.3656	0.4016
d	3.0798	3.1079	3.0522	g	0.3566	0.4570	0.4016
m	2.9849	3.1079	2.9719	ng	0.2104	0.1828	0.2410
nj	2.6900	2.7422	2.4900	aw	0.0386	0.0000	0.1606
j	2.6676	2.6508	2.7309	eu	0.0145	0.0000	0.0803
u	2.3687	2.3766	2.3293	dz	0.0124	0.0000	0.0803
aa	2.1479	2.1938	2.1687	oo	0.0113	0.0000	0.0803
z	2.1075	2.1024	2.0884	dzh	0.0019	0.0000	0.0803
c	1.5775	1.5539	1.6064	mg	0.0009	0.0000	0.0803
b	1.5741	1.5539	1.5261				

As the results show, using the add-on procedure without preselection the phonemes with a high relative frequency of occurrence in the primary set are covered rather well in the selected sentences. However, several phonemes with a low relative frequency of occurrence in the primary set don't occur at all in the selected sentences. Using the add-on procedure with preselection all phonemes occur in the selected sentences, however, the phonemes with a low relative frequency of occurrence in the primary set are "overrepresented" in the selected sentences. This phenomenon can be however easily eliminated when more sentences will be selected.

3   Sentences Selection for the Czech Read-Speech Corpus

The goal of the Czech read-speech corpus is to provide enough continuous speech material for the development and evaluation of continuous speech recognition systems for Czech. We plan to record speech from at least 100 speakers from various regions of the Czech Republic. The texts to be read are selected from several Czech newspapers and have to satisfy several requirements. Each sentence must contain at least 3 and at most 15 words and have to contain no foreign words (i.e. words which are difficult to read for Czech people) and no numbers and abbreviations (since they may not be read identically by all speakers). Each speaker will be asked to read 150 sentences, where 40 sentences of the 150 are identical for each speaker. The remaining 110 sentences are carefully selected in order to satisfy several requirements. The sets of the 110 sentences from all speakers together will be used to train a Czech speaker-independent speech recognizer, the 40 sentences will be then used to adapt the recognizer to a particular speaker.

3.1   Selection of Adaptation Sentences

Our primary intention was to select the set of 40 adaptation sentences in such a way that the set will contain all triphones with approximately identical relative frequency. However, as the experimental results in [7] and [8] showed, this requirement was satisfied not very well for phonemes and the less it will be satisfied for triphones. For that reason we changed our primary intention and decided to select the adaptation sentences in such a way that they will contain triphones according to their relative frequency of occurrence in natural speech. To do it we used the add-on procedure described in Section 2 and the score (1). There were 2 additional conditions during the selection: no two adaptation sentences had to be identical and no sentence containing a triphone occurring only in that sentence had to be selected. The former condition is quite obvious, the latter one together with requirements posed on training sentences assures that no triphone which occurs in the phonetically transcribed sentences will be missing in the training sentences. Achieved results are given in Table 2. The sentences were selected from a primary set of 24 442 phonetically transcribed sentences containing 8 223 distinct triphones. The 40 selected sentences contain 1 492 distinct triphones what is only 18.14 % of all distinct triphones occurring in the primary set. However, these 18.14 % of triphones cover about 73.66 % of the text in the primary database. Such a result can be regarded as a very good

Table 2. Results of the adaptation sentences selection

| Database | number of distinct triphones | | covered text |
	absolute	relative with regard to primary set	in the primary set
primary set	8 223	100 %	100 %
set of of adaptation sentences	1 492	18.14 %	73.66 %

one since the 40 sentences will be used, as mentioned above, to adapt a speaker-independent speech recognizer to a particular speaker and therefore they should contain mainly the triphones occurring with a very high relative frequency in the natural speech.

3.2 Selection of Training Sentences

The set of training sentences has to satisfy two main requirements. It should contain all triphones occurring in the set of phonetically transcribed sentences and it should be phonetically balanced with respect to triphones. To do it we used the add-on procedure with preselection. The sentences were selected from the same set of phonetically transcribed sentences as the adaptation sentences, however, the adaptation sentences were already eliminated from the phonetically transcribed database. It means no adaptation sentence could be selected as a training sentence.

Using the preselective procedure 1 786 so called necessary sentences were selected which contain all triphones occurring in the set of 24 442 phonetically transcribed sentences. Remaining sentences were then selected to the necessary sentences using the add-on procedure. The number of the remaining sentences that had to be selected was given by the number of speakers in such a way that the total number of training sentences (including the necessary sentences) had to be 110 times higher than the number of speakers. In contradistinction to the adaptation sentences selection, a sentence could be selected several times during the training sentences selection. However, the number of repetition of each training sentence had to be at least 3 times lower than the number of speakers.

As mentioned above, a speaker-independent speech recognizer will be trained using the set of training sentences from all speakers together. However, each speaker will read only 110 training sentences. For that reason the whole set of training sentences was divided among particular speakers in such a way that each speaker will read 110 training sentences in total, no speaker will read a sentence more than once and all speakers will read approximately equal number of sentences.

4 Conclusion

The paper deals with the problem of phonetically balanced sentences selection. Two iterative procedures have been presented which can be used to select a set of sentences that will contain phonetic events according to their occurrence in the natural speech. Both procedures have been tested on a primary set of 24 442 phonetically transcribed sentences from that a set of 40 phonetically balanced sentences was selected. In the end of the paper the use of the proposed methods for the Czech read-speech corpus design is described.

Acknowledgement. The work was supported by the Grant Agency of the Czech Republic, project no. 102/98/P085, and by the Ministry of Education of the Czech Republic, project no. VS 97159 and project KONTAKT no. ME 293.

References

1. Falaschi, A.: An Automated Procedure for Minimum Size Phonetically Balanced Phrases Selection. In: Proc. of the ESCA Workshop on Speech I/O Assessment and Speech Databases (1989) 5.10.1–5.10.4
2. Frasen, J., at al.: WSJCAM0 Corpus and Recording Description. Technical Report: CUED/F-INFENG/TR.192. Cambridge University, Engineering Department, Cambridge, UK (1994)
3. Gauvain, J.-L., Lamel, L.F., Eskénazi, M.: Design Consideration and Text Selection for BREF, a Large French Read-Speech Corpus. In: Proc. of the ICSLP (1990), 1097–2000
4. Gibbon, D., Moore, R., Winski, R. (eds.): Handbook of Standards and Resources for Spoken Language Systems. Mouton de Gruyter, Berlin New York (1997)
5. Nouza J., Psutka J., Uhlíř J.: Phonetic Alphabet for Speech Recognition of Czech. Radioengineering 4 (1997) 16–20
6. Paul, D.B., Baker, J.M.: The Design for the Wall Street Journal-based CSR Corpus. In: Proc. of the ICSLP (1992) 899–902
7. Radová, V.: Design of the Czech Speech Corpus for Speech Recognition Applications with a Large Vocabulary. In: Sojka, P., Matoušek, V., Pala, K., Kopeček, I. (eds.): Text, Speech, Dialogue. Proc. of the First Workshop on Text, Speech, Dialogue. Brno, Czech Republic (1998) 299–304
8. Radová, V., Vopálka, P., Ircing, P.: Methods of Phonetically Balanced Sentences Selection. In: Proc. of the 3rd Multiconference on Systemics, Cybernetics and Informatics to be held in Orlando, USA (1999)

Speaker Identification Using Discriminative Centroids Weighting – A Growing Cell Structure Approach

Bogdan Sabac and Inge Gavat

Polytechnic University of Bucharest, Aleea Faurei 8-11, Bucharest 78409, Romania
sbogdan@helix.elia.pub.ro and inge@helix.elia.pub.ro

Abstract. A new method of text-dependent speaker identification using discriminative centroids weighting is proposed in this paper. The characteristics of the proposed method are as follows: feature parameters extraction, vector quantization with the growing cell structures (GCS) algorithm, stochastic fine-tuning of codebooks and discriminative centroids weighting (DCW) according to the uniqueness of personal features. The algorithm is evaluated on a database that includes 25 speakers each of them recorded in 24 different sessions. All 25 speakers spoke the same phrase for 240 times. The overall performance of the system was 99.5 %.

1 Introduction

In the early works of speaker identification, feature parameters are not always considered whether they have sufficient information for verifying the identity of individuals. In the case of identifying a person in our everyday life, it is considered that we extract the person's personal features from various aspects and identify the person by integrating those extracted features. Therefore, the same concept of feature selection process will be useful in speaker identification. To realize this idea, we propose a new verification scheme based on discriminative features selection.

2 Unsupervised Growing Cell Structures

The model we propose consists of a set A of formal units. Every unit $c \in A$ has associated a n-dimensional representative vector $w_c \in R^N$. The set W of all representative vectors is the current codebook. The GCS model has a structure consisting of hypertetraehedrons (or simplices) of a dimensionality chosen in advance [1]. The vertices of the hypertetrahedrons are the neurons and the edges denote neighborhood relations. The general idea of our method is to construct the codebook incrementally by interpolating new codebook vectors from existing ones. Interpolating is always done among topologically close neighboring units. Every modification of the network, however, is performed such that afterwards

F. Jelinek, E. Nöth (Eds.): TSD'99, LNCS 1692, pp. 171–174, 1999.

the network consists solely of k–dimensional simplices again. After each interpolation the current codebook is adapted with a fixed number of vectors from the original data.

In principle the adaptation of the synaptic vectors in our model is done as earlier proposed by Kohonen: determine the best-matching unit (bmu) for the current input signal, increase matching at the bmu and its topological neighbors. The growing cell structures model follows the same basic strategy. There are however, two important differences [2]

- the adaptation strength is constant over time. Specially are used constant adaptation parameters ε_b for the bmu and ε_n for the neighboring cells, respectively.
- only the bmu and its direct topological neigbors are adapted

An adaptation step in our model can be formulated as follows:

1. Choose an input signal x from the training lot.
2. Locate the bmu.
3. Increase matching for bmu and its direct topological neighbors

$$\Delta w_{bmu} = \varepsilon_b(x - w_{bmu}) \tag{1}$$

$$\Delta w_i = \varepsilon_n(x - w_i) \qquad (\forall i \in N_c) \tag{2}$$

The symbols ε_b and ε_n are adaptation constants with $\varepsilon_b \geq \varepsilon_n$. N_c denotes the set of direct topological neighbors of a cell c.

Furthermore, at each adaptation step a local information is accumulated at the wining unit bmu:

$$\Delta E_{bmu} = \| w_{bmu} - x \|^2 \tag{3}$$

The accumulated error information is used to determine (after a fixed number of adaptation steps) where to insert new units in the network.

When an insertion is done the error information is locally re-distributed, increasing the probability that the next insertion will be somewhere else. The local error variables act as a kind of memory which lasts over several adaptation/insertion cycles and indicates where much error has occurred. Deletion of a neuron takes place if after a preset number of adaptation steps that neuron have not been a bmu. By insertion and deletion of neurons the structure is modified. The result are problem-specific network structures potentially consisting of several separate sub networks.

2.1 Stochastic Approximation

Once a codebook of the desired size has been generated it is fine tuned by stochastic approximation. This denotes a number of adaptation steps with a strength α decaying slowly:

$$\sum_{t=1}^{\infty} a(t) = \infty, \text{ but not too slowly: } \sum_{t=1}^{\infty} \alpha^2(t) < \infty \tag{4}$$

One specific sequence of parameters fulfilling the above conditions is the harmonic series

$$\alpha(t) = \frac{1}{t} \qquad or\ its\ more\ general\ form\ \alpha(t) = \frac{(c+1)}{(c+t)} \tag{5}$$

for a non-negative constant c. For the stochastic approximation phase of our algorithm the neighborhood relations are completely ignored. Every adaptation step consists thus of: presentation of an input signal x, determination of the bmu unit s and adaptation of the bmu by:

$$\Delta w_s = \alpha(t) \cdot \varepsilon_b \cdot (x - w_s) \tag{6}$$

2.2   Discriminative Centroids Weighting

The weighting coefficients are calculated according to the uniqueness of personal features in each codebook as described in 7:

$$P_w(i) = \frac{Sc_m^P(i) \cdot H_n^P(i) - Sc_m^I(i) \cdot H_n^I(i)}{Sc_m^P(i) \cdot H_n^P(i)} \tag{7}$$

if $P_w(i) < p$ then $P_w(i) = 0$, with $0 < p < 1$ \qquad\qquad where:

$P_w(i)$ represents the weight associated to the i-th centroid from the current codebook $Sc_m^{P,I}(i)$ represents the medium (m) score (Sc) obtained at the i-th centroid when trough the current codebook are passed the codebook training vectors (P) or the impostor training vectors (I).
$H_n^{P,I}$ represents the normalized (n) histogram (H) of the bmu when trough the current codebook are passed the codebook training vectors (P) or the impostor training vectors (I).

3   Speaker Identification Setup

Speech signal was sampled at 16 kHz with a 8 bit digitizer. The speech signals are analyzed with a 30 ms Hamming window shifted every 15 ms in order to extract the following parameters from each frame of speech discarding low energy speech frames: 20 mel frequency cepstral coefficients (MFCC), 20 delta mel frequency cepstral coefficients (DMFCC) calculated as polynomial expansion coefficients over speech segments of five frames in length. With the extracted feature vectors from a speaker, using the growing cell structures algorithm two codebooks are constructed and centroids weighted for that speaker. This process is repeated for all speakers in the population. The linear opinion pool has been considered in our speaker verification system for the combination of features, namely cepstrum and delta cepstrum features, in order to take the identification/rejection decision.

4 Experimental Results

The algorithm is evaluated on a database that includes 25 speakers each of them recorded in 24 different sessions. 20 speakers are selected as registered speakers and 5 as impostors speakers with the aim of testing the identification system in 'open set' mode. All speakers spoke the same phrase: "My voice is my passport" for ten times in each session. All speakers where male between 21 and 23 years old. The codebooks where constructed using the first 2 pronunciations of the recording sessions 1 to 5 for each speaker. Speaker identification experiments where performed using the fixed thresholds established in the test phase on the utterances form recording sessions 6 to 12 and also using the cohort comparisons technique.

Table 1. Speaker identification success rate as a percent for GCS and GCS + DCW

Codebook size	8	16	32	64
Overall Speaker Identification Performance (fixed threshold)				
GCS	67 %	89 %	91 %	93.2 %
GCS + DCW	92 %	96 %	97.5 %	98.7 %
Overall Speaker Identification Performance				
GCS	76 %	92 %	94 %	94.6 %
GCS + DCW	94 %	98 %	99.5 %	99.7 %

5 Conclusions

We present a new method of text-dependent speaker identification using DCW. The system is evaluated for a combination of two feature sets, employing the linear opinion pool criterion. The DCW technique improve significantly the performance of the speaker identification system. The reliability of the proposed method is also discussed with some simulation results. The overall performance of the system was 99.5 %.

References

1. B. Fritzke (1994): *Growing Cell Structures – A self-organizing Network for Unsupervised and Supervised Learning.* N.N., 7(9): 1441–1460.
2. B. Sabac, I. Gavat: Speaker Verification with Growing Cell Structures. accepted for publication at the EUROSPEECH99, 5–9 September, Budapest, Hungary.

Speech Analysis and Recognition Synchronised by One-Quasiperiodical Segmentation

Taras K. Vintsiuk and Mykola M. Sazhok

NAS Institute of Cybernetics & UNESCO/IIP International Research-Training
Centre for Information Technologies and Systems
Kyjiv 252022 Ukraine

Abstract. It is shown that the best ASR results are attained when a pre-processing is carried out synchronically with pitch. Specifically, an analysis step has to be equal to the current one-quasiperiod duration and current analysis intervals have to consist of an entire number of quasiperiods with total 45–60 ms duration. Quasi-periodicity and non-qusiperiodicity models and measures as well as their applications for the optimal segmentation of speech signals into one-quasiperiods are given and discussed. Then the ways to embed these pre-processing results into the recognition procedure are described.

1 Introduction

A lot of problems in speech signal pre-processing still await for solution. Among them there are such questions: Is it really necessary to perform speech signal pre-processing before its recognising? If so, than must it be synchronised by pitch or not? What is the analysis interval duration? And what is the analyser on the whole?

In this paper it is shown experimentally that speech signal pre-processing, if it is performed before the recognition, must be fulfilled synchronically with a current pitch period. So analysis interval bounds must match the bounds of quasiperiods, and current analysis interval duration must be in range of 10–60 ms and more.

Further, there are considered: models of the speech signal quasi-periodicity and non-periodicity, similarity measures, the algorithm for the speech signal optimal partition into quasiperiods, pointing their beginnings, the ASR procedure, that is synchronised with the one-quasiperiod speech signal segmentation.

2 Influence of Discretisation Effects, Analysis Interval Length and Analysis Step on the Recognition Accuracy

The most used analyser of a speech signal is following. At first, speech signal is divided into segments or analysis intervals with the constant step ΔT and duration $\Delta T'$. Then, such picked up speech signal segments are analysed. It is

F. Jelinek, E. Nöth (Eds.): TSD'99, LNCS 1692, pp. 175–180, 1999.

obviously, that under such approach speech analysis intervals are placed randomly relatively to the beginning of a speech signal. In [1] it is shown, that the signal amplitude spectre distinctively changes if the analysis interval is shifted even by one discrete. Thus, the results of analysis depend on discretisation effects that means the randomness of the analysis interval shift relatively to the speech signal beginning.

At the time of speech signal analysis typically the overlapped analysis intervals are used ($\Delta T < \Delta T'$), and their duration is in range of $10 - 30\,ms$. Although, it is no clarity neither theoretical nor experimental here.

To make clear how discretisation effects, analysis interval duration and step influent on the recognition accuracy the two series of experiments were made.

The recognition training and proper recognition were performed accordingly with algorithms based on the Dynamic Time Warping [1]. The training sample consisted of 500 speech signals for 100 isolated words (5 realisations per word). The test sample was similar, namely 500 realisations, exactly 5 speech signals per word.

In the first set of experiments the dependence of recognition accuracy on discretisation effects was studied. It was considered 30 different analysis interval durations $\Delta T'$ under the fixed analysis interval step $\Delta T = 15\,ms$. Since the analysis step was invariable $\Delta T = 15\,ms$ and speech signal discretisation step was equal to $\Delta t = 50\,ms$, therefore due to discretisation effects, that is by the shifting of analysis interval by different number n discretes, $n = 1 : (N-1)$, $N = \Delta T/\Delta t = 300$, it was received 299 additional test samples for each original one. Thus, actually test sample consist of $300 \times 5 \times 100 = 150,000$ word realisations.

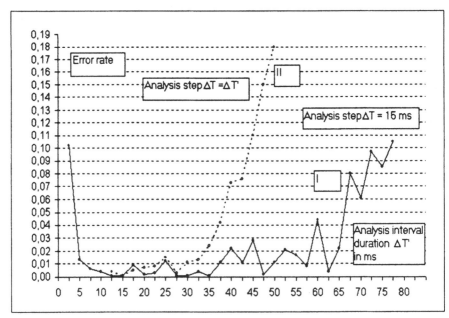

Fig. 1. The recognition error rate dependence on the analysis interval duration $\Delta T'$.

In the second set of experiments 16 different join non-overlapping analysis intervals $\Delta T = \Delta T'$ were studied. Here the number N of additional realisations was changing.

Speech signal analysis and similarity measures (both elementary and integral) used in the Dynamic Time Warping matching were based on the auto-correlation, co-variation, linear predictive, spectral, cepstral or coded descriptions [1].

In Fig. 1 the recognition error rate dependence on the analysis interval duration $\Delta T'$ under the invariable analysis step $\Delta T = 15\,ms$ (continuous Curve I) and on the same analysis interval duration $\Delta T'$ under the condition $\Delta T = \Delta T'$ (dotted Curve II) are given [2]. As a speech signal description it was used the 48-bit binary code, that was the discrete analogue of the auto-regressive spectre derivative sign on the set of 49 frequencies. As an elementary similarity measure for observed and reference elements it was used the Hamming distance [1].

Studying dependencies in Fig. 1 allows to conclude:

1) Curve I has emphatic oscillative quasi-periodical tendency with average speaker pitch period.

2) The smallest error rate ("cavities" on the Curve I) comes across when analysis interval duration $\Delta T'$ is fit by entire number of quasi-periods.

3) The best recognition accuracy is reached on the wide range of analysis interval duration from 10 to 65 ms. Sensitivity to one-quasiperiod synchronisation grows with the analysis interval increasing.

3 Speech Recognition Synchronised by One-Quasiperiodical Segmentation

3.1 Two-Level ASR System Structure

In Fig. 2 it is considered the two-level speech recognition system.

At the first level a problem of the optimal current pitch period discrimination and speech signal partition into quasi-periods is solved. It consists in finding the best quasiperiod beginnings or the best one-quasiperiod segments.

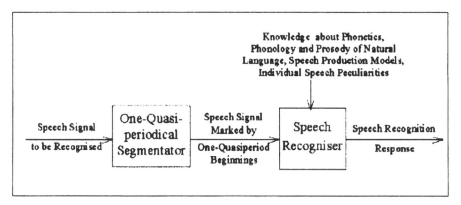

Fig. 2. Speech recognition system structure.

At the second level the input speech signal marked out by one-quasiperiod beginnings is recognised.

3.2 Optimal Signal Partition into One-Quasiperiod Segments

In [3] the algorithm for optimal speech signal partition into one-quasiperiods is described. Each hypothetical one-quasiperiodical signal segment is considered as a random distortion of previous or following one taken with an unknown multiplying factor. The problem consists in finding the best quasiperiod beginnings under restrictions on both value and changing of the current quasiperiod duration and multiplying factor.

Different elementary similarity measures for the one-quasiperiodical speech segments comparison are introduced. Dynamic programming matching procedure guarantees a choice of the best speech signal partition into quasiperiods for which the integral sum of respective elementary similarity measures for hypothesised joint quasiperiods is the largest.

Thus, the notion of one quasiperiodical segment is applied not only to properly quasi-periodical signal segments, but it is extended to any kind of one too, particularly to noise segments.

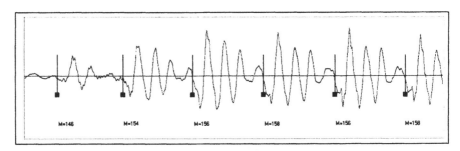

Fig. 3. One-quasiperiodical partition onto the voiced speech signal. The value of M shows the one-quasiperiod segment duration in discretes. Discretisation step Δt is equal to 50 ms.

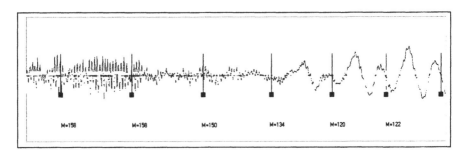

Fig. 4. One-quasiperiodical segmentation on the transitive noisy/voiced speech signal ['su].

In Fig. 3 and Fig. 4 the examples of the optimal speech signal partition into quasi-periods are given for quasi-periodical and non-periodical signals, respectively.

3.3 How to Run with Quasiperiods at the Stage of Recognition

When the speech recognition process we run with one-quasiperiodical segments by such a way. All one-quasiperiod beginnings are considered as the potentially optimal bounds of phoneme-threephones and each observed one-quasiperiodical segment is tested as a random distortion of the reference one taken from the codebook of a so-called Speaker Voice File or Passport (SVP).

The individual SVP is computed through the speaker training speech data. Such parameters are to be estimated: a set (alphabet) of typical one-quasi-periodical segments, that is a codebook; acoustical transcriptions of phoneme-threephones in names of typical one-quasiperiodical elements; intonation contours for syntagmas. Thus, SVP describes the individual phonetic-acoustical diversity and peculiarities of pronouncing.

Further we will consider the ASR for words and phrases taken from the Word/ Phrase-Book. Thereby, it will be not running with prosodic information.

Of course, the SVP refers to general linguistic and phonetic data and knowledge base for a concrete natural language. Typical one-quasiperiodical elements are chosen from the real speech training sample. They are specified by speech signal segments in time domain or by any other equivalent description, e.g. by linear predictive co-variance vector $e\left(k^1\right) \in E^1$, $k^1 \in K^1$ where K^1 is the name alphabet of typical one-quasiperiodical segments and E^1 is a set of reference elements. We interpret k^1 and $e\left(k^1\right)$ as micro-phonemes or the first level speech patterns.

An observed one-quasiperiodical segment x is compared with a reference element e by such measure of similarity like

$$g\left(x, e\right) = ln\left(1 + \left(a - e\left(k^1\right)\right)^T B\left(a - e\left(k1\right)\right)\right),\tag{1}$$

where x and B are the one-quasiperiodical segment given by its co-variance both predictive vector a and matrix B respectively.

Phoneme-threephones (PT) form the second level speech patterns. The PT is the basic phoneme that is considered under influence of neighbouring phonemes in context, they are the first which precedes and the second which follows. For each natural language there are fixed about $2,000 - 3,000$ PT. Each PT from the SVP is specified by its speaker transcriptions in the individual microelement alphabet.

Since the isolated words or phrases recognition, the orthographic text of each hypothesised word or phrase is converted into phoneme and respective phoneme-threephone transcriptions. Accordingly to the latter and referring to acoustical phoneme-threephone transcriptions, a so-called initial model signal of the word or phrase, presented by a sequence of reference microelements, is composed. Then, in time domain, the non-linear transformations of the initial model signal are performed, and results of these transformations are compared with the input signal by using the Dynamic Time Warping procedure [1].

Non-linear transformations allow to repeat or remove a microelement (typical one-quasiperiod) between neighbouring ones. More exactly, it is forbidden to repeat the same microelement more than two times, and to remove more than two microelements running. At last, the transformed one-quasiperiodic sequence has to have the same quantity of one-quasiperiodical elements as it is in the speech signal to be recognised.

Finally, running the one-quasiperiod-to-one-quasiperiod comparison accordingly with (1), the best integral matching for a hypothesised word or phrase is found.

Recognition response is a word or phrase with the best matching on a word/phrase-book.

3.4   Experimental Results

Two series of experiments were set. In the first one it was repeated the experiments described in the chapter 2. The difference was in that that analysis interval length and step were synchronised by one-quasiperiodical pre-segmentation. Any error was fixed when analysis interval duration $\Delta T'$ and step ΔT were equal to the entire number of current quasiperiods from 1 to 9 and from 1 to 3 respectively.

The second series of experiments followed the chapter 3.3 technology.

4   Conclusion

In this paper by the experimental way it is shown that robust word and phrase recognition with higher accuracy is reached when the recognition procedure is synchronised by one-quasiperiodical pre-segmentation.

It is expected a similar effect in the automatic speech understanding under taking into account a prosodic information.

References

1. T.K. Vintsiuk. Analysis, Recognition and Understanding of Speech Signals. – Kyjiv: Naukova Dumka, 1987, pp. 264 , in Russian.
2. T.K. Vintsiuk, L.I. Khomenok. Influence of Discretisation Effects, Analysis Interval Length and Step to Speech Signals Recognition Accuracy. – Proceedings, 15 All-Union Seminar "Automatic Recognition of Sound Patterns", Tallinn, 1989, pp 81–82, in Russian.
3. Taras K. Vintsiuk. Optimal Joint Procedure for Current Pitch Period Discrimination and Speech Signal Partition into Quasi-Periodic and Non-Periodic Segments. - In "Text, Speech, Dialogue", Proc. of the First Workshop on Text, Speech, Dialog – TSD'98, Brno, 1998, pp 135–140.

Spanish Phoneme Classification by Means of a Hierarchy of Kohonen Self-Organizing Maps

A. Postigo Gardón, C. Ruiz Vázquez, and A. Arruti Illarramendi

The University of the Basque Country. Faculty of Informatics.
Apdo. 649. E-20080 Donostia. (Spain)
postigo@si.ehu.es,acpruvac@si.ehu.es,acparila@si.ehu.es

Abstract. In this paper, some results of the classification of spanish phonemes by means of Kohonen Self-Organizing Maps (SOM) are presented. These results show that SOM may be very useful in the previous steps of a continous speech recognizer, as well as a valuable aid in the signal phonetic segmentation process. This intermediate classification provides a new representation of the signal such that the "phonetic" gap between inputs and outputs of the recognizer is drastically reduced, so simplifying the task of the recognizer.

1 Introduction

The different methods used in speech recognition applications take as inputs some parameters extracted from the original raw speech signal. These parameters must provide as much relevant information on the signal as possible but with the minimum possible redundancy. In order to get such "good" parameters, many transformations on the original digitized speech signal are found in literature. Almost all of them produce a change from time-domain to frequency-domain, but it still remains a large "phonetic" gap between these parameters and the speech units (phonemes either words) we are trying to recognize. If these parameters are directly used as inputs to a speech recognizer, it will have a hard work trying to find the relationship among the units to be recognized and the input parameters. This hard work can be measured in terms of a very large training time and a relatively poor recognition rate.

In this paper, the possibility of an intermediate classification of the signal parameters, intended to improve the subsequent recognition process, is analyzed. A non-supervised classifier (Kohonen self-organizing maps) has been used for this task, due to the good results in recognizing finish phonemes reported in [1,2].

2 Experimental Method

2.1 Acoustic Signal Preprocessing

Among the techniques to extract speech signal parameters, one of the most popular nowadays in speech recognition applications are cepstrum coefficients, due to

F. Jelinek, E. Nöth (Eds.): TSD'99, LNCS 1692, pp. 181–186, 1999.

the good results they provide. This is one of the reasons the work presented here is based on cepstrum coefficients. Nevertheless, the intermediate classification technique we analyze could be used with any other kind of parameters.

We have used as training and test data the spanish speech database FRASES, © of the Universidad Politécnica de Valencia, as well as the cepstrum coefficients they obtained. They obtain cepstrum coefficients as follows: 1) Antialiasing filter (7566 Hz), signal sampled at 16 kHz and digitized in 12 bits; 2) Windows of 32 ms (512 samples) overlapped every 10 ms; 3) Discrete Fourier Transform applied to obtain 256 significant coefficients. 4) Coefficients grouped in 21 bands (0 – 8 kHz), bandwidth s given by the Bark scale; 5) A cosine transformation on the logarithm of these coefficients; 6) Only the first 10 coefficients are taken into account. For the experiments presented in the following sections, we have used as training data 86 phrases, pronounced in standard spanish by a single speaker, and as test data, 50 different phrases.

2.2 Classification

First Classification Step Based on Kohonen Self-Organizing Maps. The vectors of 10 cepstrum coefficients are used as inputs to train a 200 nodes (10×20) Kohonen self-organizing map [3], obtaining a non supervised classification of speech signal units smaller than phonemes. Each map node is represented by a vector of the same characteristics as those we are trying to classify, i.e. 10 components. These representative vectors are randomly initialized. During the training phase, the euclidean distance between the particular training vector and all node representatives is computed. Then, the representative vector of the node nearest to the training one as well as its nearest neighbours are modified. This process is iteratively repeated for all the training vectors until a certain fit condition is satisfied. At that moment, each node represents all the training vectors nearest to it by means of a unique vector. In this way, the map is an associative classifier: given a test vector, the map will assign to it the representative vector of the nearest node.

Taking into account that a training vector is a coded representation of a speech signal segment of only 32 ms and that phonemes are usually largest than that (even in the worst case of plosive phonemes), there exists a high probability that a particular map node represents vectors coming from different phonemes due to their similitude. For this reason, a map node does not represent a particular phoneme but a particular utterance of the speech signal present in different phonemes.

Labelling Map Nodes. After training the map, it is possible to substitute all the training vectors of the database (in our particular case, nearly 20000 different vectors) for their corresponding node representative vector (just 200 different vectors) and directly use it as input to the speech recognizer. In this way, the training time of the recognizer can be drastically reduced due to the reduction in the number of different vectors present in training data. Nevertheless, recognition rates will not improve due to the above mentioned "mixing" effect: some speech signal parts coming from different phonemes are represented by the same node vector. For this reason, a new labelling of the nodes becomes necessary in order to avoid this "mixing" effect. The goal of this new labelling is not only to avoid that effect but, if possible, also to clear the gap between spectral vectors and phonemes, just finding a node or class identifier that reflects all the characteristics of the class.

In his work [1], Kohonen points out that, after training, the resulting map can be calibrated by using the spectra of known phonemes as input vectors. In this way, the nodes of the map are labelled by means of the particular phoneme to whom they learned to response, but in the basis of the majority voting for a number of different responses, so neglecting what we have called "mixing effect".

Taking into account that the spectral vectors we have used to train the map were already labelled by phonemes, we decided not to calibrate the map as Kohonen proposes, but just to use those phonetic labels. We define a class identifier as a new vector of 24 components, each one related to one of the 23 spanish phonemes plus silence, in order to try to preserve as much information on the phonemes as possible. In this way, recognition rates can be improved just giving to the recognizer more precise information about the degree up to which a particular node or class represents a particular phoneme. We only need to take into account not only how many coefficient vectors coming from a particular phoneme are inside a particular class, but also how far or near they are located from the class representative, so that the longer the distance, the smaller the degree in which that class represents that phoneme. Each component of the new vector represents the degree in which a particular phoneme is present in a particular class; it is given by the next expression:

$$P_{ph_i,C_j} = \cfrac{\cfrac{\#\left(V_{ph_i} \in C_j\right)}{\#V}}{\cfrac{\displaystyle\sum_{k=1} Euclidian\_distance\left(R_{C_j} - V_{ph_i}^k\right)}{\#\left(V_{ph_i} \in C_j\right)}} \tag{1}$$

where P_{ph_i,C_j} is the i component (i ranging from 1 to 24) related to phoneme ph_i of the C_j class identifying vector (j ranging from 1 to 200); $\#V$ is either the number of all the spectral vectors labelled as phoneme ph_i present in the whole training set, $\#V_{ph_i}$ (equation relative to the phoneme), or the total number of vectors present in the C_j class, $\#\left(V \in C_j\right)$ (equation relative to the class); $\#\left(V_{ph_i} \in C_j\right)$ is the number of those vectors of phoneme ph_i represented by class C_j, and R_{C_j} is the C_j class representative vector.

3   Experiments

First of all, we have trained a 10×20 map, which we call global map, with all the training vectors and then labelled each node according to the previous equation. This map can be shown in Fig. 1.a), where each node is assigned the label corresponding to the component having the maximum value. Figure 1.b) shows the same map but after having grouped the phonemes in 5 groups: vowels (i,e,a,o,u), fricatives (f,z,x,s,c,y), plosives (p,t,k,b,d,g), liquids (l,r,R) and nasals (m, n, h). The group label has been assigned according to the maximum value of the vector after summing up all the components of the same group.

```
    a a a a a o o o o o          i i i i i i i i i i
    a a a a a d o o o o          i i i i i i i i i i
    a a a a a a d d o o          i i i i i i i p i i
    a a a a a R o o o o          i i i i i r p p i i
    a a a a R R r b u u          i i i i r r r p i i
    e e e R l l u b u u          i i i r r l m p i i
    e e e r r l m m u u          i i i i r l m m i i
    e e e l l m m n u b          i i i l p m m m m p
    e e e e l d m n h b          i i i i l p m m m p
a)  e e e R l m h h g b      b)  i i i r p m m m p p
    i e y y R m m y g b          i i m m r m m m p p
    i y y r l d y y b S          i m f r r p f p p S
    i y i e g R y y S S          i f i i p p f f p p
    i i y e d d d z S S          i i i p p p f p p p
    i i y y x f k p t S          i i f f p f p p p S
    i s z t x k k k S S          i f f p f p p p S S
    s z f t x f S k S S          f f f p f f p p p p
    c s f f f S k S S t          f f f f f p p p p p
    s s s x z z S p p p          f f f f f p p p p p
    s s s z z z t S S S          f f f f f f p p p p
```

Fig. 1. Spanish phonemes classified by a "global" map. a) Label corresponds to the phoneme with the maximum value in the representative vector. b) Label corresponds to a phonetic group

We have also separately trained 5 smaller Kohonen maps (5×10), using as training set for each one only the vectors that in the initial training set are labelled as a phoneme of that particular group. These maps are shown in Fig. 2.

4   Results of Some Tests on the Maps

When a test phrase is presented to the map, a sequence of winning nodes is obtained. This sequence can be used in some different ways.

On the one hand, if the test phrase is nor labelled neither segmented, the sequence of nodes can be used as a reliable aid tool for the labelling and segmentation into phonemes, just taking into account the label corresponding to the component having the maximum value, as can be seen in Figure 3.

```
b b b b b S S p p t        h n h n n m m m m n        s s s x x f x f x x
d d b k S S t p S S        h n n m n m m m h n        s s s s z z z z f x
d d d p k S p t S p        n n m m m m m m h n        s s s s z z z z c y
d g d g k p S S S S        n h m m m m n n n n        s s s c z z f z y y
k k t t p S p k S S        h h m h n m m m n n        s s c c c s s y y y

      1 R R R R r r 1 1 1        e e e e a a a a a a
      R 1 r R r r 1 R R 1        e e e e a a a o a a
      R r 1 R r 1 r r r 1        i i e e a a o o o o
      R r 1 R R 1 1 R R R        i i e e e a u u o o
      R R r 1 1 1 r R 1 R        i i i u u a u u u u
```

Fig. 2. Maps of the 5 phoneme groups

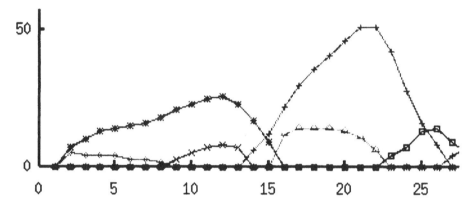

Fig. 3. The sequence of phonetic labels obtained in the test of a phrase can be used as a valuable aid for segmentation into phonemes

On the other hand, the sequence of labels could also be considered as the output of a phonetic recognizer (the SOM itself), since recognition results or phoneme-matching ratios are quite good. In order to measure these recognition rates, it is absolutely necesary to know the labels of the test phrase, to be able to determine wether the selected node label matches or not with the phoneme to be recognized. Nevertheless, these results must not be calculated in the basis of all the samples of the test phrase, but some heuristics are needed in order to "filter" in some way the presence of the "mixing" effect in the output sequence. For instance, five consecutive labels are examined and undesired ones are removed in the basis of less than three been equal. These results are presented in Table 1. As can be seen, group matching ratios are better when the group maps are used, although global recognition rate is slightly worst. In particular, results on the plosives group is much better.

Table 1. Some recognition rates

		Most appearances	Equation relative to class	Equation relative to phoneme (weighted)	idem (not weighted)	Group maps
Total rate	100.00	68.88	68.07	68.28	62.52	66.98
Plosives: ptkbdg	16.49	13.68	12.10	16.89	22.45	55.66
Nasals: mnh	7.55	56.26	55.99	55.10	29.72	53.42
Fricatives: fzsxyc	3.23	25.31	28.84	42.53	48.96	42.12
Liquids: lrR	5.70	32.08	31.84	30.43	20.80	35.37
Vowels: ieaou	39.53	85.67	85.91	84.50	79.90	80.86
Silence	27.51	94.04	91.41	90.24	80.80	67.02

5   Conclusions

Results of the classification of spanish phonemes by means of SOM have been presented. The sequence of phonetic labels obtained in the test of a phrase can be used as an aid to the phrase phonetic segmentation. SOM can be considered the first step of a phonetic recognizer. The sequence of nodes in the test can be considered as an intermediate representation that can help clearing the "phonetic" gap between spectral vectors and phonemes, just using the 24 component vectors as input to a posterior recognition step.

Acknowledgements. Thanks to the speech recogniton research group from U.P. Valencia the cession of the data base FRASES©.
This work has been partially supported by Diputación Foral de Gipuzkoa and Universidad del País Vasco.

References

1. Kohonen, T.: The "Neural" Phonetic Typewriter. IEEE Computer, March 1988. 11–22
2. Kohonen, T.: The Self-Organizing Map Proceedings of the IEEE. Vol. 78, N.9. September (1990)
3. Kohonen, T., Hynnimen, J., Kangas, J., Laaksonen, J.: The Self-Organizing Map Program Package version 3.1. SOM programming team of the university of technology. Laboratory of computer and information science. Finland. (1995)

Information Theoretic Based Segments for Language Identification[*]

Stefan Harbeck, Uwe Ohler, Elmar Nöth, and Heinrich Niemann

University of Erlangen-Nuremberg,
Chair for Pattern Recognition (Computer Science 5),
Martensstr. 3,
D-91058 Erlangen, Germany
{snharbec,niemann}@informatik.uni-erlangen.de
http://www5.informatik.uni-erlangen.de

Abstract. In our paper we present two new approaches for language identification. Both of them are based on the use of so-called multigrams, an information theoretic based observation representation. In the first approach we use multigram models for phonotactic modeling of phoneme or codebook sequences. The multigram model can be used to segment the new observation into larger units (e.g. something like words) and calculates a probability for the best segmentation. In the second approach we build a fenon recognizer using the segments of the best segmentation of the training material as "words" inside the recognition vocabulary. On the OGI test corpus and on the NIST'95 evaluation corpus we got significant improvements with this second approach in comparison to the unsupervised codebook approach when discriminating between English and German utterances.

1 Introduction

Language identification has been a field of interest for the last ten years. A wide spread method for language identification is based on the evaluation of phonotactic knowledge which is usually done by using stochastic language models [9]. The stochastic language models are trained and evaluated on phoneme sequences, which are extracted out of the speech signals using a phoneme recognizer.

In contrast we had focused on methods for language identification which require less information about the training material [2]: We need only a set of signals for each language and no additional transcription. Application on new domains and different signal quality is possible just by recording the samples within this domain and use them to train the new language identification module.

In the following paper two different approaches are described, which are both based on information theoretic units called multigrams. In the first approach the standard stochastic language model is replaced by the multigram model, in the second the acoustic units which will be used inside the recognizer will be replaced by the multigrams units.

---

[*] This work was supported by MEDAV GmbH.

F. Jelinek, E. Nöth (Eds.): TSD'99, LNCS 1692, pp. 187–192, 1999.

The paper is organized as follows: In the next section an introduction to multigrams is presented. An overview about the base line system based on codebook sequences is given in section 3. The description of the two new approaches based on multigrams follows. In section 5 experiments on a part of the OGI corpus are presented. A conclusion will be given in section 6.

2 Multigrams

Chomsky's idea about a relationship between quality of grammars and their length lead to the minimum description length (MDL) principle by Rissanen [6]. This principle can be interpreted as follows: When comparing two different grammars, the bigger one might be able to interpret every output but it is not likely to generalize well. The best theory within the MDL principle is the simplest one which adequately describes the observed data. The *quality* of a grammar can be expressed in terms of length of the grammar itself and the given observation **O**. This can be formalized by

$$G = \operatorname*{argmin}_{G' \in \mathcal{G}} |G'| + |\mathbf{O}|_{G'}, \tag{1}$$

where $\mathcal{G}$ denotes the set of all possible grammars G which describe the observation data. $|G'|$ is the shortest encoding of the grammar G and $|\mathbf{O}|_{G'}$ is the shortest encoding of the observation **O** with given knowledge of grammar G'. Every coding scheme for observations can be interpreted as a stochastic grammar and vice versa. In the *multigram* coding scheme the grammar consists of a lexicon. Every word inside the lexicon is associated with a probability that determines the relative frequency of that word. The MDL principle can be refined by

$$G = \operatorname*{argmin}_{G' \in \mathcal{G}} \sum_{w \in G'} |w|_{G'} + \sum_{o \in O} |o|_{G'}, \tag{2}$$

where $|x|_{G'}$ is the description length of x using grammar G'.

Assuming that the codewords w are chosen to minimize the total description length, the codeword length $l(w)$ is related to the a priori probability of w by $l(w) = -\log P(w)$, so the coding system defines a stochastic language model. The probability of an observation sequence **O** under the grammar G is

$$P_G(\mathbf{O}) = \sum_n P_G(n) \sum_{w_1 \ldots w_n = \mathbf{O}} P_G(w_1) \cdots P_G(w_n)$$

$$\approx \sum_n \sum_{w_1 \ldots w_n = \mathbf{O}} P_G(w_1) \cdots P_G(w_n) \tag{3}$$

Here the probability of **O** is given by summarization over the probabilities of all possible segmentations of **O** or in the context of codes over all possible representations of **O**. The factor $P_G(n)$ describes the probability for a segmentation in n segments using this grammar and will be ignored during the rest of this paper. This kind of stochastic language model is called a *multigram model*. Multigrams reflect statistical dependencies within a sequence of letters by assigning a probability $P(w)$ to a variable length block w. When thinking in terms of observation of letters in an English text, the probability of $P(\texttt{the})$ should be larger than $P(\texttt{t}) \cdot P(\texttt{h}) \cdot P(\texttt{e})$. The modeling power of this multigrams can be greatly influenced by the maximum length of w. By increasing the length, the number of parameters increases exponentially, so there is a drawback between accuracy and the robustness in parameter estimation within this model.

As reflected above the maximization of equation (3) is equivalent of minimizing the description length of the underlying grammar. The maximization is done using a variant of the EM algorithm, which is equivalent to a Baum-Welch procedure [4].

3   Base Line System

Our base line system for language identification consists of a two step process:

1. Extraction of language independent observation units which can be either codebook classes, phonemes or fenons.
2. Language dependent phonotactic modeling using n-gram models with $n = 1, 2, 3$ together with either discriminative [5,8] or usual interpolation schemes [7].

In the current system only phonotactic knowledge and no explicit knowledge on acoustic differences between languages is used. The stochastic framework is described as follows [3]: The classification of an observation $\mathbf{X}$ is done selecting the language which yields the maximum a posteriori probability according to

$$\mathcal{LS}^* = \underset{\mathcal{LS}_j}{\operatorname{argmax}} \, P(\mathcal{LS}_j|\mathbf{X}) = \frac{P(\mathbf{X}|\mathcal{LS}_j)P(\mathcal{LS}_j)}{P(\mathbf{X})} \approx P(\mathbf{S}^*|\mathcal{LS}_j)P(\mathcal{LS}_j) \quad (4)$$

with $\mathbf{S}^*$ is the best segmentation given observation $\mathbf{X}$.

4   Using Multigrams for Language Identification

In this section we describe two different kind of applications for multigrams inside our base line system.

4.1   Replacement for Language Models

The phonotactic model $P_{\mathcal{LS}_i}(\mathbf{S})$ is normally modeled by a stochastic n-gram language model and will be replaced by our multigram model with the codebook symbols as observations. Instead of calculating the probability of all possible segmentations as indicated in equation (3) only the probability of the best segmentation $s_1^* \ldots s_n^*$ is used

$$P_{\mathcal{LS}_i}(\mathbf{S}) = P_{\mathcal{LS}_i}(s_1^*) \cdots P_{\mathcal{LS}_i}(s_n^*) \quad (5)$$

4.2 Building a Fenon Recognizer

In our opinion there are two major problems when using codebook classes for language identification:

- Codebook segments do not represent phonemes so phonotactic modeling based on codebook classes is not regular
- Codebook classes are very close inside the feature space so there is a tendency for substitution among them during recognition

It makes sense to search for more phoneme equivalent and more robust segments. One method to do this is to search for acoustic homogenous regions. But phonemes are not necessarily homogenous inside feature space and every phoneme shows a special movement or trajectory inside the feature space [1] which is indicated by different codebook classes. Typically the multigram approach is used in applications for unsupervised lexicon acquisition. The observation consists of letters where the word boundaries are not available, and the task is to find regular words inside the observation. Instead of letters we observe codebook classes, and instead of searching for words we are looking for sequences of codebook classes which are hopefully similar to phonemes.

The construction of the *fenon approach* is done with the following steps:

1. Train the codebook quantizer using LBG
2. Build the multigram language model using the quantized training material as observation
3. Estimate the most probable segmentation of the training material using the multigram model
4. Choose a subset of segments inside the best segmentation as fenons
5. Label the different fenons and use this as the new transcription
6. Train an HMM based recognizer on the new transcription
7. Use the fenon recognizer to extract the best fenons on the same training data, or if available on a disjunct training material
8. Train language specific phonotactic language models based on the output of the fenon recognizer

Like inside the *codebook approach* the acoustic frontend in this version is language independent and might be extended to language dependent models in the future. Only the phonotactic frontend represents language specific knowledge. The fenons do not have to represent only phonemes but are also able to represent common words like functional words which occur very often inside the training corpus.

5 Experiments

In our experiments we used the languages German and English of the OGI corpus. As training set the training plus as validation annotated utterances are used (1 hour 20 minutes per language). As test either the test material annotated utterances (30 minutes per language) or the official NIST database was used (20 minutes per language). For comparison we evaluated in our first experiment the standard *codebook approach* and also used a supervised trained phoneme recognizer for language identification.

Table 1. Recognition rates of language identification using different approaches for two languages on the OGI corpus evaluated on 10 and 30 seconds of speech.

Method	OGI test set		NIST test set	
	10	30	10	30
Codebook approach	79	81	84	90
Phoneme approach	84	91	86	98
Multigrams	73	84	82	90
Fenons	76	87	87	98

As shown in Table 1, the phoneme recognizer is the best on both sets when observing 30 second utterances. Comparing only the unsupervised trained approaches, the use of the fenon recognizer reduces the error rate of the *codebook approach* by 30 percent on the OGI test set and by 80 percent on the NIST test set which was even as good as using a supervised trained phoneme recognizer. When comparing the recognition rates on the 10 second utterances, the *codebook approach* is better than the fenon recognizer only on the OGI test set. So the use of fenons or phonemes seems to work especially on longer sentences. When the *multigram model* replaces the standard n-gram model the recognition rates drops down significantly on the 10 second sentences. On the 30 second sentences of the OGI test set the use of multigrams is better than using n-grams. One reason might be the artificial boundaries which are inserted into the observations when splitting the utterances into 10 second utterances. Also, there is no method to prevent over-adaptation to the training data as it is done inside the n-gram models.

6   Conclusion and Outlook

In this paper two new methods were proposed which are based on the information theoretic multigram models. These multigrams are developed to get a model for building a lexicon from scratch similar to language acquisition. Using these models as a replacement for standard n-gram models does not improve the recognition. But it might be promising to combine both modeling schemes e.g. inside a neural network or train them using discriminative methods in the future.

Nevertheless, the use of multigrams for finding semi-phonemes or fenons is quite promising as it increases recognition rate on the used test corpora, especially when observing long sentences and it is also as good as the supervised phoneme recognizer.

References

1. L. Deng. A stochastic model of speech incorporating hierarchical non-stionarity. *TIEEE*, 1(4):471–474, 1993.
2. S. Harbeck, E. Nöth, and H. Niemann. Multilingual Speech Recognition. In *Proc. of the 2nd SQEL Workshop on Multi-Lingual Information Retrieval Dialogs*, pages 9–15, Pilsen, April 1997. University of West Bohemia.
3. S. Harbeck, E. Nöth, and H. Niemann. Multilingual Speech Recognition in the Context of Multilingual Information Retrieval Dialogues. In *Proc. of the Workshop on TEXT, SPEECH and DIALOG (TSD'98)*, pages 375–380, Brno, September 1998. Masaryk University.
4. S. Harbeck and U. Ohler. Multigrams for language identification. In *Proc. European Conf. on Speech Communication and Technology*, page to appear, Budapest, Hungary, 1999.
5. U. Ohler, S. Harbeck, and H. Niemann. Discriminative training of language model classifiers. In *Proc. European Conf. on Speech Communication and Technology*, Budapest, Hungary, 1999.
6. J. Rissanen. *Stochastic complexity in statistical inquiry*. Singapure World Scientific, 1989.
7. E.G. Schukat-Talamazzini, F. Gallwitz, S. Harbeck, and V. Warnke. Interpolation of maximum likelihood predictors in stochastic language modeling. In *Proc. European Conf. on Speech Communication and Technology*, pages 2731–2734, Rhodos, Greece, 1997.
8. V. Warnke, S. Harbeck, E. Nöth, H. Niemann, and M. Levit. Discriminative estimation of interpolation parameters for language model classifiers. In *Proc. Int. Conf. on Acoustics, Speech and Signal Processing*, volume 1, pages 525–528, Phoenix, USA, 1999.
9. M.A. Zissman. Comparison of four approaches to automatic language identification of telephone speech. *IEEE Transactions on Speech and Audio Processing*, 4(1): 31–44, January 1996.

Fast and Robust Features for Prosodic Classification*

Jan Buckow, Volker Warnke, Richard Huber, Anton Batliner, Elmar Nöth, and
Heinrich Niemann

University of Erlangen-Nuremberg,
Chair for Pattern Recognition (Computer Science 5),
Martensstr. 3,
D-91058 Erlangen, Germany
{buckow,warnke,huber,batliner,noeth,niemann}@informatik.uni-erlangen.de
http://www5.informatik.uni-erlangen.de

Abstract. In our previous research, we have shown that prosody can be
used to dramatically improve the performance of the automatic speech
translation system VERBMOBIL [5,7,8]. In VERBMOBIL, prosodic information is made available to the different modules of the system by annotating the output of a word recognizer with prosodic markers. These
markers are determined in a classification process. The computation of
the prosodic features used for classification was previously based on a
time alignment of the phoneme sequence of the recognized words. The
phoneme segmentation was needed for the normalization of duration
and energy features. This time alignment was very expensive in terms
of computational effort and memory requirement. In our new approach
the normalization is done on the word level with precomputed duration
and energy statistics, thus the phoneme segmentation can be avoided.
With the new set of prosodic features better classification results can
be achieved, the features extraction can be sped up by 64%, and the
memory requirements are even reduced by 92%.

1   Introduction

The aim of the VERBMOBIL project is to develop a system that translates spontaneous human-to-human speech from a source to a destination language [5].
During this translation process prosodic information is used at various stages
[8]. In VERBMOBIL the output of a word recognizer is structured as a word hypotheses graph (WHG). Every edge represents a word hypothesis and every path
through the graph a possible acoustic–phonetic interpretation of the observed
utterance. The edges in the graph are marked with start and end time, thus
making it possible to determine the corresponding segment of the speech signal.

---

* This work was funded by the German Federal Ministry of Education, Science, Research and Technology (BMBF) in the framework of the VERBMOBIL Project under
Grant 01 IV 102 H/0. The responsibility for the contents lies with the authors.

F. Jelinek, E. Nöth (Eds.): TSD'99, LNCS 1692, pp. 193–198, 1999.
© Springer-Verlag Berlin Heidelberg 1999

In order to make prosodic information available, each edge in the WHG is enriched with probabilities for prosodic events. The probabilities are determined in a classification process. For each word hypothesis, prosodic features are extracted from the speech signal and used as input to a multi layer perceptron (MLP) for each prosodic event. The output of the MLP can be interpreted as *a-posteriori* probability [3].

In our previous experiments, a time alignment of the phoneme sequence of the recognized words was necessary to perform a phone intrinsic normalization of energy and duration features. A phone intrinsic normalization is important because individual phonemes are affected very differently by a change in speaking-rate or loudness [2,6,1]. The time alignment was by far the most expensive operation in terms of computational effort and memory requirement.

In this paper, we present a new set of prosodic features. Phone intrinsic variations are taken into account without the need to perform a time alignment of the phoneme sequence. All that is required is the duration of each word hypothesis. The phone intrinsic normalization is done on the word level with the help of precomputed duration and energy statistics. The new features are described in Section 2. We show that with the new set of features we achieve better results for all prosodic classes that are distinguished in the VERBMOBIL system. These results are detailed in Section 3.

2   Feature Extraction

Aim of the extraction of prosodic features is to compactly describe the properties of the speech signal which are relevant for the detection of prosodic events. Prosodic events, such as phrase boundaries and phrase accents, manifest themselves in variations of speaking-rate, energy, pitch, and pausing. The exact interrelation of these prosodic attributes is very complex. Thus, our approach is to find features that describe the attributes as exactly but also as compactly as possible.

At each edge of the WHG, not only the current edge (i.e. the current word interval) is used for feature extraction but also intervals containing several words. These intervals from the beginning of word f to the end of word t are referred to by $I_{(f,t)}$. Intervals that we use are e.g. $I_{(-2,-1)}$ or $I_{(-1,0)}$. At the end of the word *"not"* in the utterance shown in Figure 1 the Interval $I_{(-2,-1)}$ e.g. denotes

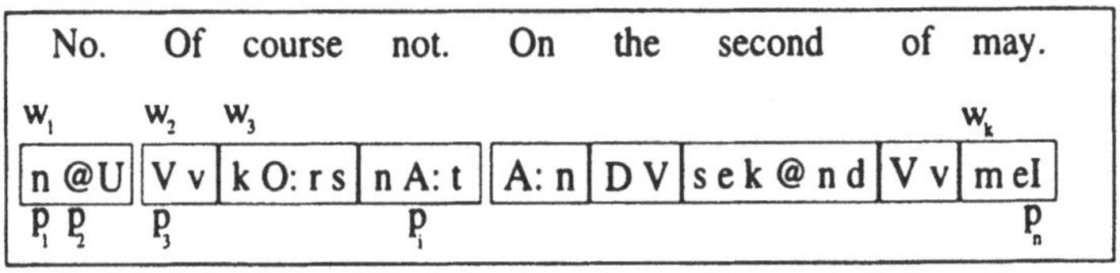

Fig. 1. Utterance *"No. Of course not. On the second of May."* with the phoneme sequence in SAMPA notation.

the time interval from the beginning of the word *"Of"* to the end of the word *"course"*.

Each of the features that we used in our experiments (see Section 3) corresponds to an interval as described above. The pause features are easily extracted: These are simply the duration of *filled pauses* (e.g. "uhm", "uh", ...) and *silent pauses*. Energy and pitch features are based on the short term energy and F0 contours. Duration features should capture variations in speaking-rate and are based on the duration of speech units. A normalization of energy, duration, and pitch features can be performed in order to take phone intrinsic variations and the optional use of prosodic marking into account.

2.1   Features Describing Contours

As mentioned above, energy and pitch features are based on the short-term energy and F0 contour, respectively. Some of the features that are used to describe a pitch contour are shown in Figure 2. Additionally, we use the mean and the median as features (not shown in the figure).

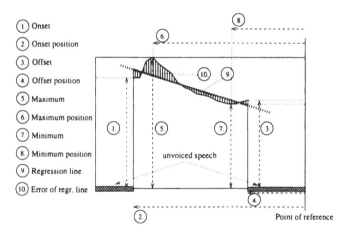

Fig. 2. Example of features used to describe a pitch contour.

2.2   Duration Features on the Phoneme Level

In our previous experiments, a time alignment was performed and $\tau_{duration}$ was computed according to Equation 1 (with $F = duration$, I being some interval and $\#I$ denoting the number of units u in the interval I). The units u are phonemes in this case.

$$\tau_F(I) = \frac{1}{\#I} \sum_{u \in I} \frac{F(u)}{\mu_{F(u)}} \tag{1}$$

$$\zeta_F(J, I) = \frac{1}{\#J} \sum_{u \in J} \frac{F(u) - \tau_F(I)\mu_{F(u)}}{\tau_F(I)\sigma_{F(u)}} \tag{2}$$

Thus, $\tau_{duration} = \frac{1}{\#I} \sum_{u \in I} \frac{duration(u)}{\mu_{duration(u)}}$ is a measure of how much faster or slower the phonemes in the interval I were spoken compared to their mean duration. This value $\tau_{duration}(I)$ was subsequently used to compute the measure $\zeta_{duration}(J, I)$ (see Equation 2) that we included in our feature vector as normalized speaking-rate for interval J. This value $\zeta_{duration}(J, I)$ is a measure of how much faster or slower the interval J of the utterance was spoken compared to the interval I. This measure takes into account phone intrinsic dependencies as well as the optional use of prosodic marking. The standard deviation $\sigma_{duration(u)}$ and the mean of the duration $\mu_{duration(u)}$ have been computed previously on a large training database.

2.3   Duration Features on the Word Level

A major disadvantage of the approach described in Section 2.2 is the necessity to determine the phoneme intervals. In our feature extraction module the computation of the phoneme intervals requires 92 % of the total computation time and 64 % of the total memory needed. Therefore, one would prefer to do a normalization on the word level. But for most words w there is not enough training data to get reliable estimates for the $\mu_{F(w)}$ and $\sigma_{F(w)}$. Equation 2 can be interpreted as a transformation of a feature with mean $\tau_F(I)\mu_{F(X)}$ and standard deviation $\tau_F(I)\sigma_{F(X)}$ to a feature with mean 0 and standard deviation 1. If we assume that the $F(u)$ are independent random variables then $\sigma^2_{F(u_1)} + \sigma^2_{F(u_2)} = \sigma^2_{F(u_1)+F(u_2)}$ (see e.g. [4]). Thus, we can compute the mean $\mu_{F(w)}$ and the standard deviation $\sigma_{F(w)}$ for a word $w = (p_1, p_2, p_3, \ldots, p_n)$ with phonemes p_i as shown in Equations 3 and 4 (if $F(w) = F(p_1) + F(p_2) + \ldots F(p_n)$).

$$\mu_{F(w)} = \sum_{i=1}^{n} \mu_{F(p_i)} \tag{3}$$

$$\sigma_{F(w)} = \sqrt{\sigma^2_{F(p_1)+F(p_2)+\ldots F(p_n)}} = \sqrt{\sum_{i=1}^{n} \sigma^2_{F(p_i)}} \tag{4}$$

In case of $F = duration$ this means that if we assume the durations of the phonemes are independent random variables then the word duration statistics can be deduced from the phoneme duration statistics. Thus, if during recognition a normalization on word level has to be performed according to Equations 1 and 2 then either word duration statistics $\mu_{F(w)}$ and $\sigma_{F(w)}$ can be used if reliable estimates exist or the estimates can be deduced according to Equations 3 and 4.

2.4   Energy Features

In order to describe the short-term energy contour we used only a subset of the features that are shown in Figure 2 because not all of them provide useful information (e.g. onset and offset). Furthermore, we included normalized energy in our feature vector; the same normalization as described in Section 2.3 can be applied here, i.e. $F = energy$ has to be used in Equations 1 and 2.

3   Experiments and Results

In order to evaluate our new feature set we performed several experiments.

1. On a subset of the German VERBMOBIL corpus, we compared the memory requirements and the computation time of the old and the new feature extraction methods. For this experiment, we used a set of 95 pitch, duration, pause, and energy features.
2. On the prosodically labeled German and English subsets of the VERBMOBIL corpus we performed classification experiments for all prosodic events that are used in the system, i.e. phrase boundaries, phrase accents, sentence mood, irregular boundaries, and emotion.

The results of the first experiments are shown in Table 1. As can be seen, the extraction of features could be sped up by a factor of more than 12, while at the same time the memory requirements were reduced almost by a factor of three.

Computation Time		Memory Requirement	
old features	new features	old features	new features
216 min	17 min	73 MByte	26 MByte

Table 1. Computation time and memory requirement of the old and new feature extraction methods on 112 min of speech

In Table 2 the recognition results for phrase boundary and phrase accent recognition are displayed ($\mathcal{RR}$ is the absolute, $\overline{\mathcal{RR}}$ the relative recognition rate; see Equations 5 and 6). The recognition did improve, even though the old feature set consisted this time of 276 features based on word, syllable and syllable nuclei intervals, whereas the new feature set comprised only 105 word based features.

$$\mathcal{RR} := \frac{\text{\# correct classified patterns}}{\text{\# all patterns}} \tag{5}$$

$$\overline{\mathcal{RR}} := \frac{1}{\text{\# classes}} \sum_{c \in classes} \frac{\text{\# correct classified patterns of class c}}{\text{\# all patterns of class c}} \tag{6}$$

	English		German	
	old features	new features	old features	new features
$\overline{\mathcal{RR}}$ boundary	84.0	89.0	84.0	84.7
$\mathcal{RR}$ boundary	86.0	88.5	85.6	86.0
$\overline{\mathcal{RR}}$ accent	77.0	81.4	81.2	81.7
$\mathcal{RR}$ accent	75.0	81.0	80.9	81.0

Table 2. Recognition results for phrase boundaries and phrase accents recognition.

4 Conclusion and Further Work

In our experiments we have shown that our new word based features have at least as much discriminative power as the old features that were based on words, syllables, and syllable nuclei. With the new normalization, recognition results could be improved for all prosodic events. Furthermore, the memory requirements could be reduced by 64 % and computation times even by 92 %.

In the experiments described in this paper, we have always used an entire utterance for normalization, i.e. in Equation 1 interval I was always an entire utterance. This is a disadvantage if long utterances have to be dealt with. In further experiments we are going to investigate if smaller context sizes can be used.

References

1. A. Batliner, A. Kießling, R. Kompe, H. Niemann, and E. Nöth. Tempo and its Change in Spontaneous Speech. In *Proc. European Conf. on Speech Communication and Technology*, volume 2, pages 763–766, Rhodes, 1997.
2. M. Beckman. *Stress and Non-stress Accent*. Foris Publications, Dordrecht, 1986.
3. C.M. Bishop. *Neural Networks for Pattern Recognition*. Oxford University Press, NY, 1995.
4. I.N. Bronstein and K.A. Semendjajew. *Taschenbuch der Mathematik*. Verlag Harri Deutsch, Thun und Frankfurt/Main, 24 edition, 1989.
5. T. Bub and J. Schwinn. Verbmobil: The Evolution of a Complex Large Speech-to-Speech Translation System. In *Int. Conf. on Spoken Language Processing*, Volume 4, pages 1026–1029, Philadelphia, 1996.
6. Andreas Kießling. *Extraktion und Klassifikation prosodischer Merkmale in der automatischen Sprachverarbeitung*. Berichte aus der Informatik. Shaker Verlag, Aachen, 1997.
7. R. Kompe, A. Kießling, H. Niemann, E. Nöth, A. Batliner, S. Schachtl, T. Ruland, and H.U. Block. Improving Parsing of Spontaneous Speech with the Help of Prosodic Boundaries. In *Proc. Int. Conf. on Acoustics, Speech and Signal Processing*, Volume 2, pp. 811–814, München, 1997.
8. Ralf Kompe. *Prosody in Speech Understanding Systems*. Lecture Notes for Artificial Intelligence. Springer–Verlag, Berlin, 1997.

A Segment Based Approach for Prosodic Boundary Detection*

Volker Warnke, Elmar Nöth, Heinrich Niemann, and Georg Stemmer

Universität Erlangen-Nürnberg,
Lerhstuhl für Mustererkennung (Informatik 5),
Martensstr. 3,
D-91058 Erlangen, Germany
warnke@informatik.uni-erlangen.de
http://www.mustererkennung.de

Abstract. Successful detection of the position of prosodic phrase boundaries is useful for the rescoring of the sentence hypotheses in a speech
recognition system. In addition, knowledge about prosodic boundaries
may be used in a speech understanding system for disambiguation. In
this paper, a segment oriented approach to prosodic boundary detection is presented. In contrast to word oriented methods (e.g. [6]), it has
the advance to be independent of the spoken word chain. This makes
it possible to use the knowledge about the boundary positions to reduce search space during word recognition. We have evaluated several
different boundary detectors. For the two class problem 'boundary vs.
no-boundary' we achieved an average recognition rate of 77 % and an
overall recognition rate up to 92 %. On the spoken phoneme chain 83 %
average recognition rate (total 92 %) is possible.

1 Introduction

State–of–the–art speech understanding systems use different knowledge sources
to process on spoken utterances. In the VERBMOBIL speech-to-speech translation
system [8] prosodic boundary information is used for disambiguation of phrase
boundaries. For example the word chain *Of course not on Friday* may have the
two different meanings:

> *1. Of course not ! on Friday.* vs. *2. Of course ! not on Friday.*

Currently the prosodic boundary classifier depends on the output of a word recognizer [5]. If boundary information would be available during the word recognition task, the search space of the word recognizer could be reduced. Thus we
have to develop a boundary classifier, that does not depend on information from

---

* This work was funded by the German Federal Ministry of Education, Science, Research and Technology (BMBF) in the framework of the VERBMOBIL Project under
Grant 01 IV 102 H/0 and by the DFG (German Research Foundation) under contract
number 810 939-9. The responsibility for the contents lies with the authors.

F. Jelinek, E. Nöth (Eds.): TSD'99, LNCS 1692, pp. 199–202, 1999.

the word recognizer. In this paper we present two segment based approaches for prosodic boundary classification.

2   Data

The VERBMOBIL-database contains spontaneous–speech dialogs of German, English, and Japanese speakers. For each utterance, a basic transliteration is given containing the spoken words, the lexically correct word form, pronunciation, and several labels for (filled) pauses and non–verbal sounds. In addition to this basic transliteration, large parts of the corpus are annotated with supplemental labels, such as prosodic (B) and syntactic-prosodic (M) phrase boundaries, dialog act boundaries (D), phrase accents (A), and dialog act classes (DA) [3,1].

For the experimental evaluation we use the subset of the VERBMOBIL-database, labeled with the prosodic B boundaries. It consists of 118 minutes. 790 turns are used for training and 64 for testing.

3   Experiments and Results

All classifiers described in the following attempt to distinguish the prosodic events that mark prosodic boundaries from all other acoustic events that mark no boundary. Those include normal speech and also irregular phrase boundaries, like hesitations and interruptions; i.e. detection of the prosodic boundaries in speech data is viewed as a sequence of classification steps.

We investigated into two major types of classifiers. The first type works equidistantly, after each 10 milliseconds the classifier decides, whether to detect a boundary or not. The second type works in a non-equidistant way, it uses segments of variable length to incorporate durational modeling. In successive order, each segment gets mapped to one of the two classes 'boundary' or 'no-boundary'.

3.1   Fixed Length Segments

The equidistant approach uses Gaussian distribution densities to model the acoustic correlates of the boundaries. The Gaussian distributions are estimated on fixed length segments. In the training data, no information about the extent of the acoustic correlate of a prosodic boundary is given. For the robust supervised estimation of the Gaussian distributions we have to determine in advance, which of the fixed length segments belong to an acoustic correlate of a boundary or not. We use the following heuristic: All segments within the time interval between the end of the word at the prosodic boundary and the beginning of the next word belong to the acoustic correlate of a boundary, i.e. all pauses and non–verbals after a prosodic boundary are used to estimate the Gaussian distribution of the corresponding class. During classification a post processing step is used to reduce the false detection rate: If successive segments were classified as 'boundary', only the middle segment of the sequence is marked as 'boundary', the remaining segments are marked as 'no-boundary'.

We considered different segment lengths between 40 and 160 msec. In order to achieve a robust estimation of the covariance matrices, Karhunen-Loève transformation is applied to each segment to reduce its dimension. The resulting feature vectors we investigated have a dimension between 2 and 80. The best results were achieved using segments with a duration of 160 msec and a dimension of 10; the detection rate is 44 % at an insertion rate of 151 %. That is equivalent to a precision of 23 %.

3.2   Variable Length Segments

The non-equidistant approach is motivated by the n-gram classifier [4,7] for boundary detection as described in [6]. In [6] the spoken word chain of the training data is labeled with 'boundary' and 'no-boundary' symbols. A stochastic language model is estimated on the resulting symbol chain. Classification is done with the Bayes rule by computing the a-posteriori probabilities for the occurrence of a 'boundary' or a 'no-boundary' symbol, given the recognized word sequence. We examine, if this method may be applied to chains of symbols other than words. The difficulty is to find a symbol representation, that contains enough of the information about the boundaries, while it must be as simple as possible to ensure that the boundary detector can be used as a fast preprocessing module. We took two major types of symbol representation into account. The first uses unsupervised learning for symbol generation, while the second uses phone models, that were trained by supervised learning.

For all experiments that are described in the following, a bigram stochastic language model has been used. Our first experiments in unsupervised generation of symbols used the codebook classes of a vector-quantizer for symbol representation. This led to disappointing results (average boundary recognition rate: 70 %, total 75 %).

Better recognition rates can be achieved by incorporating durational variability of the symbols and adding more selectivity to the segment models. For this purpose we used fenones [2] to represent the symbols. A fenone recognizer can be looked upon as a recognizer for subword units, but it is trained unsupervised. If the number of fenones is small (< 10), this corresponds to a phonetic category recognizer (nasals, fricatives, ...), if the number is about 40–200, this corresponds to a phone recognizer. Each fenone has a duration between 30 and 80 msec. The fenone model is a simple linear HMM with one or three looped states and Gaussian output densities. The fenone recognizer uses a bigram language model. The fenone codebook was designed in two steps. The first step consists of clustering the training data with the LBG-algorithm into a fixed number of partitions. In the second step, subsequent equal codebook symbols in the training data get merged. The resulting variable-length symbols are the fenones. We considered different sizes of fenone codebooks between 7 and 120 symbols. The experiments resulted in an average boundary recognition rate of 77 % (total 83 %) on a codebook size of 15 fenones.

We got the best results, when we used a phone recognizer to convert the feature vector sequence into a symbol sequence. The phone recognizer has a lexicon of 62 phones and three different pauses. The phone sequence was used for the polygram classifier as input symbol sequence. This approach achieved an average recognition rate of 77 % (total 92 %). Evaluation of the accuracy of the phone recognition resulted in the very bad value of 35 %. In order to show that further improvement of boundary detection can be achieved by using a better phone recognizer, we applied the polygram classifier to the spoken phone sequence (100 % accuracy). A much better boundary detection was the result: An average recognition rate of 89 % together with a total recognition rate of 90 %.

4 Conclusion and Further Work

We have shown that successful recognition of prosodic phrase boundaries is possible without using the spoken word chain. Further improvements may be achieved by using a better phone recognizer.

Our future work is to combine the boundary detector with a word recognizer. We will evaluate the influence of the information about the boundary positions on the word recognition rate.

References

1. J. Alexandersson, B. Buschbeck-Wolf, T. Fujinami, M. Kipp, S. Koch, E. Maier, N. Reithinger, B. Schmitz, and M. Siegel. Dialogue Acts in VERBMOBIL-2 – Second Edition. Verbmobil Report 226, 1998.
2. L. R. Bahl, J. R. Bellegarda, P. V. de Souza, P. S. Gopalakrishnan, D. Nahamoo, and M. A. Picheny. A New Class of Fenonic Markov Word Models for Large Vocabulary Continuous Speech Recognition. *Proceedings International Conference on Automatic Speech and Signal Processing*, pages 177–180, 1991.
3. A. Batliner, R. Kompe, A. Kießling, M. Mast, H. Niemann, and E. Nöth. M = Syntax + Prosody: A syntactic–prosodic labelling scheme for large spontaneous speech databases. *Speech Communication*, 25(4):193–222, 1998.
4. F. Jelinek. Self–organized Language Modeling for Speech Recognition. In A. Waibel and K.-F. Lee, editors, *Readings in Speech Recognition*, pages 450–506. Morgan Kaufmann Publishers Inc., San Mateo, California, 1990.
5. Andreas Kießling. *Extraktion und Klassifikation prosodischer Merkmale in der automatischen Sprachverarbeitung*. Berichte aus der Informatik. Shaker Verlag, Aachen, 1997.
6. Ralf Kompe. *Prosody in Speech Understanding Systems*. Lecture Notes for Artificial Intelligence. Springer–Verlag, Berlin, 1997.
7. E. G. Schukat-Talamazzini. *Automatische Spracherkennung – Grundlagen, statistische Modelle und effiziente Algorithmen*. Vieweg, Braunschweig, 1995.
8. W. Wahlster, T. Bub, and A. Waibel. Verbmobil: The Combination of Deep and Shallow Processing for Spontaneous Speech Translation. In *Proc. Int. Conf. on Acoustics, Speech and Signal Processing*, volume 1, pages 71–74, München, 1997.

Speech Recognition and Syllable Segments

Ivan Kopeček

Faculty of Informatics, Masaryk University
Botanicka 68a, 602 00 Brno
Czech Republic
kopecek@fi.muni.cz
http://www.fi.muni.cz/~kopecek/

Abstract. Speech recognition based on the syllable segment is discussed in the paper. Special emphasis is put on the methods for estimation of syllable boundaries and methods of comparison of syllable segments by DTW algorithms. The method of characteristic function for estimating of the boundaries between syllable segments is introduced. This method is also utilized in the proposed DTW algorithm for syllable segments.

1 Introduction

The syllable based approach to speech processing (see e.g. [1,2,3,5]) is motivated by the fact that syllables create perceptually and acoustically coherent form and that they also represent the basic prosodic units.

The feeling of syllable boundaries, although very strong in most cases, is subjective and in many cases not unique. This forces us to define an independent subset of syllables, i.e. the subset that is able to generate the suprasegmental structures of the language and is uniquely defined. We must, however, take into consideration that syllables are not always identical to the corresponding segments of phonetic phrases, especially for higher speech rates.

In what follows we discuss approaches to syllable segments and propose a method for automatic segmentation and a DTW algorithms that utilize the syllable segment form.

2 Syllables and Syllable Segments

Phonetics gives no exact specification of syllables. The feeling of syllable boundaries, although usually very strong, is subjective and often not unique.

A possible way how to solve this problem is to define a subset that can generate the language (or, better to say, suprasegmental structures) and is uniquely defined. This is complicated by the fact that syllables are not always identical to the corresponding segments of phonetic phrases. For instance, the Czech word *odstranit* (banish, get off, take away) standardly syllabified *od-stra-nit* may have basic phonetic forms *ot-stra-nit*, *o-ctranit* or *oc-tra-nit* depending on the accuracy of the pronunciation and the speech rate. (Neither second nor third form

F. Jelinek, E. Nöth (Eds.): TSD'99, LNCS 1692, pp. 203–208, 1999.

is consistent with the standard syllabification and moreover both the segments "o" and "oc" do not respect the fact that "od" is a prefix).

Another reason that complicates the determination of an appropriate set of syllable segments is the necessity to respect the coarticulation between the adjacent syllables and to keep the number of segments in reasonable limits.

Fortunately, the coarticulation effect is in most cases reduced on the boundaries of the adjacent syllables in comparison to the inner parts of syllables) and syllabic onsets are mostly realized in standard form (see [4,8]). Also acoustical intensity often decreases at the syllable boundaries.

On the other hand there are also situations where the coarticulation between adjacent syllables is more strong and should not be neglected.

To manage these problems, special syllable segments were proposed and analyzed (L-syllables, CVS-syllables, and other types – see [6,10,1]).

Roughly speaking, L-syllables are segments that start at the boundary of a vowel/consonant couple (or at the beginning of speech), and end at the boundary of next vowel/consonant couple (with an exception when boundary is the last one in a connected portion of speech; then L-syllable ends at the end of speech). The right-hand side of the transient part of the L-syllable is removed to ensure coarticulation compatibility and simultaneously reasonable number of the segments. CVS-syllables and other types of syllable segments can be viewed as complementary segments to L-syllables.

3   Determination of the Boundaries Between Segments

For the speech recognition based on syllable segments the determination of boundaries between segments represents a crucial task.

The semi-automatic segmentation procedure for syllable based speech synthesizer *Demosthenes* ([6,10]) was based on detecting local maximums of the *function of sonority decrease*. Provided $v_i = (v_{i,1}, v_{i,2}, ..., v_{i,n})$ is the acoustic vector in the time i, the function of sonority decrease is based on the value

$$D(i) = \sum_{j=k}^{l} c_j(i)$$

where

$$c_j(i) = max\{v_{i,j} - v_{i-\delta,j}, 0\}.$$

The indices k, l correspond to the vowel formants frequency region, δ is the length of the used difference.

The estimation of the syllable segment boundaries for speech synthesis may be however strongly supported by the estimation of the segment length, which for the speech recognition more is complicated, not so precise and not so confident.

On the other hand, the determination of the segment boundaries must be fully automatic and as reliable as possible. For this purposes we propose the following method which is in fact based on the same idea as the method mentioned above. The main difference lies in more complex and more precise determination of the segment boundary characteristics.

Let $w_1, ..., w_n$ be differences of the acoustic vectors corresponding to characteristic boundaries of the segments. We assume that these values are obtained by choosing typical boundaries that approximately cover all possible boundary regions.

Consider the function

$$X(v) = max\{A(v, w_1), ..., A(v, w_n)\},$$

where

$$A(r, s) = | (r, s) | / (\| r \| \| s \|),$$

(r, s) is the scalar product of the vectors u, v and $\| r \|$ denotes the Euclidean norm of the vector r. (We can express the function $A(r, s)$ also in the form $A(r, s) = 1 - \| r - s \|^2 / 2$.)

Let us denote

$$W_X = \{w_1, ..., w_n\}.$$

W_X will be said *set of characteristic vectors* and the function $X(v_I)$ *characteristic function*. The vector for which the function $\{A(v, w_1), ..., A(v, w_n)\}$ attains its maximum will be *called critical vector*.

The characteristic function $X(v_i)$ has the following properties:

1. $0 \leq X(v) \leq 1$;
2. $X(v) = 1 \Rightarrow v \in W$;
3. $X(v)$ is a continuous function.

In comparison to the function of sonority decrease this characteristic function possesses the following advantages, that are important for implementation of the method in a speech recognition system:

- Since the value of the function is 1 for all elements of the set W, we can expect that the values of the function in boundary points will lie in a vicinity of 1. (The boundary value of the function of sonority decrease depends on the type of the boundary and on the acoustic intensity.)
- By defining the set W we can easily define and also modify the type of the segments we want to use.
- By determining of the syllable boundaries we also obtain characteristics of the corresponding segments.

Let us denote by R the set of the points where at least one of the following conditions is fullfilled:

1. the function of differences of the normed acoustic vectors attains its maximum on a predefined interval;
2. the function of the acoustic intensity is for the first time (in a time interval) lower than a predefined constant;
3. the function of the acoustic intensity; attains its minimum on a predefined interval.

The scheme of the segment boundary estimation algorithms based on the characteristic function can be now described as follows:

1. Choose critical values ϵ and δ;
2. Choose minimal segment length σ;
3. Move pointer to the beginning;
4. Find next point i which belongs to the set R;
5. If the maximum is less than δ repeat the previous step;
6. If $X(v_i) \geq 1 - \epsilon$ you have found the boundary point;
7. otherwise go to 4.

Let us mention that for further processing of the detected syllable segment the right-hand part of the vowel - consonant transient zone should be removed in the case that the transient zone contains the right hand side syllable boundary. The reason is to decrease the coarticulation effect of the adjacent syllable on the vocalic nucleus.

4   DTW Algorithms for Syllable Based Approach

Of course, standard DTW (=*Dynamic Time Warping*) algorithms as well as most of the heuristics and variants can be successfully used also for the syllable based speech speech recognition. Nevertheless, it seems to be natural and useful to take advantage of the special form of the segments and also exploit the characteristic function (if we use it for syllable boundaries estimating).

In what follows, we will restrict our considerations to the most frequent standard syllable segments that contain the syllabic onset and the syllabic nucleus. For such segments the dynamical transient part consonant-vowel is a characteristic feature of the syllable. This feature can be localized, quantified and characterized by means of the characteristic function $Y(v_i)$ which is defined by the set W_Y of characteristic vectors. The characteristic vectors are obtained in consonant-vowel transient part points where the function of acoustic vector changes is maximal.

Assume that our database of the segments contains for each segment u (and its identification) also the value k of the point of the consonant-vowel transient part in which the characteristic function is maximized, the corresponding critical vector and the value $Y(u_k)$. Now, consider two segments $u_1 = (u_{1,1}, u_{1,2}, ..., u_{1,m})$ with the corresponding transient point k_1 and critical vector w_1, and $u_2 = (u_{2,1}, u_{2,2}, ..., u_{2,n})$ with the corresponding transient point k_2 and critical vector w_2,. Suppose $m \geq n$. Then we can take

$$\| u_1 - u_2 \|_Y = c_1 \| w_1 - w_2 \| + c_2 \sum_{i=1}^{m} \| u_{1,i} - u_{2,t(i)} \|$$

where the DTW transformation function $t(i)$ can be approximated by the quadratic function determined by the following equations:

$$t(1) = 1; \qquad t(k_1) = k_2; \qquad t(m) = n;$$

(c_1 and c_2 denote the weight coefficients of the method). The transformation function t can be also taken as a starting approximation for other types of DTW algorithms.

5   Statistical Approach to Syllable Based Speech Recognition of the Czech Language; Prosody

Basic statistics of the Czech frequency syllables and the phonetic analysis can be found in [15,12]. The following statistics of the syllable types frequency in Czech (C=consonant, V=vowel) is illustrative: V – 12,8 %, VC – 2,8 %, KV – 45,3 %, CVC – 17,9 %, CCV – 9,5 %, CCVC - 8,8 %; other eleven types of syllables appear with negligible frequency ([12]). A statistics of syllable segments (L-syllables etc.) can be found in [1]. A statistics of the syllable unigrams, bigrams and trigrams based on the analysis of the corpora DESAM and ESO (see [14]) was done in the work [13].

One of the main contemporary aims of the corpora analysis at Faculty of Informatics Brno is to create a syllable based language model of Czech. We expect that this will bring an essential asset to the syllable based speech recognition of Czech.

Prosodic factors play important role in pronunciation and therefore the investigation of prosodic attributes and the of the prosodic governing mechanism is one of the most important topic of speech recognition.

The advantage of the methods based on syllables is in close connection of prosody and syllables as basic prosodic segments. In Czech, the lexical stress appears usually very regularly, typically on the first syllable in a word. This can be also utilized for detecting word boundaries in continuous speech recognition. One of the first tasks to manage the prosody and to utilize the advantage of syllable segments is to develop reliable method for lexical stress determination.

6   Conclusions

The experiments show that the proposed method give good results. The future work will be aimed to further improving the method, to the statistical approach and to the prosody aspects.

Acknowledgement. The author is grateful to K. Pala for reading the draft of the paper and valuable comments. The research has been partially supported by the Czech Ministry of Education under the Grant VS97028 and by the Grant Agency of the Czech Republic under the Grant 201/99/1248.

References

1. Batušek, R.: Statistics of the Syllable Segments for Speech Synthesis of the Czech Language; Proceedings of the First Workshop on Text, Speech and Dialogue - TSD'98, 1998, pp. 153–158.
2. L. Josifovski, D. Mihajlov, D. Gorgevik: Speech Synthesizer Based on Time Domain Syllable Concatenation; Proceedings SPECOM'97, Cluj-Napoca, 1997, pp. 165–170.
3. G. Doddington: Syllable Based Speech Processing; WS97 Project Report, Research Notes No. 30, J. Hopkins University, 1997.

4. S. Greenberg: Speaking in Shorthand - A Syllable-Centric Perspective for Understanding Pronunciation Variation; Proceedings of the workshop Modeling Pronunciation Variation for Automatic Speech Recognition, 1998, pp. 47–56.
5. S. Greenberg: A Syllable-Centric Framework for the Evolution of Spoken Language; Commentary on MacNeilage, P. The frame/content theory of evolution of speech production. Brain and Behavioral Sciences, 21, 518.
6. I. Kopeček: Speech Synthesis Based on the Composed Syllable Segments; Proceedings of the First Workshop on Text, Speech and Dialogue - TSD'98, 1998, pp. 259–262.
7. I. Kopeček: Automatic Segmentation into Syllable Segments; Proceedings of First Int. Conference on Language Resources and Evaluation, 1998, pp. 1275–1279.
8. I. Kopeček: Syllable Segments in Czech ; Proceedings of the XXVII. Mezhvuzovskoy naucznoy konferencii, Vypusk 10, St. Petersburg, March 1998, pp. 60–64.
9. I. Kopeček, K. Pala: Prosody Modelling for Syllable-Based Speech Synthesis; Proceedings of the IASTED Conference on AI and Soft Computing, 1998, pp 134–137.
10. I. Kopeček: Syllable Based Speech Synthesis; Proceedings of the 2nd International Workshop SPECOM'97, Cluj-Napoca, 1997, pp. 161–165.
11. I. Kopeček: Speech Synthesis of Czech Language in Time Domain and Applications for Visually Impaired; Proceedings of 2nd SQEL Workshop, Pilsen, 1997, pp. 141–145.
12. H. Kučera, G.K. Monroe: A Comparative Quantitative Phonology of Russian, Czech and German; New York, 1968.
13. H. Malková: Statistical Evaluation of the Text Corpora with Respect to Syllable Segments (in Czech); Diploma Thesis, Faculty of Informatics, MU Brno, 1999.
14. K. Pala, P. Rychlý, P. Smrž: DESAM – Annotated Corpus for Czech. Proceedings of SOFSEM'97, Springer–Verlag, 1997.
15. Z. Palková: Phonetics and Phonology of Czech (in Czech); Charles University, Prague, 1994.
16. L. Shastri, S. Chang, S. Greenberg: Syllable Detection and Segmentation Using Temporal Flow Neural Networks; in Proceedings of the International Congress of Phonetic Sciences, San Francisko, 1999.

Text Preprocessing for Czech Speech Synthesis

Robert Batůšek and Jan Dvořák

Faculty of Informatics, Masaryk University, Brno, Czech Republic
{xbatusek,jdvorak}@fi.muni.cz,
http://www.fi.muni.cz/laf

Abstract. Some algorithms for preprocessing of some more complicated parts of Czech texts are presented in the paper. Preprocessor uses some rules taking into account word context. Some other problems with transcription of such complicated parts of text are discussed as well: influence of the speech rate and prosodic phrases segmentation. All the presented algorithms are implemented in the TTS system Demosthenes ([5,6]).

1 Introduction

There are several possible approaches how to realize phonetic transcription, for instance pronunciation treatises, expert rule-based systems, trained rule-based systems or neural networks (see [3]). Expert rule-based systems are probably the most used. Such systems usually use rules in the form

$$A \rightarrow B/L\_R : C, \tag{1}$$

where A is a letter chain being transcribed, B is the result of the transcription[1], $L(R)$ means the chain of letters in left (right) context and C is a condition that must hold to apply the rule.

However, some phonetic rules using only letter context are not sufficient to make transcription of some more complicated parts of text as acronyms, abbreviations, numbers or non-alphabetic characters. Clearly, text containing these elements have to be preprocessed. We propose to use rules similar to (1) for preprocessing. These rules formally look like phonetic rules

$$A \rightarrow B/L\_R @E, \tag{2}$$

but meanings of their components are different. A, B are words; for the output word B is not used a phonetic alphabet but the common alphabet. L and R stand for word contexts. Of course, contexts may not contain only concrete words but also some equivalence classes of words can occur here. Frequently used classes are for example numbers, numerals, prepositions, but also whitespace,

---

[1] It is assumed that we have already designed a phonetic alphabet which allows us to take down phonetic (or at least phonemic) form of a language. In the paper we will use SAMPA phonetic alphabet (see [10]) for phonetically transcribed text.

F. Jelinek, E. Nöth (Eds.): TSD'99, LNCS 1692, pp. 209–214, 1999.

punctuation,[2] etc. Our experiments with various Czech texts show that the sizes of both left and right contexts need not exceed 5 words.

Instead of *conditions* we rather use some *expressions*. There can be multiple expressions assigned to each rule. Expressions are used, for instance, to

- Assign a grammatical category to the expanded word. Category can be either directly specified or acquired from a word in the context.
- Change the order of processing words or skip some words. For instance, "$20" should be expanded as "twenty dollars".
- Assign an attribute to a word which is used when testing next words (and therefore the word currently processed is then a part of the left context).
- Specify a speech rate levels in which the rule will be applied (see Section 3).
- Specify a weight of a rule. Rules are ordered in the descending order according to their weights and are tested in this order until the matching context is found or until all the rules concerning a given abbreviation are tested. Default weight value is 0.

Clearly, this set of functions is easily extensible.

2   Phonetic Transcription Preprocessing Based on Rules using Word Context

2.1   Expanding Acronyms

There are three basic possibilities how to expand an acronym:

1. It is read as a common word – "NATO"→"nato", "ODA"→"oda" (Civil Democratic Aliance – a Czech political party). To decide whether it is possible to pronounce an abbreviation as a common word the number and the positions of syllabic phonemes (vowels and syllabic consonants l and r) are analyzed.
2. It is spelled – "NATO" → "en á té ó", "ODA" → "ó dé á".
3. It is expanded to the expression from which it was abbreviated – e.g. "NATO" → "severoatlantická aliance", "ODA" → "občanská demokratická aliance". In addition, for some frequently used abbreviations "abbreviated" expansions exist (e. g. when expanding "USA" the last word "americké" is usually omitted). In our dictionary the full expansions of some frequently used acronyms are included only.

To decide which type of the expansion will be used the speech rate must be taken into account (see Section 3).

---

[2] In fact, these symbols are not words, but we will use the term *word* for them for the sake of simplicity.

2.2 Expanding Abbreviations

Some abbreviations have more possible transcriptions (and also meanings) depending on a word context. For example, abbreviation "př." should be transcribed as "příklad" (example) in the context "př. 1.1" or "před" (before) in the context "př. n. 1." (B. C.). Transcription of the mentioned above abbreviation can be accomplished using the following rules:

př→před/[$NUMBER|$NUMERAL|"st"^WS"."$WS]_^WS"."$WS"n"^WS"."
@weight(3)

př→příklad/_^WS"."

where $NUMBER, $NUMERAL and $WS stand for equivalence classes of the word contexts (number, numeral, whitespace), symbol "^" stands for negation and "@weight(3)" denotes the weight of a rule.

2.3 Expanding Special Characters

Special characters like % or @ are in fact non-lexical abbreviations of some widely used words in written text. Punctuation marks are another example of characters that cannot be simply omitted during transcription.

Transcription of the special characters is in general more ambiguous than transcription of abbreviations. (See e. g. symbol "-": "1998-99", "bude-li" and "10 - 9 = 1".) In our approach each non-alphabetic character is analyzed and one of the following three possibilities is chosen:

- to elide it · "bude-li" → "bude li" (if it will), "v 10:30" → "v deset třicet" (at half past ten)
- to let it unchanged for further processing (e. g. determining some prosodic attributes of a clause) – "Co?" (What?)
- to expand it using some word rules – "10–20" → "... deset až dvacet" (from ten till twenty)

2.4 Expanding Numbers

We distinguish five basic types of numbers: *integer numbers, floating point numbers, ordinal numbers, times of the day* and *dates.* For each number type the particular expansion algorithm is used. In what follows we describe only the expansion of integer numbers as it is the basic algorithm and the other number reading algorithms are derived from that.

Firstly, the number is grouped into segments consisting of three digits beginning from the right. Appropriate place value of each segment is determined and the word corresponding to this value is loaded from a configuration file. The segment is then expanded respecting gender and number of the word. Segments consisting of zeros only are skipped. When the whole number starts with the segment containing only the digit "1" the word "one" is omitted. (In Czech it is more natural to say "tisíc" – thousand – than "jeden tisíc" – one thousand – in the beginning of the number.) When the size of the number exceeds a threshold it is considered not to be a representation of a basic number and it is read digit by digit (typically bank account numbers).

Numbers Containing Spaces Numbers larger then 1 000 are usually separated by spaces for better readability. It can cause problems when trying to recognize the type of such a number. Numbers in expressions like "tel. 11 22 33" or "PSČ 602 00" (postal code) we should read separately, but as one number in the expression "11 222 333 Kč". Our algorithm detects the length of number segments divided by spaces and when the length is 3, it reads all the numbers as one (larger) number.

2.5   Declination

All the expanded expressions mentioned in the previous sections (acronyms, abbreviations, numbers and special symbols) have to be correctly declined. We have to expand "Kč" into "*koruna*" (crown) in the expression "1 Kč" but into "*korun*" (crowns) in the expression "9 Kč". The exact solution of the problem of declination should use syntactic analysis of Czech sentence. For our current purposes a partial solution is sufficient that declines words only when correct declination can be assigned without any doubt, e. g.:

- Correct case is given by a preposition. This rule should be used only together with prepositions that are connected with the only case.
- An abbreviation that follows a number takes a case of the number (typically a currency symbol).
- The gender and number of the number segment (see Section 2.4) is given by the digit place name.
- The gender of a number is defined by the gender of the following word.

3   Dependence of a Phonetic Transcription to the Speech Rate

Higher speech rate leads sometimes to different phonetic transcription. We propose to define some levels of the speech rate and corresponding sets of (both phonetic and word) rules. One rule can belong to more sets. Thus some levels are assigned to each rule in which this rule could be applied. For example, transcription of the abbreviation "OSN" (UN) is realized using the following rules:
OSN → ''Organizace Spojených Národů''/_ @level(1,1)
OSN → /_ @level(2) @spell(cz),
where the expression @level(low,high) assigns a range of levels to a rule in which the rule is applied (parameter **high** can be omitted). We have implemented three levels of transcription that could be characterized as follows:

1. *Accurate pronunciation* A set of the phonetic rules corresponding to relatively slow and very accurate pronunciation is applied. Acronyms are fully expanded if their expansion is present in the lexicon.

2. *Faster but correct pronunciation* When some letter chain allows more forms of the phonetic transcription the rule is added to the corresponding set which has the shortest output from among those rules that don't violate correct Czech pronunciation. For the other letter chains the only existing rule is used. Acronyms are usually spelled at this level. (e. g. "odstranit" (to remove) → "otstran'it" instead of "ot-stran'it").

3. *Abbreviated pronunciation* This kind of pronunciation is used to accomplish the maximal speech rate still preserving comprehensibility. For each letter chain the rule is added to the corresponding set which has the shortest output even if the correct Czech pronunciation doesn't allow it. Acronyms are read as other words whenever it is possible, otherwise they are spelled, e.g. "království" (kingdom) → "kra:lostvi:", "přijd'te" (come) → "pr'it'te".

4 Dependence of the Prosodic Phrases Segmentation on the Text Preprocessing

Segmentation to the prosodic phrases   (called also prosodic clauses or rhythm units) is an often   discussed problem in current speech synthesis research (see [3,4]). The problem was also studied for the Czech language in [2,6,8].

In [6] is a proposal of an algorithm based on works the cited above using so called sentence constituents. Each clause is parsed into some sentence constituents and each boundary between two neighboring words is characterized to be either prone, neutral or not susceptible to match the boundary between prosodic phrases. Clearly, boundaries inside sentence constituents are usually not susceptible to form the boundary between prosodic phrases.

Their algorithm, however, doesn't consider sentences with numbers or abbreviations. Expansion of these parts of text can, however, change the segmentation of text to prosodic phrases. Let us take, for example, a sentence:
Sametová revoluce začala 17. 11. 1989.
(The Velvet Revolution started at November 17, 1989.)

This sentence has to be expanded to
"Sametová revoluce — začala sedmnáctého listopadu — tisíc devět set osmdesát devět."

(Boundaries between prosodic phrases are marked by the — character.) Without exact knowledge of the expansion the date in the end of the sentence could be characterized as to contain three words instead of seven and thus it would be likely to create only one sentence constituent.

It follows from the previous sections that we cannot a priori know how more complicated texts will be expanded. We thus propose the following structure of modules that have to produce a phonetic representation of a text:

1. Preprocessing of the text including transcription of abbreviations, numbers and special characters.

2. Sentence constituents detection and marking the boundaries between the neighboring words.
3. Finding the prosodic phrases boundaries (based e. g. on the approach designed in [6]).
4. Real phonetic transcription of the marked text. The phonetic transcription must already take into account prosodic phrase boundaries because some phonetic phenomena don't act on these boundaries (i. e. voicing assimilation in Czech).

5   Conclusions

Our algorithms for expanding acronyms, abbreviations, numbers and special characters that appear in common Czech text enhance the quality of the phonetic transcription module of speech synthesizers and thus the quality of the synthesized speech. Preprocessor based on word context rules expands correctly most of these parts of text. Integration of syntactic, semantic and pragmatic analysis of sentence which is currently under discussion should help to correct the cases where our rules fail.

Acknowledgement. The authors are grateful to Ivan Kopeček for reading a draft of the paper and to Marie Krčmová for valuable comments concerning Czech phonetics.

References

1. Batůšek, R.: Statistics of the Syllable Segments for Speech Synthesis of the Czech Language. Proceedings of TSD'98, Brno, 1998.
2. Daneš, F.: Intonace a věta ve spisovné češtině (Intonation and Sentence in Czech, in Czech). Czech Academy of Science, Prague, 1957.
3. Dutoit, T.: An Introduction to Text-to-Speech Synthesis. Kluwer Academic Publishers, Dortrecht/Boston/London, 1997.
4. Kompe, R.: Prosody in Speech Understanding Systems. Springer, 1997.
5. Kopeček, I.: Speech Synthesis Based on the Composed Syllable Segments, Proceedings of TSD'98, Brno, 1998.
6. Kopeček, I., Pala, K.: Prosody Modelling for Syllable-Based Speech Synthesis. Proceedings of the IASTED Conference on AI and Soft Computing, 1998, pp. 134–137.
7. Kuznetsov, V. I., Sherstinova, T. Y.: Phonetic Transcription for the System of Speech Synthesis and Recognition. Proceedings of TSD'98, Brno, 1998.
8. Palková, Z.: Fonetika a fonologie češtiny (Czech Phonetics and Phonology, in Czech). Charles University, Prague, 1994.
9. Psutka, J.: Komunikace s počítačem mluvenou řečí (Communication with Computer Using Speech, in Czech). Academia Press, Prague, 1995.
10. SAMPA www page: http://www.phon.ucl.ac.uk/home/sampa/home.

MLPs and Mixture Models for the Estimation of the Posterior Probabilities of Class Membership

Alexei V. Ivanov and Alexander A. Petrovsky

Computer Engineering Department at the Belarusian State University of Informatics
and Radioelectronics,
6, P. Brovky str, 220027, Minsk, Belarus
alwork@glasnet.ru, palex@it.org.by

Abstract. In this paper the MLP and Gaussian mixture model approaches to the estimation of the posterior probability of class membership in the task of phoneme identification are analyzed. The paper discuss differences between the described methods altogether with discussing advantages and drawbacks of each method. Based on this analysis several ways of the joint employment of the models are proposed.

1 Introduction

The phoneme identification problem is a part of the general speech recognition task. It can be stated as a static pattern classification problem in which our goal is to assign different degrees of belief that the observed short period of a spoken utterance belongs to various phonemic classes. This information can be further used in some kind of temporal models of speech (for instance in various modifications of hidden Markov models) to generate general decisions about a pronounced phrase. These degrees of belief are usually described with the help of posterior probabilities $P(C|X)$ of a phonemic class C, given the observed input X. In this paper we compare two semi-parametric methods of estimation of the $P(C|X)$, based on application of neural network algorithms:

1. Direct estimation of the posterior probabilities through the maximum likelihood parameter optimization of the multi-layer perceptrons.
2. Two stage estimation procedure through modeling the class conditional distributions of the input data with the help of a mixture of kernel functions of some kind and further estimation of the posteriors with the help of Bayes formula.

2 MLP Training

Multi-layer perceptrons consist of several layers of neurons. Each neuron has a sigmoid transfer function and weighed connections to each (in a general case)

F. Jelinek, E. Nöth (Eds.): TSD'99, LNCS 1692, pp. 215–218, 1999.

neuron of the previous layer (or components of the input vector in the case of the first neuron layer) and an additional bias connection:

$$y_j = \frac{1}{1 + exp\{-\sum_i x_i w_{ij} + b_j\}}. \tag{1}$$

Let us consider that the training procedure is done with the help of a number of input vectors X, labeled with 1-from-C coding, so that for each training example coming from the class C_j, j-th component of the label vector equals unity, while all the other components are zeros.

It is proved [1,2,3,4] that for various parameter optimization strategies (such as gradient decent) with minimization of the Mean Square Error function or Cross-Entropy Error function and the backpropagation technique used to compute derivatives of the error function with respect to each of the free parameters, the trained network estimates posterior probabilities of class membership $P(C|X)$ directly.

3   Mixture Models

The core idea of this method is in the approximation of the generally unknown probability density with a mixture of some functional forms. Gaussian (2) is a common choice for such functional form due to a number of useful properties and relatively small number of free parameters needed to be defined. In some realizations [5] even greater simplification is achieved through applying additional constraints on the form of a covariance matrix.

$$f(X) = \frac{1}{(2\pi)^{d/2} \| \Sigma \|^{1/2}} exp\left\{ -\frac{1}{2}(X - \vec{\mu})^T \Sigma^{-1}(X - \vec{\mu}) \right\}, \tag{2}$$

in which X – input feature-vector, $\vec{\mu}$– mean vector, Σ– covariance matrix.

For each of the phonemic classes we can assign a mixture of M Gaussians to represent a class conditional likelihood of the data:

$$p(X|C) = \sum_{i=1}^{M} p(X|J_i)p(J_i|C). \tag{3}$$

And using the maximum likelihood Expectation-Maximization algorithm [1] we can optimize parameters $(\vec{\mu}, p(J|C)$, components of matrix $\Sigma)$ of each of the Gaussians. As in the case of the MLP, the target distribution can be modeled with an arbitrary accuracy.

The obtained estimation of the class conditional likelihoods can be further used in the Bayes formula to get an estimation of the posterior probabilities of class membership:

$$P(C|X) = \frac{p(X|C)P(C)}{p(X)} = \frac{p(X|C)P(C)}{\sum_i p(X|C_i)P(C_i)}. \tag{4}$$

Class priors $P(C)$ can be estimated directly from the training data.

4   The Difference Between Models

The basic difference between the two described approaches is in the form of the representation of the posterior probability of class membership.

The MLP creates a distributed representation; in other words, typically many hidden units contribute to each network output value. As a result, training procedures are highly non-linear and have problems with stacking on the local minima. Hidden units in the MLP compute a monotonic function of the weighed sum of the inputs and thus their activation values are constant along the $(d-1)$-dimensional planes, parallel to the decision hyperplane in the d-dimensional input space. It is schematically demonstrated in Fig. 1 by strait lines separating different classes. MLP models estimate the posteriors directly and if the class priors in the training and testing sets are different such models require additional compensation at the recognition time [2].

Mixture models generate a local representation, the activation of hidden units in the equivalent RBF net are determined by the Mahalanobis distance from the mean vector and thus only few hidden units will contribute to each output value of the model. In order to achieve approximation accuracy equivalent to that of the MLP, mixture models require more free parameters. The hidden unit activations in the equivalent RBF network are constant on the $(d-1)$-dimensional hyperellipsoids around the mean value (Fig. 1). Parameters in the mixture models can be determined with a less computational effort, especially with additional constraints on the form of covariance matrix. Free parameters of the mixture models have precise meanings unlike those of the MLP.

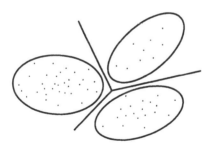

Fig. 1. Schematic illustration of the differences in representation in the MLP and mixture models.

5   Usage of the Combinations of the Models

Although the above techniques are used to estimate the posterior probabilities of class membership, they employ very different methods of approximation. From this perspective one can get some advantage by using both models during the training process.

5.1 Outliers

In practical applications there is some amount of erroneous information in the training set, resulting, for instance from wrong phonemic labeling or wrong calculations of the phoneme boundaries. This problem is known as an "outliers problem". Erroneous data can easily destroy valid discrimination boundaries. This problem can be overcome by employing joint training of both models as follows:

1. Estimation of the parameters of the mixture model, calculating class conditional likelihoods for every data point in the training set.
2. Using these likelihoods as parameters governing the learning rate in the MLP on-line training or excluding some amount of the least probable data points from the training set for the MLP.

This procedure would create a model which is "suspicious" of the data falling outside the most probable region for each class.

5.2 Committee of the Networks

As estimators of the posterior probability both models can produce a committee in order to increase an approximation accuracy, or can be employed into a mixture of experts by a third network, trained to distinguish which approximation gives better results in each particular region of the input space.

6 Conclusion

It is accepted to consider two mentioned models as alternative ways of the class posterior calculation. But the drawn comparison shows that while estimating posterior probabilities they provide essentially different representation. This fact givess us a reason to believe that joint employment of these models can increase performance in classification.

References

1. C. Bishop "Neural Networks For Pattern Recognition" Clarendon Press, Oxford 1995, pp. 482.
2. H. Bourlard, N. Morgan "Connectionist Speech Recognition, A Hybrid Approach" Kluwer Academic Publishers, Boston, Dordrecht, London 1994, pp. 312.
3. B. Ripley "Pattern Recognition and Neural Networks" Cambridge University Press, 1996.
4. R. M. Golden "Mathematical Methods for Neural Network Analysis and Design" The MIT Press, 1996.
5. Y. Bengio "Neural Networks for Speech and Sequence Recognition" International Thomson Computer Press, 1996.

A Simple Spanish Part of Speech Tagger for Detection and Correction of Accentuation Error

S. N. Galicia-Haro, I. A. Bolshakov, and A. F. Gelbukh

Laboratorio de Lenguaje Natural, C.I.C., I.P.N.
Av. Juan de Dios Bátiz. AP 75-476, CP 07738, Mexico City, Mexico
{igor, gelbukh, sofia}@pollux.cic.ipn.mx

Abstract. One of the most frequent kind of typographic errors specific to Spanish is connected with accentuation, namely, with omission of an obligatory stress mark or insertion of a superfluous one. If such an error transforms one word to another existing one, the latter cannot be detected by usual spell-checkers, since some context analysis is necessary. A simple procedure is proposed for this task. It relies on (1) some simple heuristics that determine linear context and (2) on a small list of pairs of words that differ only in accentuation mark. This idea is applied to numerous nouns or adjectives like *número* that pass to quasi-homonymous personal verb forms if they lose their stress marks.

1 Introduction

The usual spell-checkers consider each word out of its context. With such a strategy only those typographic and orthographic errors may be detected by the spell-checker, that change an existing word to a senseless letter string. As to the errors converting one existing word to another existing one, the words with them stay unnoticed and make the text in essence ungrammatical. For their reliable correction we propose some simple heuristic method to solve this problem.

In Spanish there is a peculiar source of typographic errors which are fully undetectable if the words are taken out of context. These errors are connected with Spanish accentuation rules. For example, the phrases *este artículo tiene ...* 'this article has ...', or *las páginas siguientes ...* 'the following pages ...' would be considered correct by a spell-checker even if the underlined words lost their stress marks: **este articulo tiene ...* **'this I join has ...'*, **las paginas siguientes ... *'the you paginate following ...'*. Indeed, words *articulo*, *numero*, and *paginas* are true Spanish verb forms, though fully unacceptable in the mentioned contexts.

Meanwhile, this type of errors is quite usual for foreigners. One of the authors, has made more than 60 accentuation errors in his first paper in Spanish [1], and only half of them were detected by the spell-checker of Word for Windows, version 6 [2]. Newer versions of Word for Windows use a powerful grammar checker to detect such errors, though this seems to be a task rather for a simple enough spell-checker, since for the user this is an ordinary typo.

F. Jelinek, E. Nöth (Eds.): TSD'99, LNCS 1692, pp. 219–222, 1999.
© Springer-Verlag Berlin Heidelberg 1999

2   Linguistic Information

All error-prone Spanish words with stress marks may be divided to several non-intersecting groups. One group contains those nouns or adjectives that pass to personal verb form, if the stress mark is omitted. At the first stage of our study, the quasi-homonymous pairs "noun - verb" were gathered through the manual search in large Spanish dictionaries, such as of Academy of Madrid [3] and Anaya Group [4]. It gave us about 150 pairs then a special program was written for automatic extracting from large electronic dictionaries. Thus the full amount of pairs reached nearly 300.

The common features of the pairs are the following: 1) The pairs are Spanish words of middle statistical ranks, 2) An accentuated counterpart is a specific word form of a noun, an adjective or a vocable combining two homonyms, 3) A non-accentuated counterpart is a specific personal form of a verb in singular. So the components of each pair correlate as a word form to another word form. 4) Each accentuated form, independently of its part of speech is characterized with a specific combination of number and gender.

In our method we employ a kind of a part-of-speech tagger, that can set only one mark: "possible adjective or noun", leaving all the other words unmarked. Good part-of-speech taggers always take into account the linear context for disambiguation part-of-speech homonymy [5]. Fortunately, Spanish presents good conditions for detecting the noun or adjective context and there is grammatical agreement between the nouns, adjectives and even articles.

3   The Algorithm

The algorithm is based on a procedure described in the next section, that for a given noun or adjective $\widetilde{\omega}_0$ can determine whether the immediate 4-word context is suitable for it. We do not have such a procedure for verbs though. Then we use a technique used by some style checkers [6]: instead of checking the current situation in the text, a hypothesis is form about a possible error of the text, and check this hypothesis. The hypothesis is incompatible with the original user's text. If the hypothesis looks reasonable, a possible error is reported.

The algorithm scans the text. Each word is looked up in the two lists: the list of accentuated words, and the automatically compiled list of their non-accentuated counterparts. The characteristics of the found word, namely its gender and number, are retrieved from the list. Let the word under consideration be ω_0, and its immediate linear context be ω_{-1}, ω_1, ω_2, such that the word order is ω_{-1}, ω_0, ω_1, ω_2.

Then the work of the algorithm depends on the list in which the word was found: 1) If the word was found in the accentuated list, it is considered to be a noun or adjective and the suitability of the immediate context is checked; the variable $\widetilde{\omega}_0$ described there is set to the word ω_0 itself. If the context is *not* suitable for a noun or adjective, a possible error is reported. 2) If the word was found in the non-accentuated list, it is considered to be a verb. Since we cannot check the context for a verb, a hypotheses is considered that the true intended word was the corresponding accentuated noun. The variable $\tilde{\omega}_0$ is set to this accentuated counterpart, and the context is checked for this hypothetical word. If the context is suitable for it, the hypothesis is accepted and a possible error is reported.

4   Linear Context

For the algorithm, a procedure is necessary to check for the word 0 that is supposed to be a *noun* or *adjective* in the form of already known *gender* and number, whether a specific 4-word immediate linear context ω_{-1}, ω_1, ω_2, such that the word order is ω_{-1}, ω_0, ω_1, ω_2, is suitable for a noun or adjective with these gender and number. Here $\tilde{\omega}_0$ is either the current word considered by the ω_2 main algorithm, or its accentuated counterpart.

Let Preps be the list of all simple (one-word) prepositions, Preps = {*a, de, con, por, sin, en, sobre, para, ...*}, and Dets be the list of quasi-determinatives that depends on the gender and number of ω_0 according to the following table:

	Singular	*Plural*
Masculine	un, el, este, ese, aquel, mi, tu, su, al, del, buen, mal, primer, gran	unos, los, estos, esos, aquellos, mis, tus, sus, buenos, malos, primeros, grandes
Femenine	una, la, esta, esa, aquella, mi, tu, su, buena, mala, primera, gran	unas, las, estas, esas, aquellas, mis, tus, sus, buenas, malas, primeras, grandes

Let us use the notation $u\tilde{\,}v$ for grammatical agreement of words u and v in gender and number, i.e., for the fact that the first word form has or could have the same gender and number as the second one. The word $\tilde{\omega}_0$ is considered to be a noun or adjective properly used in the given context, and thus the word ω_0 is considered likely to be an error, if any of the following four conditions is satisfied:

1. $\omega_{-1} \in Dets \cup Preps$, or
2. $\omega_{-1}\tilde{\,}\tilde{\omega}_0$ or
3. $\omega_1\tilde{\,}\tilde{\omega}_0$ or
4. $\omega_1 \in$ {más, menos}& $\omega_2\tilde{\,}\tilde{\omega}_0$

The tests should be carried out in the given order, that helps to cope with the combinations like el *número* y el *género* gramaticales, where the agreement is more difficult to check. Since the gender and number of $\tilde{\omega}_0$ are known, to check the agreement in the conditions 2 to 4, it is enough to check the hypothesis that the corresponding word ω_{-1}, ω_1, ω_2 is compatible with the hypothesis about its number and gender.

5   Experimental Results

The algorithm was realized in a complete program, consisting of 27 subprograms in Pascal, including the scanner of the input text and the dialog with the user for interactive correction of the reported errors in the text.

The program was applied to several unprepared Spanish texts. Before processing, all the stress marks were removed from the text, and the text was corrected by Word for Windows, version 6. This guaranteed that the text contained only the errors that could not be detected out of context. To find such remaining errors, our program was applied. In the text [1] consisting of 9 pages, there were 35 such errors not detectable by a usual spell-checker. As much as 32 of them, or 91.4%, were detected by the program. Only one of the missed ones corresponded to quasi-homonyms from the list: *no es practica, the correct form being no es práctica; two other were connected with the other words.

6   Conclusions

A simple method is proposed to detect and correct one very common type of errors in Spanish text, namely, absent or misplaced accentuation marks. Such errors arise under the following circumstances: 1) Typos or orthographic errors of native speakers, as well as "simplified" Spanish spelling, 2) Errors made by foreigners, and 3) Problems of technical nature, such as the absence of the special Spanish keyboard or problems of different encoding of the accentuated letters under different operating systems.

Because of these reasons, some large text corpora, especially the ones collected from Internet, have significantly large fragments with the accented marks totally or partially lost. Due to informal genre of such texts the use of full-featured grammar checkers is not effective for them; this makes the suggested simple approach attractive for processing of large corpora. Also, the ideas similar to the described method can be used for simple detection of other types of errors, especially those of agreement in number and gender.

References

1. Bolshakov, I.A.: El modelo morfólogico formal para sustantivos y adjetivos en el español. Computacion y Sistemas, 1 (1996).
2. Word for Windows95. User's Guide. MicrosoftCorp. (1995).
3. RAE Diccionario de la lengua Española. Real Academia Española, Edición en CD-ROM (1996).
4. Diccionario del Español contemporáneo. Grupo ANAYA, http://www.anaya.es.
5. Cutting, D., et al.: A Practical Part-of-Speech Tagger. In: Proceedings of the Third Conference on Applied Natural Language Processing. Trento, Italy (ACL) (1992).
6. Ashmanov, I.: Grammar and Style Corrector for Russian Texts (in Russian). In: Proc. Of International Workshop on Computational Linguistics and its Applications, Dialogue-95, Kazan, Russia (1995).

Slovene Interactive Text-to-Speech Evaluation Site – SITES

Jerneja Gros, France Mihelič, and Nikola Pavešić

University of Ljubljana, Faculty od Electrical Engineering,
Tržaška 25, SI-1000 Ljubljana, Slovenia
nejka@fe.uni-lj.si,
http://luz.fe.uni-lj.si

Abstract. The Slovene Interactive Text-to-Speech Evaluation Site (SITES) was built according to standards for interactive speech synthesiser comparison sites as set by COCOSDA (International Committee for the Co-ordination and Standardization of Speech Databases and Assessment Techniques for Speech Input/Output) and the LDC (Linguistic Data Consortium).

SITES aims to give the interested listeners a thorough and honest impression of the current text-to-speech (TTS) system and provides valuable feedback about strong and weak points of the system. The SITES web site enables to evaluate the S5 Slovene TTS system either interactively or off-line by sending the synthesized speech file to a given e-mail address. We implemented various standard text selection methods and set up rules for construction as Semantically Unpredictable Sentences for the Slovene language.

The evaluation web site has the capability to accept arbitrary input text, and returns a speech file. A CGI script first reads the user's form input. When the user submits the form, the script receives the form data as a set of name-value pairs, which is parsed. In the CGI script, the TTS system is called with the parameters specified by the user. The TTS system generates a temporal audio file which is sent back to the user.

1 Introduction

Text-to-speech synthesis (TTS) enables automatic conversion of any available textual information into its spoken form. Until recently, when TTS systems were presented on the Internet or on text-to-speech workshops and conferences, it was usually done in form of prepared demonstrations. This inevitably led to great uncertainty concerning the true quality of the systems.

Therefore, the International Committee for the Co- ordination and Standardization of Speech Databases and Assessment Techniques for Speech Input/Output (COCOSDA) http://www.itl.atr.co.jp/cocosda/, together with the Linguistic Data Consortium (LDC) set the standards of a Interactive Speech Synthesizer Comparison Site http://www.ldc.upenn.edu/ltts/. The specific purpose of the LDC/COCOSDA TTS website is to help users:

F. Jelinek, E. Nöth (Eds.): TSD'99, LNCS 1692, pp. 223–228, 1999.

- to find interactive TTS websites: sites where you type in your own text, instead of listening to speech specially prepared by vendors for demonstration purposes,
- to select useful test text from a wealth of text corpora made available by the LDC.

We continue the paper with a brief introduction into Slovene text-to-speech synthesis. Then we go on describing the Slovene Interactive TTS Evaluation Site (SITES) and the text selection procedures. For the time being, the SITES site can be accessed on a local computer; by summer 1999 it will be available on `http://sites.fe.uni-lj.si`.

2   Slovene Text-to-Speech Synthesis

In the Laboratory of Artificial Perception at the University of Ljubljana, we started on text-to-speech synthesis in 1994. In the following year we presented the first PC-based TTS system for the Slovene language [1]. We used it as a reference system for further improvements. In 1998 another diphone based Slovene TTS system developed at the Jožef Stefan Institute in Ljubljana was presented [2]. Speech files produced by this system on a limited application domain can be obtained via e-mail from the following address `http://www-ai.ijs.si/~ema/welcome-s.html`.

In the recent version of our TTS system S5 we implemented a novel procedure for determination of prosodic parameters, added a pronunciation dictionary and improved text normalization [3].

The S5 TTS system has already been implemented in different applications:
- S5 as a stand-alone TTS system (reading machine) [1],
- S5 providing the spoken output in the SQEL speech recognition and dialogue system for automatic airline timetable retrieval [4] and
- S5 integrated into a special application - HOMER, a reading system for the blind and partially sighted people, combined with optical character recognition devices and voice control [5].

The input text is transformed into its spoken equivalent by a series of modules. A grapheme-to-phoneme or -to-allophone module produces strings of phonetic symbols based on information in the written text. A prosodic generator assigns pitch and duration values to individual phones. Final speech synthesis is based on diphone concatenation using TD-PSOLA [6]. The quality of the synthesized speech was assessed in terms of intelligibility and naturalness of pronunciation. Additionally, various aspects of the synthetic speech production process were tested. The assessment results of the Slovene TTS system are given and discussed in [7].

2.1   Grapheme-to-Allophone Transcription

Input to the S5 system is unrestricted Slovene text. It is translated into a series of allophones in two consecutive steps. First, input text normalization is performed. Next, word pronunciation is derived, based on a user-extensible pronunciation dictionary and letter-to-sound rules. The dictionary covers over 16.000 most frequent inflected word forms. In case where dictionary derivation fails, words are transcribed using automatic lexical stress assignment and letter-to-sound rules [3].

2.2 Prosody Generation

Prosody generation in S5 consists of four phases: intrinsic duration assignment, extrinsic duration assignment, modeling of the intra word F0 contour and assignment of a global intonation contour.

A large quantity of Slovene speech material was recorded and processed in order to extract the relevant prosodic parameters for Standard Slovene speech. In addition to the usual measurements, the impact of speaking rate on syllable duration and duration of phones was studied in a number of ways. An extensive statistical analysis of lengthening and shortening of individual phones, phone groups and phone components, like closures or bursts was performed, the first of the kind for the Slovene language [8].

Similarly to [9], our two-level duration model first determines the words' intrinsic duration, taking into account factors relating to the phone segmental duration, such as: segmental identity, phone context, syllabic stress and syllable type: open or closed syllable. Further, the extrinsic duration of a word is predicted, according to higher-level rhythmic and structural constraints of a phrase, operating on a syllable level and above. Finally, intrinsic segment duration is modified, so that the entire word acquires its predetermined extrinsic duration. It is to be noted, that stretching and squeezing does not apply to all segments equally. A method for segment duration prediction was developed, which adapts a word with an intrinsic duration to the determined extrinsic duration, taking into account how stretching and squeezing apply to the duration of individual segments [8].

Since the Slovene language has been defined as a pitch accent language [10], special attention was paid to the prediction of tonemic accents for individual words. First initial vowel fundamental frequencies were determined according to [10], creating the F0 backbone. Each stressed word was assigned one of the two tonemic accents, characteristic for the Slovene language. The acute accent is mostly realized by a rise on the posttonic syllable, while with the circumflex the tonal peak usually occurs within the tonic. Five typical F0 patterns were chosen from the variety of F0 patterns. Finally a linear interpolation between the defined F0 values was performed.

We used a relatively simple approach for prosody parsing and the automatic prediction of Slovene intonational prosody which makes no use of syntactic or semantic processing, but rather uses punctuation marks and searches for grammatical words, mainly conjunctions which introduce pauses [11].

2.3 Diphone Concatenation

Once appropriate phonetic symbols and prosody markers are determined, the final step within S5 is to produce audible speech by assembling elemental speech units. This is achieved by taking into account computed pitch and duration contours, and synthesizing a speech waveform. The TD-PSOLA diphone concatenative technique was applied [6].

3 SITES

The Slovene Interactive Text-to-Speech Evaluation Site (SITES) was built according to the standards set by the LDC/COCOSDA TTS Evaluation site, based on [12]. It aims to give the interested listeners a thorough and honest impression of the current system and to provide valuable feedback about strong and weak points of our system.

The SITES web site enables to evaluate the S5 Slovene TTS system either interactively or off-line (by sending the synthesized speech file to a given e-mail address). The SITES Web site has the capability to accept arbitrary input text, and returns a speech file. A CGI script first reads the user's input. When the user submits the form, the script receives the form data as a set of name-value pairs, which is parsed. In the CGI script, the TTS system is called with the parameters specified by the user. The TTS system generates a temporal audio file which is sent back to the user.

The following general options concerning the voice type and speech quality can be adjusted:

- speaking rate: normal, fast, slow
- file type format: au, wav, aiff, pcm
- frequency: 8 kHz, 11.025 kHz, 16 kHz
- quality: 8 bit u-law, 8bit linear, 16 bit linear and
- download file type: .tar.gz, .tar.Z, .tar.zip, .tar.

3.1 Text Selection

The available text material covers a range of challenges for the TTS system. Three text types are used: newspaper text (former daily newspaper Slovenec, 264.736 words), fiction (The Bible, 152.212 words) and semantically unpredictable sentences.

The evaluator has the opportunity to either type in his own text or the text material is created by one of the standard automated methods, based on a chosen text corpus:

- enter your own text in the appointed area and press the Submit the request button,
- select a sample from the listed text collections using the Collections and Selection Method buttons and then press the Submit the request button.

3.2 Text Selection Methods

From the fiction and newspaper text one sentence is selected according to one of the 4 following text selection methods as proposed by COCOSDA:

1. random selection. This method simply randomly selects a sentence from the entire text corpus.

2. minimum word frequency based selection. This method selects short sentences made up entirely of common words. These sentences should pose no problems for the TTS system. This method involves the following steps:

- determine number of occurrences (frequency) of each word in the text corpus,
- for each sentence, determine the frequency of the least frequent word,
- sort sentences in descending order by least frequent word frequency,
- randomly select from the top 1, 5, or 10 sorted list.

3. overall word frequency based selection. This method selects longer sentences with many high-frequency words, although they may contain some rarer words as well. Selected sentences are more taxing. This method involves the following steps:

- determine number of occurrences (frequency) of each word in the corpus,
- for each sentence, add the log frequencies of all its words,
- sort sentences in descending order by log frequency sum,
- randomly select from the top 1, 5, or 10 sorted list.

4. overall trigram frequency based selection. This method uses successive letter triples as the basic unit, but is otherwise the same as overall word frequency based selection. The sentences tend to be long, and may contain several rare words. However, the phoneme combinations are common. Selected sentences are more taxing, in particular for the dictionary and pronunciation rule components of systems. This method involves the following steps:

- determine number of occurrences (frequency) of each trigram in the corpus,
- for each sentence, add the log frequencies of all its trigrams,
- sort sentences in descending order by log frequency sum,
- randomly select from the top 1, 5, or 10 sorted list.

3.3   Semantically Unpredictable Sentences

Semantically Unpredictable Sentences (SUS) are used in intelligibility tests at sentence level. In order to reduce contextual semantic information given by the sequence of single words, not concrete sentences but basic sentence structures are the raw material. The filling of each component of these structures is done by random selection of words out of a pre-defined list of possible candidates. These lists contain only mono-syllabic, and exceptionally disyllabic, words.

A set of basic pattern structures for constructing Slovene Semantically Unpredictable Sentences was built, similar to the five basic syntactic structures of the SAM project [13,14]. The rules were of the following type: Subject – Verb Adverbial, Subject Transitive Verb – Object, etc. The word lists were constructed from the MULTEXT- EAST lexicon [15] since it provides also the basic morphosyntactic descriptions of a word form.

3.4   Conclusion

The outlines of the SITES Slovene Interactive Text-to-Speech Evaluation Site were presented. We hope the casual evaluators will give some useful comments on the quality of the synthetic speech.

References

1. Gros, J., Pavešić, N., Mihelič, F: Text-to-speech synthesis: A complete system for the Slovenian language. Journal of Computing and Information Technology 5 (1997) 11–19.
2. Šef, T., Dobnikar, A., Gams, M., Grobelnik, M.: Slovenski govor na internetu. Proceedings of the Conference Language Technologies for the Slovene Language ISJT'98. Ljubljana Slovenia (1998) 60–64.
3. Gros, J.: Samodejno pretvarjanje besedil v govor. PhD Thesis. University of Ljubljana (1997).
4. Pavešić, N., Mihelič, F., Ipišić, I., Dobrišek, S., Gros J., Pepelnjak, K.: Spoken queries in European languages (SQEL) project: an overview of the Slovenian information retrieval system for flight services. Proceedings of the International Conference on Software in Telecommunications and Computer Networks SoftCOM'98. Split Dubrovnik Croatia Bari Italy (1998) 69–74.
5. Dobrišek, S., Gros, J., Mihelič F., Pavešić, N.: HOMER: a voice driven reading system for the blind. Proceedings of the COST 254 Workshop. Ljubljana (1998).
6. Moulines E., Charpentier, F.: Pitch-Synchronous Waveform Processing Techniques for Text-to-Speech Synthesis Using Diphones. Speech Communication 9 (1990) 453–467.
7. Gros, J., Mihelič, F., Pavešić, N.: Speech Quality Evaluation in Slovenian TTS. Proceedings of the LREC98. Granada Spain (1998).
8. Gros, J., Pavešić, N., Mihelič, F: Speech timing in Slovenian TTS. Proceedings of the EUROSPEECH'97. Rhodes Greece (1997) 323–326.
9. Epitropakis, G., Tambakas, D., Fakotakis, N., Kokkinakis, G.: Duration modelling for the Greek language. Proceedings of the EUROSPEECH'93. Berlin Germany (1993) 1995–1998.
10. Srebot-Rejec, T.: Word Accent and Vowel Duration in Standard Slovene: An Acoustic and Linguistic Investigation. Slawistische Beitraege. Verlag Otto Sagner 26 (1988).
11. Sorin, C., Laurreur D., Llorca, R.: A Rhythm-Based Prosodic Parser for Text-to-Speech Systems in French. Proceedings of the XIth ICPhS. Tallin Estonia (1987) 125–128.
12. van Bezooijen, R., van Heuven, R.: Assessment of Synthesis Systems. Handbook of Standards and Resources for Spoken Language Systems. Eds. D. Gibbon, R. Moore. Mouton de Gruyter. Berlin (1997) 481–563.
13. Grice, M.: Syntactic structures and lexicon requirements for Semantically Unpredictable Sentences in a number of languages Proceedings of the ESCA Workshop on Speech Input/Output Assessment and Speech Databases. Noordwijkerhout (1989) 1.5.1– 1.5.4.
14. Benoit, C.: An intelligibility test using semantically unpredictable sentences: towards the quantification of linguistic complexity. Speech Communication 9 (1990) 293–304.
15. East meets West: A Compendium of Multilingual Resources, Eds. T. Erjavec, A. Lawson, L. Romary. CD-ROM produced and distributed by TELRI (1998)

Developing HMM-based Recognizers with ESMERALDA

Gernot A. Fink

University of Bielefeld, Faculty of Technology,
P.O. Box 100131, 33501 Bielefeld, Germany
Tel.: +49 521 106 2935, Fax: +49 521 106 2992
gernot@techfak.uni-bielefeld.de

Abstract. ESMERALDA is an integrated environment for the development of speech recognition systems. It provides a powerful selection of methods for building statistical models together with an efficient incremental recognizer. In this paper the approaches adopted for estimating mixture densities, Hidden Markov Models, and n-gram language models are described as well as the algorithms applied during recognition. Evaluation results on a speaker independent spontaneous speech recognition task demonstrate the capabilities of ESMERALDA.

1 Introduction

For building successful speech recognition systems currently Hidden Markov Models (HMM) are the technology of choice (cf. [7,8,13]). However, for a complete recognizer many additional problems have to be addressed ranging from signal processing to statistical language modeling. The application of the estimated models during recognition is also not at all trivial – especially for large vocabulary tasks.

In the literature numerous approaches for statistical model building and decoding have been proposed. The goal of the ESMERALDA[1] system is to put together a tractable set of conceptually simple yet powerful techniques in an integrated development environment.

The architecture of the system is organized around the incremental recognizer as shown in Fig. 1. Separate modules for estimating mixture density models in conjunction with HMMs and for building n-gram language models are provided. The approaches adopted will be described in the following section. For completeness also selected methods for signal processing and feature extraction are available. The incremental recognizer that applies all the statistical models and additionally also grammatical restrictions during recognition will be presented in Section 3. In order to demonstrate the capabilities of ESMERALDA evaluation results on a speaker independent spontaneous speech recognition task will be given in Section 4.

---

[1] Environment for Statistical Model Estimation and Recognition on Arbitrary Linear Data Arrays

F. Jelinek, E. Nöth (Eds.): TSD'99, LNCS 1692, pp. 229–234, 1999.

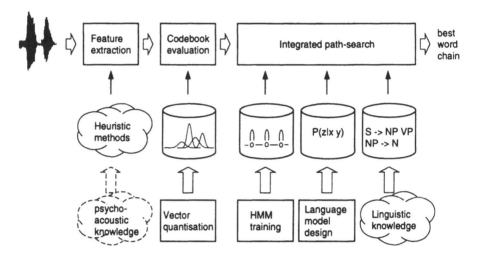

Fig. 1. Overall architecture of the ESMERALDA system

2 Statistical Model Design

2.1 Mixture Densities

For HMM systems using Gaussian mixture models for emission probability density calculation a good initialization of the Gaussians is of fundamental importance. ESMERALDA provides a variant of the LBG-algorithm for vector quantizer design [10]. Homogeneous regions within the feature space are described not only by their centroid vector but using full- or diagonal-covariance Gaussians.

Starting with some initial set of Gaussians each intermediate codebook is optimized alternating between classification and parameter adaptation steps. In order to reach the desired number of codebook classes selected models are split and the new codebook is again optimized. As even today this method is computationally extremely expensive the parallel processing on a cluster of workstations is supported using the DACS distributed programming environment [4].

2.2 Hidden Markov Models

The methods provided by ESMERALDA are organized around a core set of tools for handling structured HMMs. Elementary models specifying a certain type of allowed state transitions are built from individual state definitions that carry all statistical model parameters. From these elementary models more complex HMMs can be constructed by using a declarative specification language — a feature that was inspired by the ISADORA system [13]. The use of the configuration language makes it possible to implement most of the HMM structures for e.g. different types of sub-word units that are known from the literature. Model parameters can be initialized on labelled data and optimized by applying the standard Baum-Welch re-estimation algorithm (cf. [7, pp. 152–158]).

For most recognition tasks it is crucial to find a good balance between precision of models and robustness of parameter estimates. A conceptually simple and very effective method to at least partly solve this problem automatically is the use of *state clustering* which was first proposed in [9, pp. 103–106] and is now widely used (e.g. [1,14]). In ESMERALDA model states without own parameters can be defined that are merely alias names referring to some base state. For tri-phone sub-word units the base states could be the states of the mono-phone models while initially all tri-phone states are alias names. During the training phase parameter updates are assigned to the most specific entities i.e. the alias states. Though for the majority of them the estimates will not be robust enough for good models they can serve as a basis for a similarity measure calculated between aliases of a common base state. Using a greedy algorithm similar alias states are grouped into clusters for which individual new parameter carrying states are created that are subsequently used by the tri-phone models.

2.3 Statistical Language Models

Especially for large vocabulary recognizers language model restrictions are of fundamental importance in order to achieve high recognition accuracy. In combination with HMM-based systems statistical n-gram models currently are the most widely used technique (cf. [8, pp. 57–78]). Within this framework the hardest problem to solve is that of so called *unseen events* for which no useful probability value can be assigned without additional effort. Therefore, ESMERALDA provides tools for estimating arbitrary n-gram models and offers standard methods for redistributing probability mass and smoothing distributions in order to assign robust non-zero probability estimates to unseen n-grams (cf. [3]).

3   The ESMERALDA Recognizer

The statistical models set up by the tools described in the previous section are applied to signal decoding by the ESMERALDA recognizer. In order to be able to handle complex statistical models and large vocabulary tasks the system is designed and moderately optimized for efficiency.

3.1   Time-Synchronous Search

The ESMERALDA recognizer processes input signals in a cascade of modules for feature extraction, statistical mixture decoding, emission probability calculation, HMM search, and finally language model search. For HMM decoding the well known *beam search* algorithm (cf. [8, pp. 81–84]) is applied. The recognition lexicon is represented as a lexical tree [11] resulting in a substantially improved efficiency of the search process.

In order to combine language model scores with acoustic scores in a tree-based recognition system it is, however, necessary to start copies of the search tree at every word end reached as the identity of the successor word is not known before a leaf node of the tree is reached. The ESMERALDA recognizer uses a technique similar to the original method proposed in [11] but developed independently from [12] where tree copies are indexed by their starting time frame and not by the predecessor word. This method has the advantage that the necessary number of tree copies is less dependent on the vocabulary size and mostly determined by the average length of the acoustic word models used.

All calculations from feature extraction to language model search are carried out strictly time-synchronously i.e. all restrictions are applied in combination frame by frame [14]. This method sets the ESMERALDA recognizer apart from multi-pass search strategies as can be found in e.g. [1,6]. Therefore, the system can not exploit the benefits of a model hierarchy with increasing complexity. However, the incremental generation of recognition results is only possible with a time synchronous search strategy.

3.2 Incremental Processing

In order to be able to produce recognition results for an utterance while the user is still speaking i.e. the end of the input signal is not yet reached an incremental processing strategy was developed [5]. It is especially advantageous in multi-modal systems that combine speech with e.g. image processing or gesture recognition as the results of different signal analysis systems can early be combined and used to produce feedback to the user.

As at the signal frame currently processed the final recognition results can never be judged from competing solutions it is necessary to allow for some delay in time until stable hypotheses are established in the past of the search space. We choose a rather simple method that uses a configurable time delay from the current frame to search for stable results and put these out incrementally. Experiments showed that with a delay of 2 seconds and above identical results are achieved on different tasks as with the traditional non-incremental method. However, decreasing the delay does not only lead to a faster reaction of the recognizer by also increases the error rate. A good compromise between a moderately increased error rate and short reaction time was found at a delay of 0.75 seconds [5].

3.3 Combining Statistical and Declarative Language Models

Though statistical language modeling is a well established technique especially in combination with HMMs it is effective only if large amounts of training data are available. As this need not be the case for a given scenario it is desirable that declarative language restrictions set up by experts can be applied in an equally robust way to enhance recognizer performance.

The ESMERALDA recognizer is capable of applying the constraints of a context-free grammar – which is required to be LR(1) – in conjunction with statistical n-gram models [15]. In experiments on several smaller domains considerable improvements could be achieved by combining a domain specific grammar with statistical language models estimated on limited data.

When combining a speech recognition system with an interpretation component the use of a grammar offers another important advantage over purely statistical systems. As within the grammar the structure of certain domain specific language constituents is defined structured recognition hypotheses can be generated. If the structure definitions used during recognition are well matched with the linguistic knowledge base very efficient and robust understanding systems can be built [2].

4 Results

In order to demonstrate the capabilities of ESMERALDA we carried out experiments on the German VERBMOBIL appointment scheduling domain using the official 1996 evaluation guidelines. On over 33 hours of spontaneous speech we trained a 5336 word recognition system and evaluated it on the independent test set of approximately 41 minutes of speech.

First, we compared our language modeling approach to the results obtained by using the official bi- and tri-gram models provided by Philips Research Labs, Aachen, Germany. In order to handle unseen n-grams robustly we applied shift-discounting and back-off smoothing of the distributions. The perplexities of 64.4 for bi-grams and 49.2 for tri-grams keep up well with the official results of 64.2 and 48.9 respectively.

Our feature extraction calculates 12 mel-frequency cepstral coefficients (cf. [13, p. 60]) with dynamic mean normalization (cf. [16]) and one energy coefficient every 10 ms together with the smoothed first and second order derivatives. On this 39 dimensional feature space we set up a semi-continuous HMM system [7] using tri-phone sub-word units with a final codebook of 943 full covariance Gaussians. During recognition the official bi-gram language model was used and no parameter adaptation algorithms were applied. While running at 5 times real-time on a DIGITAL AlphaStation 500/500 the system achieves a word error rate of only 20.1% – a figure which is among the best results published on this task.

5 Conclusion

In this paper ESMERALDA[2] was presented –– an integrated development environment for building statistical recognition systems. It offers a collection of powerful state-of-the-art techniques for calculating mixture density models, training structured Hidden Markov Models, and estimating statistical n-gram language models together with an efficient incremental recognizer. The evaluation result presented in the previous section demonstrate that competitive performance on non-trivial recognition tasks can be achieved using the ESMERALDA system.

---

[2] ESMERALDA is still in experimental state and mainly used for internal research purposes. However, it runs on several UNIX-type operating systems and a Linux version is used at the Centre for Autonomous Systems, Kungliga Tekniska Högskolan, Stockholm. Those interested in using ESMERALDA for their own research should contact the author.

Acknowledgement. The author would like to thank Franz Kummert for many fruitful discussions during the development of ESMERALDA.

References

1. J. Billa, T. Colhurst, A. El-Jaroudi, R. Iyer, K. Ma, S. Matsoukas, C. Quillen, F. Richardson, M. Siu, G. Zvaliagkos, and H. Gish. Recent experiments in large vocabulary conversational speech recognition. In *Proc. Int. Conf. on Acoustics, Speech and Signal Processing*, Phoenix, Arizona, 1999.
2. H. Brandt-Pook, G. A. Fink, S. Wachsmuth, and G. Sagerer. Integrated recognition and interpretaion of speech for a construction task domain. In *Proc. 8th Int. Conf. on Human-Computer Interaction*, München, 1999. to appear.
3. M. Federico, M. Cettelo, F. Brugnara, and G. Antoniol. Language modelling for efficient beam-search. *Computer Speech & Language*, 9:353–379, 1995.
4. G. A. Fink, N. Jungclaus, H. Ritter, and G. Sagerer. A communication framework for heterogeneous distributed pattern analysis. In *Proc. Int. Conf. on Algorithms And Architectures for Parallel Processing*, pages 881–890, Brisbane, 1995.
5. G. A. Fink, C. Schillo, F. Kummert, and G. Sagerer. Incremental speech recognition for multimodal interfaces. In *Proc. 24th Annual Conference of the IEEE Industrial Electronics Society*, pages 2012–2017, Aachen, September 1998.
6. T. Hain, P. C. Woodland, T. R. Niesler, and E. W. D. Whittaker. The 1998 HTK system for transcription of conversational telephone speech. In *Proc. Int. Conf. on Acoustics, Speech and Signal Processing*, Phoenix, Arizona, 1999.
7. X. Huang, Y. Ariki, and M. Jack. *Hidden Markov Models for Speech Recognition.* Edinburgh University Press, Edinburgh, 1990.
8. F. Jelinek. *Statistical Methods for Speech Recognition.* MIT Press, Cambridge, MA, 1997.
9. K.-F. Lee. *Automatic Speech Recognition: The Development of the SPHINX System.* Kluwer Academic Publishers, Boston, 1989.
10. Y. Linde, A. Buzo, and R. Gray. An algorithm for vector quantizer design. *IEEE Trans. on Communications*, 28(1):84–95, 1980.
11. H. Ney, R. Haeb-Umbach, B. Tran, and M. Oerder. Improvements in beam search for 10000-word continuous speech recognition. In *Proc. Int. Conf. on Acoustics, Speech, and Signal Processing*, volume 1, pages 9–12, San Francisco, 1992.
12. S. Ortmanns, H. Ney, F. Seide, and I. Lindam. A comparison of time conditioned and word conditioned search techniques for large vocabulary speech recognition. In *Proc. Int. Conf. on Spoken Language Processing*, pages 2091–2094, Philadelphia, 1996.
13. E. G. Schukat-Talamazzini. *Automatische Spracherkennung.* Vieweg, Wiesbaden, 1995.
14. V. Steinbiss, H. Ney, X. Aubert, S. Besling, C. Dugast, U. Essen, R. Haeb-Umbach, R. Kneser, H.-G. Meier, M. Oerder, and B.-H. Tran. The Philips research system for continuous-speech recognition. *Philips Journal of Research*, 49(4):317–352, 1996.
15. S. Wachsmuth, G. A. Fink, and G. Sagerer. Integration of parsing and incremental speech recognition. In *Proc. of the European Signal Processing Conference*, volume 1, pages 371–375, Rhodes, 1998.
16. M. Westphal. The use of cepstral means in conversational speech recognition. In *Proc. European Conf. on Speech Communication and Technology*, volume 3, pages 1143–1146, Rhodes, 1997.

Large Vocabulary Speech Recognition for Read and Broadcast Czech*

W. Byrne[1], J. Hajič[2], P. Ircing[3], F. Jelinek[1], S. Khudanpur[1], J. McDonough[1], N. Peterek[2], and J. Psutka[3]

[1] Johns Hopkins University, Baltimore MD, USA
[2] Charles University, Prague, Czech Republic
[3] University of West Bohemia in Pilsen, Czech Republic

Abstract. We describe read speech and broadcast news corpora collected as part of a multi-year international collaboration for the development of large vocabulary speech recognition systems in the Czech language. Initial investigations into language modeling for Czech automatic speech recognition are described and preliminary recognition results on the read speech corpus are presented.

1 Introduction

In this paper we describe data collection efforts to develop an infrastructure for the development of Czech language automatic speech recognition (ASR) systems. Our goal is to gather a rich variety of spoken language material so that Czech language speech processing systems can be developed for research and eventually for other purposes. Our initial collection is a Czech read speech corpus intended primarily for initial and baseline experiments. In addition, Czech language Voice of America (VOA) broadcasts are also being recorded and transcribed. These databases are being collected in the early stages of a three year research program. As such, only preliminary investigations intended to asses the effectiveness of known techniques will be reported here. Our ultimate research plans are more ambitious, however. Unlike English and other languages for which high performance ASR is available, Czech is a highly inflected, relatively free word order language. It has been observed elsewhere that this leads to significant rates of out-of-vocabulary words (OOVs) even with very large language models [1]. It is our goal to address this problem directly by incorporating pronunciation modeling, morphological modeling, and parsing early in the recognition process [2,3,4,5]. The corpora and preliminary experiments described here are initial steps towards these goals.

* This work was supported by the projects KONTAKT No. ME293, No. VS96151 and No. VS97159 of the Ministry of Education in Czech Republic and by NSF Grants No. IIS-9810517 and No. IIS-9820687.

F. Jelinek, E. Nöth (Eds.): TSD'99, LNCS 1692, pp. 235–240, 1999.

2   Czech Language Speech Corpora

2.1   CUCFN Read Speech Corpus

Our initial effort in data collection under this project is the Charles University Financial News Corpus (CUCFN). The first stage of this corpus consists of recordings of read economic news taken from the Ceskomoravsky Profit Journal. Speech read by fluent Czech speakers was recorded in quiet conditions at 22 KHz with 16 bit resolution. Recording were made simultaneously with both a HMD410 Sennheiser head-mounted, close-talking microphone and a desk-mounted K6 Sennheiser microphone. Speech from 29 male speakers and 23 female speakers has been collected and verified. Most subjects were native speakers of common Czech, except for some speakers with marked regional accents from North Moravia and South Moravia. There was also one native Russian speaker and one native Macedonian speaker. Each speaker read a randomly selected section of 100 news sentences as well as 40 common 'enrollment' sentences for use in adaptation and normalization experiments. The first stage of corpus contains a total of 7280 sentences yielding slightly more than 17 hours of speech. On average each utterance contains 17.0 words and has a duration of 9.1 seconds. This read speech corpus is meant to serve as a well-defined environment for use in developing new Czech speech recognition systems.

A second stage of data collection is currently ongoing. General news stories from the Lidové Noviny newspaper [6] read by 50 additional speakers are being recorded.

2.2   CZBN Broadcast News Corpus

We are also collecting a Czech language Broadcast News corpus. Satellite transmissions of Voice of America broadcasts are being recorded by the Linguistic Data Consortium (LDC) and transcribed at the University of West Bohemia according to protocols developed by LDC for use in Broadcast News LVR evaluation. The recordings span the period February 8 through May 4, 1999. The corpus consists of 46 recordings of 30 minute news broadcasts yielding a total of 23 hours of broadcast material. Portions of the shows containing music, speech in music, or other non-speech material are marked, but speech found during those periods is not transcribed. Based on an initial assessment of the transcribed material, this yields approximately 19:30 minutes of transcribed material from each 30 minute broadcast. The average sentence has a duration of 8.6 seconds and contains 18.3 words.

Collection and transcription has begun only recently, hence no ASR systems yet exist. This corpus will be a focus of one of the projects to be undertaken at the 1999 NSF Language Engineering Workshop held at Johns Hopkins University.

3   Czech Language Pronunciation Dictionary

We use the phonetic mappings presented by Psutka [4,5] to derive pronunciations for large vocabulary speech recognition experiments. Our phone set consists of 38 phonemes. Several phones in the original phone set [4] were discarded because the words in which they occur were infrequently observed in the acoustic training data (stage 1 of the CUCFN data). As described [4], pronunciations

Table 1. Modifications to Czech phonetic alphabet.

Description or example	Number of occurrences	Replacement
voiced counterpart of "c"	13	"c"
voiced counterpart of "ch"	10	"ch"
pneumatika	9	"e u"
tramvaj	3	"m"
long "o"	49	"o"

are derived from lexical forms directly from spelling with supplemental rules of transformation. The rules have the form $A \rightarrow B/C\_(D|E)$, which indicates that letter A is transformed to letter B when preceded by C and followed by either D or E. This yields an intermediate representation which is then mapped directly to the phone set. These rules capture coarticulation, context effects, and changes in voicing due to context.

A difficulty in the use of these rules is their treatment of words of non-Czech origin. For example the rule $t \rightarrow t'/ * \_(i|i')$ transforms *ticho* to *t'icho*, and the rule $n \rightarrow \tilde{n}/ * \_(i|i')$ transforms *nic* to *ñic*. However, both these rules when applied to the word *administrativa* yield *admiñistrat'iva*, which should be pronounced as *administrativa*. As these foreign words are observed they are added o a list of exceptions. However pronunciation of non-native words remains a problem.

4   Czech Language Modeling

Statistical language modeling for Czech poses several interesting challenges not encountered in some of the more widely studied languages such as English or French. The usual N–gram approach to language modeling runs into two main hurdles when used for Czech. The highly inflected nature of Czech causes a proliferation of word types if words are viewed as white-space-separated groups of characters without attaching any further meaning to them. At the same time Czech also has a nearly-free word order, further compounding the gathering of good statistics of word collocation.

The highly inflected nature of Czech is clearly evident in the CUCFN corpus described in Section 2.1 and used for experiments reported here. The 16.5 million word language model (LM) training portion of the corpus contains 415 K distinct words. The 64 K most frequent words, a typical vocabulary size for English automatic speech recognition systems, cover only 95 % of the training tokens and the coverage by the 100 K most frequent words increases to only 96.4 %. Of the 415 K distinct words 241 K appear only once or twice in the corpus. Thus a fairly high incidence of out of vocabulary (OOV) words can be expected on any realistic recognition task if each word form is treated as a distinct word in the language model.

A portion of about 19 K words was held out from the CUCFN collection and used as a test set in experiments reported here. It contains about 7 K distinct words, and even for a 415 K word vocabulary extracted form the LM training corpus, we find that 1.6 % of the tokens in the test set are not seen in training! For more realistic recognition vocabulary sizes, this OOV rate is even higher: 30 % for a 5 K vocabulary, 22 % for 10K, 18 % for 15 K, ..., 6.7 % for 63 K. This should be compared with typical OOV rates of 1–2 % for a 64 K–word vocabulary encountered on the English *Wall Street Journal* (WSJ) corpus, and 3–4 % for a 64 K–word vocabulary encountered on the Spanish *Broadcast News* corpus. While a detailed analysis of the incidence of OOVs due to new inflected forms as opposed to, say, new proper nouns *etc.* remains to be performed, Czech does appear to have a higher rate of inflected forms than languages like Spanish.

It is thus clear that a brute-force expansion of the vocabulary is not the ideal approach for Czech language modeling. We are investigating several alternative solutions to this problem, including class based language models using automatically derived morphological word classes, in our ongoing work. In this paper, however, we only report results with word N–gram language models using either a small vocabulary of 5 K words for pilot experiments or a large vocabulary of 63 K words for more realistic evaluations.

The small (5 K word vocabulary) speech recognition system was built primarily as a development step in order to validate the proper engineering of component modules such as the acoustic models, the dictionary and the language model. To circumvent the 30 % OOV rate of the 5 K–word vocabulary for the pilot experiments, which could make it difficult to pinpoint faulty modules in case of high recognition error rates, it was decided to close the LM vocabulary on the test set. All (4500) words in the test set reference transcriptions which did not appear in the list of 5 K most frequent words in the training corpus were added to the 5 K word vocabulary, resulting in a 9.5 K-word vocabulary.

A bigram and a trigram language model were trained for the pilot experiments with the 9.5 K word vocabulary and the 16.5 million word corpus. A bigram and a trigram language model were also estimated for the large vocabulary experiments with the 63 K word vocabulary from the same corpus.

N–gram Order	9.5 K Vocab (0 % OOVs)	63 K Vocab (6.7 % OOVs)
1gram	5895	2579
2gram	1829	737.5
3gram	1664	657.4

Table 2. Test set perplexity for various vocabulary sizes and N–gram order

Table 2 shows the perplexity of the test set under various language models. Note that the perplexity of the English WSJ corpus, which is comparable in terms of genre (newspaper text) and content (financial news), is in the range of 180–220 for a bigram model and 110-160 for a trigram model, depending on vocabulary size, amount of training text, *etc.*

The relatively high perplexity of the models for the 9.5 K vocabulary perhaps merits some explanation: when we close the 5 K vocabulary by including 4500 additional words which occur in the test set, we end up estimating a model which "knows" that these words are infrequent even though they have made an appearance in a word list otherwise comprising the most frequent words. As a result, they often get an even smaller probability than what a *uniform distribution* would assign to a set of 9.5 K words.

5   Acoustic Modeling

Initial large vocabulary speech recognition experiments have been conducted on the CUCFN corpus using HTK, the hidden Markov model toolkit [7]. The recognition system is based on a continuous density HMM trained on approximately 10 hours of read speech from 22 speakers taken from a subset of the first stage of the CUCFN corpus. The speech features parameterization employed in training and test are mel-frequency cepstra, including both delta and delta-delta sub-features; cepstral mean subtraction is applied to all features on a per utterance basis. Triphone state clustering was carried out using broad acoustic phonetic classes similar to those used for English. The final model comprises approximately 2,800 state clusters each with 12 mixture Gaussian observation distributions.

5.1   CUCFN Development Test Set Results

A portion of the first stage of the CUCFN data corpus was set aside as a development test set. The test set contained speech by 7 male and 7 female speakers. The test set consisted of 1399 sentences containing 19376 words in total.

The 9.5 K word bigram language model of Section 4 was used for initial recognition experiments on this test set. Results are reported in Table 3. As discussed earlier, this was done only as initial validation of the model and is reported here for interest only.

A second experiment was performed with a fair language model. The 63 K word bigram model of Section 4 was used in an initial recognition pass. Lattices produced using the ATT large vocabulary decoder [8] were then used with the 63 K word trigram for further acoustic rescoring. As can be seen in comparison with the initial language model cheating experiment, the increased language model size and the OOVs significantly reduce accuracy. Overall we found that performance was relatively consistent across speakers with the exception of the worst subject, a female native speaker of Russian.

Table 3. Word error rates on the CUCFN development test set.

	Language Model	OOV Rate	Word Accuracy
Type	Vocabulary Size (words)	(%)	(%)
2gram	9,500	0.0	78.66
2gram	63,000	6.7	69.88
3gram	63,000	6.7	71.33

6   Conclusion

We have described a project for the collection of spoken Czech language databases and have presented results of preliminary language modeling and speech recognition experiments. These experiments suggest that well-known acoustic modeling techniques should work well for large vocabulary Czech language speech recognition. However the relatively free word order of Czech and number of inflected word forms pose difficulties for language modeling.

Acknowledgement. Satellite news broadcast recordings were done under contract by the Linguistic Data Consortium, Philadelphia, PA, USA. We gratefully acknowledge use of the large vocabulary decoder provided by M. Riley and F. Pereira of ATT and the use of the SRI Language Modeling Toolkit provided by A. Stolcke of SRI. Language modeling data has been provided by the Lidové Noviny Publishers, Prague, Czech Republic.

References

1. P. Geutner, M. Finke, P. Scheytt: Adaptive Vocabularies for Transcribing Multilingual Broadcast News. ICASSP. 1998.
2. J. Hajič: Morphology Unification Grammar. PhD Thesis. Charles University. Prague. 1984.
3. J. Hajič: Building a Syntactically Annotated Corpus: The Prague Dependency Treebank. In: Issues of Valency and Meaning. Karolinum Press. Charles University. Prague. 1998. p. 106-132.
4. J. Psutka: Communication with Computer by Speech. Academia, Prague 1995, 287pp. (in Czech)
5. J. Nouza, J. Psutka, J. Uhlíř: Phonetic Alphabet for the Recognition of Czech. Radioengineering, vol.6, no.4, pp.16-20, 1997.
6. http://www.lidovenoviny.cz
7. S. Young et al.: The HTK Book. Entropic Inc. 1999
8. M. Mohri, M. Riley, D. Hindle, A. Ljolje, F. Pereira: Full Expansion of Context-Dependent Networks in Large Vocabulary Speech Recognition. ICASSP 1998.

Rules for Automatic Grapheme-to-Allophone Transcription in Slovene

Jerneja Gros, France Mihelič, and Nikola Pavešić

University of Ljubljana, Faculty od Electrical Engineering,
Tržaška 25, SI-1000 Ljubljana, Slovenia
nejka@fe.uni-lj.si,
http://luz.fe.uni-lj.si

Abstract. An extensive rule set for grapheme-to-allophone conversion of Slovene texts has been defined and evaluated. Another rule set has been developed for pronunciation of names. The efficiency of both *S5* rule sets was compared to the one of the *Onomastica* rule set on two manually transcribed test data sets.

A performance test applied on the *S5* pronunciation dictionary showed error rates of about 30 % in the stress assignment and consequently in the phonetic transcription. In case stress assignment and the transcriptions of graphemic /e/ and /o/ in stressed syllables had been marked in advance a transcription success rate of nearly 100 % was achieved both on names and on standard words with the *S5 names* rule sets and the *S5 standard words* rule set, respectively.

1 Introduction

One of the important aspects in text-to-speech synthesis is the grapheme-to-phoneme conversion of the input text. Although there are many different approaches to this conversion, this contribution focuses on a combined use of pronunciation dictionary and context dependent pronunciation rules in the *S5* Slovene text-to-speech system (SQEL Slovene Text-to-Speech System) [1].

Input to the *S5* text-to-speech system is arbitrary Slovene text. It is translated into a series of allophones in two consecutive steps.

First, input text normalization is performed. The normalized text is passed to the pronunciation module, which uses a range of dictionary and rule based methods to assign a pronunciation to each word. These methods are arranged in order of accuracy, with dictionary lookup as the most accurate. They are discussed in detail and evaluated in the following sections.

2 Pronunciation Dictionary

The *S5* pronunciation dictionary was originally developed with the intention of capturing as large a percentage of words in continuous running text as possible. This was achieved by a statistical analysis of many different types of Slovene

F. Jelinek, E. Nöth (Eds.): TSD'99, LNCS 1692, pp. 241–247, 1999.

texts, mainly fiction, newspaper and journal articles (in total nearly 1.000.000 words).

It was found that that about 16.000 most frequent words accounted for approaching 90% of all words used in the text corpus [1]. These words were used as the core of the dictionary. They were automatically transcribed by the $S5$ grapheme-to-allophone rules. Later their pronunciation was manually corrected.

The $S5$ pronunciation dictionary covers over 16.000 most frequent inflected Slovene word forms, collocations and proper names. For words that are pronounced differently for different parts of speech the most frequent pronunciation is chosen.

3   Lexical Stress Assignment

In case where dictionary derivation fails, input words must be transcribed automatically. First, the lexical stress position has to be determined, a quite frustrating task given that in Slovene lexical stress can be located almost on any syllable obeying no specific rules. The stressed syllable in Slovene may form the ultimate, the penultimate or the preantepenultimate syllable of a polysyllabic word.

The automatic stress assignment algorithm we developed is to a large extent determined by (un)stressable affixes, prefixes and suffixes of morphs, based upon observations of linguists [3]. For words which do not belong to these categories, the most probable stressed syllable is predicted using the results we obtained by a statistical analysis of stress position depending on the number of syllables within a word [1].

The analysis was performed on the $S5$ pronunciation dictionary where lexical stress positions had been manually inserted. According to the results of the statistical analysis disyllabic, trisyllabic and quadrosyllabic words are most likely to be stressed on the penultimate syllable.

4   Grapheme-to-Allophone Rules

Finally, a collection of over 150 context-sensitive and context-free grapheme-to-allophone rules translates each grapheme string into a series of allophones.

The rules are accessed sequentially until a rule that satisfies the current part of the input string is found. The transformation defined by that rule is then performed, and a pointer is incremented to point at the next unprocessed part of the input string, and so on until the whole string has been converted.

The context free rules are rare and they include a one- to-one correspondence, two-to-one correspondence and one-to-two correspondence.

Each context-sensitive rule consists of four parts: the left context, the string to be transcribed, its right context and the phonetic transcription. A number of writing conventions has been adopted in order to keep the number of rules relatively small and readable. The left and the right context may contain wild characters describing larger phonetic sets, e.g. '$' stands for consonants, - for white spaces.

The rules for consonants are rather straightforward, while those for vowels must handle vowel length and the variant realizations of the orthographic /e/ and the orthographic /o/ in stressed syllables. Typical context-sensitive rules have the following structure:

left context	grapheme string	right context	phonetic string
$	/er/	-	[@r]
=	/n/	k	[N]

The above rules state that the word final /er/ preceded by a consonant is transcribed as [@r] (e.g. /gaber/ into [*ga:.b@r]), and that any /n/ followed /k/ is transcribed into [N] ([N] is the allophone of [n] when followed by /k/ or /g/, e.g. in /anka/ – [*a:N.ka]).

Our initial rule set produced in 1995 [4] was based on various observations of expert linguists, e.g. [3] and other basic rule sets for Slovene grapheme-to-allophone transcription [5,6].

The initial set of rules has been undergoing continuous refinement ever since and resulted 169 rules of the *S5 standard words* rule set. Rules for pronunciation corrections of words according to the left and to the right context of the word were also added.

In the recent years, telecommunication applications of speech synthesis have increased in importance, e.g. automatic telephone directory inquiry systems. Names of locations (cities, streets, etc.) and proper names cannot be mentally reconstructed from the context when listening to the messages, correct name pronunciation is required [7]. *S5 standard words* rules developed for a standard Slovene vocabulary do not lead to satisfactory results when applied on names. Therefore, additional 'name-specific' rules were added to the *S5 standard words* rule set resulting in the *S5 names* rule set.

5 Evaluation

One way of insuring accurate pronunciation for many telephone applications is to create huge dictionaries containing all the names needed. This was one of the motivations behind the LRE and COP European funded projects *Onomastica* that aimed to produce machine- readable phonetic dictionaries for names in various European languages.

In 1997, the COP Onomastica CD-ROM was published and distributed by ELRA [8]. The CD-ROM contains manual transcriptions of 283.449 Slovene names marked by an expert phonetician (all of them Quality Band I) and also a set of 125 rules for Slovene grapheme-to-allophone transcription.

In our test we have compared the transcription efficiency of three rule sets:

– the *Onomastica* rule set (125 rules),
– the *S5 standard words* rule set (169 rules) and

- the *S5 names* rule set (174 rules)

   on two manually transcribed pronunciation dictionaries:

- the *Onomastica* pronunciation dictionary consisting of 233.919 names; for names with multiple pronunciations only the more probable one (the first one) was taken; company names were excluded from the test input data,
- the *S5* pronunciation dictionary consisting of 14.640 most frequent Slovene words.

On the *Onomastica* pronunciation dictionary the *Onomastica* rule set and the *S5 names* rule set were compared. Since the *Onomastica* rule set must be adapted to pronunciation of names it was not entirely fair to compare its performance with the *S5 standard words* rule set on the second pronunciation dictionary. We would have taken the 88 rules set for standard words mentioned in [9] if it were publicly available.

5.1   Data Preparation

The *Onomastica* rule set and pronunciation dictionary use the broad phonetic level of transcription as opposed to the narrower one applied in the *S5* case. Therefore, the *S5* rule sets and the *S5* pronunciation dictionary had to be generalized to larger phonetic groups.

The Slovene *Onomastica* rule set does not contain any rules on stress mark assignment. Furthermore, the rules for graphemic /e/ and /o/ are rather vague when it comes to [e] and [E] or [o] and [O] disambiguation in a stressed syllable [8]:

<table>
<tr><td>

```
*       E       *
1
e(e2)
*       E       *
1
E(e3)
```
</td><td>

Does the first rule apply to a /E/ in a stressed syllable? Not all /E/ in stressed syllables are transcribed as [e], many often are transformed in to [E].

As reported in [9], the stress mark position was inserted manually. We assume that the /e/ and /o/ in stressed syllables were transcribed by hand as well.
</td></tr>
</table>

Therefore, in the evaluation only names where stress position was automatically correctly detected by our stress assignment rules were used. [O]-[o] and [E]-[e] transcriptions were regarded as equal due to the lack of disambiguation capabilities in the *Onomastica* rule set.

The resulting sizes of both test data sets were:

- *Onomastica* pronunciation dictionary: 68.384 words,
- *S5* pronunciation dictionary: 8.480 words.

6   Results

The phonemisation errors were determined by comparing the automatic transcription outputs with the manual pronunciation dictionary transcriptions.

An initial count of the results on the *Onomastica* pronunciation dictionary showed that 996 of the 68.384 names were incorrectly output using the *S5 names*

Table 1. Number of false transcriptions and transcription success rate on the *Onomastica* pronunciation dictionary

rule set	No. of false transcriptions	transcription success rate [%]
Onomastica	9051	86.8
S5 names all errors	996	98.5
S5 names revised errors	20	99.9

rules set, i.e. a success rate of 98.5% (Table 1). A closer examination of the errors revealed that they could be grouped into various categories.

There were 919 words among the 996 errors where word final grapheme /l/ was manually transcribed as [w] (/nikol/ – [ni.*ko:w], /stil/ – [*sti:w] or /gril/ – [*gri:w]). We agree with these transcriptions in only 15 cases (/stol/ – [s*tO:w]). Everywhere else final /l/ should be transcribed as [l].

In 68 cases the diphthongs /ai/ or /ei/ were manually transcribed as [ai][Ei] and as [aj][Ej] by the *S5 names* rules, which is just a matter of transcribing convention. In 5 cases /ue/ was manually transcribed as [wE] and as [ue] by the *S5 names* rules. The double consonant /kk/ was manually transcribed as [kk] 8 times (/brankka/ – [b*ra:nk.ka]) and as [k] by the *S5 names* rules, which is closer to the real pronunciation.

Given this revised interpretation of the data, the results show only 20 errors out of 68.384 input words, which is a 99.6% success rate. When transcribing the *S5* pronunciation dictionary using the *S5 standard words* rule set only 3 words were transcribed incorrectly (Table 2), i.e. a 99.9% success rate.

These raw results do not, however, properly reflect the accuracy of the transcription rules since it has been assumed that lexical stress assignment and [e]/[E] and [o]/[O] disambiguation have been done manually. Therefore, the *S5 standard words* rule set was tested once again on the *S5* pronunciation dictionary, this time with the full allophone set.

For 70.6% of the test words the generated allophone sequence matched the manual transcription exactly. In 15.1% of the words, the stressed syllable was incorrectly predicted which resulted in an incorrect phonetic transcription. In the rest of the text, [e]/[E] and [o]/[O]) were confused in 12.2% of the words; other pronunciation errors were found in 2.1% of the rest of the test set.

Table 2. Number of false transcriptions and transcription success rate on the *S5* pronunciation dictionary

rule set	No. of false transcriptions	transcription success rate [%]
Onomastica	2139	74.8
S5 standard words	3	99.6

Provided over 80% of the input text can be found in the pronunciation lexicon, only 5% of the words would be processed incorrectly. In other words, in an average text of 1.000 words (including of course many repetitions of the more common words) 950 words would be expected to be correct assuming that the obtained corpus frequencies can be said to reasonably predict the actual frequencies encountered by a text-to-speech system.

7   Conclusion

An extensive rule set for grapheme-to-allophone conversion of Slovene texts has been defined and evaluated. Another rule set has been developed for pronunciation of names. Rules for stress position prediction are described as well.

A performance test applied on the *S5* pronunciation dictionary showed error rates of about 30% in the stress assignment and consequently in the phonetic transcription. In case stress assignment and the transcriptions of graphemic /e/ and /o/ in stressed syllables had been marked in advance a transcription success rate of nearly 100% was achieved both on names and on standard words with the *S5 names* rule set and the *S5 standard words* rule set, respectively. The *Onomastica* rule set produced a lower transcription success rate.

A slight improvement of stressed /e/ and /o/ transcription may be obtained by using statistical or pronunciation-by-analogy approaches [10] combined with a suffix stripper proposed in [11]. However, there are innumerable cases where also this would fail, e.g. as for Slovene words /leto/,/teto/,/četo/,/čelo/ and /pelo/, with the transcriptions [*le:.tO], [*tE:.tO], [*tSe:.tO], [*tSE:.lO] and [*pe:.lO]. The only way to ensure proper word transcription in Slovene is to use the derivative morphology approach based on a lemmatized dictionary of canonic Slovene forms along with all inflection and exception rules.

References

1. Gros, J.: Samodejno pretvarjanje besedil v govor. PhD Thesis. University of Ljubljana (1997).
2. Lindstrom, A., Ljungquvist, M., Gustafson, K.: A modular architecture supporting multiple hypotheses for conversion of text to phonetic and linguistic entities. Proceedings of the EUROSPEECH93. Berlin (1993) 1463–1466.
3. Toporišič, J.: Slovenska Slovenica (Slovene Grammar). Založba Obzorja Maribor (1991) (in Slovene).
4. Gros, J., Pavešić, N., Mihelič, F.: A text- to-speech system for the Slovenian language. Proceedings of the EUSIPCO'96, Trieste (1996) 1043–1046.
5. Hribar, J.: Sinteza umetnega govora iz teksta. MSc Thesis. University of Ljubljana (1984).
6. Weilguny, S.: Grafemsko-fonsmeki modul za sintezo izoliranih besed slovenskega jezika. MSc Thesis. University of Ljubljana (1993).
7. Belhoula, K., Kraft, V., Rinscheid, A., Ruehl, H.W.: Extension of a TTS system to rule-based pronunciation of names. Proceedings of the CRIM/FORWISS Workshop on Progress and Prospects of Speech Research and Technology. Munich (1994) 249–251.

8. ONOMASTICA-Copernicus Database. CD-ROM. EU Project COP-58. Distributed by the European Language Resources Association. ELRA (1997).
9. Kačič, Z.: Definiranje leksikona izgovarjav lastnih imen za slovenski jezik. Proceedings of the ERK'98 Conference. Portorož Slovenia (1998) 185–188.
10. Dedina, M.J., Nusbaum, H.C.: Pronounce: a program for pronunciation by analogy. Computer Speech and Language **5** (1991) 55–64.
11. Mannell R., Clark, J.E.: Text-to-speech rule and dictionary development. Speech Communication **6** (1987) 317–324.

Speech Segmentation Aspects of Phone Transition Acoustical Modelling

Simon Dobrišek, France Mihelič, and Nikola Pavešić

University of Ljubljana, Faculty of Electrical Engineering,
Laboratory of Artificial Perception, Tržaška 25, SI-1000 Ljubljana, Slovenia,
simond@fe.uni-lj.si,
http://luz.fe.uni-lj.si/

Abstract. The paper presents our experiences with the phone transition acoustical models. The phone transition models were compared to the traditional context dependent phone models. We put special attention on the speech signal segmentation analysis to provide a better insight into certain segmentation effects when using the different acoustical models. Experiments with the HMM-based models were performed using the HTK toolkit, which was extended to provide proper state parameter tying for the phone transition models. All the model parameters were estimated on the GOPOLIS speech database. The annotation confusions concerning two-phone speech units are also discussed.

1 Introduction

A training process of an HMM-based speech recognition system incorporates an alignment of a speech model with a speech signal. The alignment results in a certain segmentation of the speech signal. Speech model parameters are usually estimated using an iterative training algorithm. The new estimation of the model parameters is therefore strongly affected by the segmentation obtained by the current model parameters.

If the acoustical modelling of speech is based on the phone models then even a slight change of the segmentation can change the estimation of the model parameters considerably. This happens due to the positions of the segment borders which are placed in the non-stationary transition parts of the signal. On the other hand, if the acoustical modelling of speech is based on the phone transition models, the segment borders are expected to be placed in the relatively stationary signal regions, and therefore the model parameters should not be affected that much by the change of segmentation.

We decided to investigate the differences in the acoustical modelling using traditional phone models and phone transition models.

2 Segmentation Analysis

A speech signal segmentation produced by a given acoustical model characterises this model in comparison the other models. A question that arises here is how to compare different signal segmentations and how to present the analysis results.

F. Jelinek, E. Nöth (Eds.): TSD'99, LNCS 1692, pp. 248–251, 1999.

One possible solution is to extend the well-known problem of alignment of two strings of symbols to the problem of alignment of two sequences of labelled signal segments. We propose a variant of the string edit distance where the primitive cost function is composed of the primitive edit cost function and the additional distance function of a pair of signal segments. One of the most obvious distance functions to choose is

$$d_s(s1, s2) = \frac{|t_{b1} - t_{b2}| + |t_{e1} - t_{e2}|}{(t_{e1} - t_{b1}) + (t_{e2} - t_{b2})},$$

where t_{bi} and t_{ei} assign the beginning and end time $(t_{ei} > t_{bi})$ of the two signal segments $(i = 1, 2)$. If the returned value is below 1.0 then the two segments $(s1, s2)$ overlap at least for a short period of time.

The proposed variant of the primitive cost function returns the total number of edit operations at the optimal alignment which is comparable to the Levenshtein distance. It can be shown that the mentioned segmental-based string edit distance gives phonetically more consistent recognition score statistics, when comparing it to the traditional Levenshtein distance based speech recogniser assessment.

For all the pairs of segments in the segmental-based string edit distance which are declared to be a match or a substition, we can additionally define a function f_s which returns some information about how much the two segments are shifted relatively to each other. An example of such a function would be

$$f_s(s1, s2) = \frac{t_{b1} - t_{b2} + t_{e1} - t_{e2}}{(t_{e1} - t_{b1}) + (t_{e2} - t_{b2})}.$$

The presented segmental-based string edit distance in combination with the above function f_s provides a useful tool for obtaining some interesting speech signal segmentation statistics. All the segmentation histograms in the paper were generated using the described approach.

3   Speech Unit Annotation Confusions

Biphones and diphones are both two-phone speech units which have the ability to capture co-articulatory effects. Biphones are understood to be just left or right context dependent (mono)phones. On the other hand, diphones represent the transition regions that strech between the two "centres" of the subsequent phones.

These two speech units are obviously different when we consider them together with the speech signal segmentation. The biphone acoustical models should produce similar signal segmentation as the (mono)phone acoustical models. On the other hand, the segmentation produced by the diphone models is expected to be shifted in time when we compare it to the segmentation produced by the biphone models.

If we observe biphones and diphones only from the aspect of phonetic transcriptions, without considering any signal segmentation, than it can be easily shown that the difference between these two speech units is not that obvious any more.

For an example, let us observe the phonetical transcriptions of an isolated spoken command *Left!*. The possible different canonical transcriptions of the uttered command are the following:

	established annotation	optional annotation
left biphones:	sil sil-l l-eh eh-f f-t t-sil	sil sil~l l~eh eh~f f~t t~sil
right biphones:	sil+l l+eh eh+f f+t t+sil sil	sil~l l~eh eh~f f~t t~sil sil
diphones:		sil~l l~eh eh~f f~t t~sil

Without changing the definition of biphones the symbols '-' and '+' can be replaced by any other symbol, even by the symbol '~', which we use to annotate the transition nature of diphones. Let us assume that we allow, if required, splitting of the beginning or end silence segment into two parts that can be associated with the two subsequent speech units. It can be seen that all the above transcriptions that use the symbol '~' can be associated with signal segmentations which have a diphone transition nature.

4   Acoustical Modelling

Continuing the discussion from the previous section an interesting question arises about what signal segmentation can we expect when dealing with biphone and diphone acoustical models, respectively.

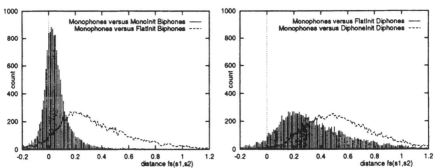

Fig. 1. The biphone and diphone segmentation histograms.

Several experiments were performed using the GOPOLIS speech database [2] and the HTK toolkit [4] which was extended to allow proper state parameter tying for the phone transition models as well. All the HMM models were built using the traditional approach.

First, we observed the differences in the signal segmentation produced by the monophone initialised biphone models and the flat started biphone models. After the initial training session which took approximately the same number of iterations through the whole available training speech data for both differently initialised models with approximately the same number of parameters, the segmentation statistics was generated. The manual signal segmentation was compared to the segmentation produced after the forced alignment procedure.

From the histograms in Figure 1 it can be seen that the monophone initialised biphone models tend to preserve the segmentation given by the monophone models. However, the flat started biphone models produce a signal segmentation which is of a rather phone transition nature. Therefore one will conclude that the flat started biphone models are actually diphone models.

The diphone models were initialised from the diphone inventory of the first Slovenian text-to-speech system [3]. The flat started diphone models performed similary to the flat started biphone models.

The comparison between the diphone and biphone models was extended to the comparison between the bi-diphone and triphone models, where bi-diphones are just context dependent diphones. We encountered similar segmentation effect and confusion problems with the symbolical representation of these two speech units.

All the acoustical models mentioned in the paper were also incorporated into a continuously spoken word recogniser [1]. An interesting observation was that the diphone models had achieved considerably higher average log likelihood values per signal frame even when compared to the triphone models. Both models had approximately the same number of parameters. That was a good sign that indicated a high word recognition score. And the score was indeed higher for the diphone models [1].

5   Conclusions

The main conclusion would be that all even-numbered-phone speech unit models tend to produce signal segmentations which correspond to the transition signal regions. On the other hand, the monophone and odd-numbered-phone speech unit models tend to produce segmentations which correspond to the phone nuclei signal regions. This means that it is probably not wise to freely combine model parameters of these different speech units.

References

1. Dobrišek, S. (1999). Analysis and Recognition of Phones in Speech Signals. *Ph.D. Thesis in preparation*, (In Slovenian). University of Ljubljana, Faculty of Electrical Engineering, Ljubljana Slovenia.
2. Dobrišek, S., Gros, J., Mihelič, F., and Pavešić, N. (1998), Recording and labelling of the GOPOLIS Slovenian speech database. *Proc. 1st Int. Conf. on Language Resources & Evaluation*, Vol. 2, ESCA, pp. 1089–1096.
3. Gros, J., Pavešić, N., Mihelič, F. (1997), Text-to-Speech Synthesis: A Complete System for the Slovenian Language. *Jurnal of Computing and Information Technology*, Vol. 5(1), pp. 11–19.
4. Young, S., Odell, J., Ollason, D., Vatchev, V., and Woodland, P. (1997), *The HTK Book*. Cambridge University, Entropic Cambridge Research Laboratory Ltd.

Context Dependent Phoneme Recognition

Dušan Krokavec

Technical University of Košice,
Faculty of Electrical Engineering and Informatics,
Department of Cybernetics and Artificial Intelligence
Letná 9/B, SK 042 00 Košice, Slovak Republic
krokavec@ccsun.tuke.sk

Abstract. The paper present one way of contextual information incorporation into continuous speech recognition systems and an application of the neural network (multilayer perceptron) to context dependent phoneme recognition. An overview of the problem is given and a survey of these techniques considered from the point of training set creation and the hybrid system interconnection is presented.

1 Introduction

In general, in a continuous speech recognition systems it is necessary to use subword models and a pronunciation dictionary for each item in the vocabulary. By specifying the pronunciation of a word in terms of a string of phonemes the task is reduced to the phoneme string probability estimation and the searching of the most probable word string from all possible phoneme strings. The most popular tools for this task are hidden Markov models (HMM).

The phonemes are dependent on contextual variables such as speaking rate, intonation, co-articulation, i.e. on the phonetic and acoustic context. To achieve speech recognition at the highest levels of performance means making efficient use of contextual information. The use of multilayer perceptrons allows a large window of speech signal to be used directly for context dependent phoneme recognition.

Multilayer perceptrons (MLP) with enough hidden units can in principle provide arbitrary mappings between network input and output. The elements of the network weights (the parameter set Θ) can be trained to associate a desired output vector with an input vector generally via the error back–propagation algorithm, using a steepest descent procedure to iterative minimize a cost function in their parameter space. To advantages, that make them particularly attractive for automatic speech recognition, belong:

- The parameters of the output classes of the MLP (representing speech units or hidden Markov model outputs) are trained to optimize each class parameters while maximizing the discrimination between the correct class and the rival ones (discriminant learning) and to produce the *posterior* probability class given the acoustic data. This is in contrast to the likelihood criterion used for HMMs parameter learning.

F. Jelinek, E. Nöth (Eds.): TSD'99, LNCS 1692, pp. 252–257, 1999.

– Neural network can incorporate and find optimal combination of multiple constrains for classification as well as contextual dependent input frame sets.

While purely connecting input/output formalism of neural networks is not very well suited to solve the sequence mapping problem in speech recognition, HMMs provide a reasonable structure for representing sequence of speech units. From this point of view one good application for MLP is to provide the local distance measure for HMM.

2   Phoneme Posterior Probability

The task in classification problem is to assign an input vector $\mathbf{x}$ to one of M classes $\{C_i, i = 1, 2, \ldots, M\}$. In stochastic classification the properties of the classes are characterized by the conditional densities $f(\mathbf{x}|C_i)$, $i = 1, 2, \ldots, M$ and the classes occur with known a priori probability $P(C_j)$, $i = 1, 2, \ldots, M$.

The cost function using in neural network phoneme classification was the squared-error

$$\sum_{i=1}^{M} [y_i(\mathbf{x}) - d_i)]^2 \tag{1}$$

where $y_i(\mathbf{x})$, $i = 1, 2, \ldots, M$ are outputs of the neural network and d_i, $i = 1, 2, \ldots, M$ are desired outputs.

The cost expectation over the all training set is

$$\int_{\Omega} \sum_{j=1}^{M} \sum_{i=1}^{M} [y_i(\mathbf{x}) - d_i]^2 f(C_j|\mathbf{x}) f(\mathbf{x}) d\mathbf{x} = E\{\sum_{j=1}^{M} \sum_{i=1}^{M} [y_i(\mathbf{x}) - d_i]^2 f(C_j|\mathbf{x})\}, \tag{2}$$

where $E\{.\}$ is the expectation operator. Expanding the summation of bracketed expression yields

$$\sum_{j=1}^{M} [y_i(\mathbf{x}) - d_i]^2 f(C_j|\mathbf{x}) =$$
$$= y_i^2(\mathbf{x}) \sum_{j=1}^{M} f(C_j|\mathbf{x}) - 2y_i(\mathbf{x}) \sum_{j=1}^{M} d_i f(C_j|\mathbf{x}) + \sum_{j=1}^{M} d_i^2 f(C_j|\mathbf{x}) = \tag{3}$$
$$= y_i^2(\mathbf{x}) - 2y_i(\mathbf{x}) E\{d_i|\mathbf{x}\} + E\{d_i^2|\mathbf{x}\},$$

where $E\{d_i|\mathbf{x}\}$ and $E\{d_i^2|\mathbf{x}\}$ are the conditional expectation of d_i, and d_i^2, respectively. The cost expectation is than

$$E\{\sum_{j=1}^{M} \sum_{i=1}^{M} [y_i(\mathbf{x}) - d_i]^2 f(C_j|\mathbf{x})\} =$$
$$= E\{\sum_{i=1}^{M} [y_i^2(\mathbf{x}) - 2y_i(\mathbf{x}) E\{d_i|\mathbf{x}\} + E\{d_i^2|\mathbf{x}\}]\} =$$
$$= E\{\sum_{i=1}^{M} [y_i(\mathbf{x}) - E\{d_i|\mathbf{x}\}]^2\} + E\{\sum_{i=1}^{M} [E\{d_i^2|\mathbf{x}\} - E^2\{d_i|\mathbf{x}\}]\} = \tag{4}$$
$$= E\{\sum_{i=1}^{M} [y_i(\mathbf{x}) - E\{d_i|\mathbf{x}\}]^2\} + E\{\sum_{i=1}^{M} D\{d_i|\mathbf{x}\}\},$$

where $D\{d_i|\mathbf{x}\}$ is the conditional variance of d_i. Since the second expectation term is independent of the network output, and the first expectation term is the mean-squared-error between the pair of the network output and the conditional expectation of desired output, the squared-error cost function minimization by the network estimate the conditional expectation of desired outputs

$$E\{d_i|\mathbf{x}\} = E\{\sum_{j=1}^{M} d_i f(C_j|\mathbf{x})\} = f(C_i|\mathbf{x}), \tag{5}$$

which are the Bayesian probabilities.

3   Scaled Likelihoods

If the multilayer perceptron input is provided with acoustic context

$$\mathbf{x}_l = \mathbf{u}_{l-H}^{l+H} =_{l-H}, \mathbf{u}_{l-H+1}, \ldots, \mathbf{u}_l, \ldots, \mathbf{u}_{l+H-1}, \mathbf{u}_{l+H} \tag{6}$$

the output values will estimate the probability distribution over classes conditioned on the input acoustic context frame set

$$g_j(\mathbf{u}_l, \Theta) = P(C_j|\mathbf{x}_l), \quad \forall j. \tag{7}$$

Since the network outputs approximate Bayesian probabilities, $g_j(\mathbf{u}_l, \Theta)$ is an estimate of

$$P(C_j|\mathbf{x}_l) = \frac{P(\mathbf{x}_l|C_j)P(C_j)}{P(\mathbf{x}_l)} \tag{8}$$

which implicitly contains the a priori class probability $P(C_j)$.

In an neural network it is generally not possible to make the output probability constrain equal to one by locally normalization. Scaled likelihoods $P(\mathbf{x}_l|C_j)$ for use as emission probabilities in hidden Markov models during decoding can be obtained by dividing the probability distribution over classes conditioned on the input acoustic context frame set (output of the network) by the a priori class probabilities $P(C_j)$, which leads to a formulae

$$\frac{P(C_j|\mathbf{x}_l)}{P(C_j)} = \frac{P(\mathbf{x}_l|C_j)}{P(\mathbf{x}_l)} \tag{9}$$

The a priori probabilities are estimated by the relative frequency of phonemes as observed in the training set.

During recognition, the scaling factor $p(\mathbf{x}_l)$ is a constant for all classes (do not change the classification) and during classification a priori class probability is a multiplicative factor (it is possible to vary them without retraining).

4   Complete Labeled Training Set

In the basic approach the neural network is decoupled from the hidden Markov model during training, i.e. the HMM and MPL are only combine for recognition. MPL is trained separately to classify frames of speech into a set of phonemes. After training the MPL estimates the posterior probability of the different phonemes $P(C_j|\mathbf{x}_l)$, given a symmetric context inputs $\mathbf{x}_l$.

The problem of this approach is that the complete labeling (i.e. the phoneme segmentation) for a training set is required in order to train the network. In speech recognition the complete labeling corresponds to knowing which phoneme each particular observation $\mathbf{u}_l$ is associated with. In this case each observation has a label and sequence of labels is as long as the observation sequence.

When dealing with incomplete labeling (i.e. the phonetic transcription) the sequence of observation is associated with a shorter sequence of labels and the label of each individual observation is unknown. In continuos speech recognition it corresponds to unknown time boundaries between the phonemes while correct string of phonemes is known (spoken words are known in the training set).

If only the incomplete labeling is available, the iterative procedure using current network and current parameter set of HMM must be used to obtain the complete labeled training set. First step iteration of the complete labeling is obtained by performing a Viterbi alignment for each training sequence based on the current results of neural network and HMM. This iteration is used for training the networks as a desired labeling in the second step. This procedure is repeated until the labeling does not change anymore. The initial labeling for network training is output of the HMM for the incomplete labeled training set.

5   Experimental Setup

The context dependent phoneme recognition was realized in the configuration of a match neural network for the 36 phoneme task and the network was fully connected MLP with sigmoid output functions, and one hidden layer with 200 hidden units. The cost function for classification was the squared–error expectation between actual outputs and desired output of the network, the elements of the network weights was setup via the error back–propagation algorithm, using a steepest descent procedure to a cost function space.

It is known that a symmetric context input increases the accuracy substantial only for models without hidden units [11]. When using hidden units in the match network the accuracy increases, however, at the cost of a much larger number network parameters. Contrary to the models without the hidden layer a larger context than two frames does not improve significantly the global recognition properties for network with hidden units but local accuracy (for individual phonemes) can be increased using up to four context frames. From this point of view the input layer was adapted for symmetric context 1 up to 4 frames, where in speech segmentation a 24-dimensional feature vector based on the Mel–cepstrum and their derivatives was preferred, and the standard resolution (16 bits) and sampling frequency (16 kHz) have been used within the 10 ms frame duration.

The observation vectors were normalized (to zero mean and unit variance) in order to speed up training of neural network. For the discrete HMM there was used a codebook of 186 prototype feature vectors.

The incomplete labeled training set was derived from the corpus of 5400 domain depending training sentence from 100 speakers, developed in the the

COPERNICUS 94 Project No. 1634 SQEL - Spoken Queries in European Languages [3,8], where one sentence from each of the speakers was used. Using the ISADORA software package for acoustic modeling there were determined the parameters of phoneme hidden Markov models and initial labeling for network training. The above mentioned iterative procedure was used to obtain the complete labeled training set.

The results are reported in the form of the recommended TIMIT test set. The Table 1 shows how the total number of errors is distributed on phoneme substitutions, insertions and deletions.

Table 1. Class recognition accuracy

Substitutions	Insertions	Deletions	Total error	Accuracy
8.0 %	14.4 %	3.0 %	25.4 %	74.6 %

The neural network training phase for this limited complete labeled database was 4 days for Sun SPARCstation 20 processor.

The effect of using hidden units in the match neural network was proven, too. It was seen that a few hidden units improve the accuracy on the test set and more than 220 hidden units can drops below the best results obtained by the model with 200 hidden units.

6   Concluding Remarks

The paper present some background material on contextual information in continuous speech recognition systems and an application of the neural network (multilayer perceptron) to context dependent phoneme recognition. An overview of the problem is given and a survey of these techniques considered from the point of training set creation and the hybrid system connection is presented. The contribution is concerned as well with some aspects of feature extraction and the phoneme recognition experiments based on the domain depending database, developed in the frame of the COPERNICUS 94 Project No. 1634 SQEL. A hybrid system for phoneme probability estimation is described in the context of methods for neural networks implementation, where was obtained that the hybrid system can be efficiently learned to distribute recognition task between different outputs of the neural network.

References

1. Boulard, H., Wellekens, C.J.: Links Between Markov Models and Multilayer Perceptrons. IEEE Transactions on Pattern Analysis and Machine Intelligence, **12** (1990) 1167–1178.
2. Krokavec, D.: Neural Network for Phoneme Probability Estimation. In: Proceedings of the Conference on Artificial Intelligence in Industry AIII '98. High Tatras, Slovak Republic (1998) 182–191.

3. Krokavec, D., Ivanecký, J.: Slovak Spoken Dialog System. In: Proceedings of the Conference on Artificial Intelligence in Industry AIII '98. High Tatras, Slovak Republic (1998) 447–456.
4. Krokavec, D., Saxa, J.: Recursive Neural Network for Phoneme Recognition. In: Proceedings of the 3rd Conference on Digital Signal Processing DSP'97. Herlany, Slovak Republic (1997) 5–8.
5. Levin, E.: Hidden Control Neural Architecture Modeling of Nonlinear Time Varying Systems and Its Application. IEEE Transaction on Neural Networks, 4 1994 109–116.
6. Lipman, R.P.: Review of Neural Networks for Speech Recognition. Neural Computation, 1 1989 1–38.
7. Matoušek, V.: Phonetic Segmentation Method for the Continuous Czech Speech Recognition. In: Proceedings EUROSPEECH '93 (1993) 713–717.
8. Matoušek, V., Oceliková, J., Krutišová, J., Mautner, P.: A Dialog Module for the Information Retrieval System. In: Proceedings of the Conference on Artificial Intelligence in Industry AIII '98. High Tatras, Slovak Republic (1998) 469–478.
9. Niles, L.T., Silverman, H.F.: Combining Hidden Markov Models and Neural Network Classifiers. Proceedings ICASSP '90 1990 417–420.
10. Richard, M.D., Lippmann, R.P.: Neural Network Classifiers Estimate Bayesian a Posteriori Probabilities. Neural Computation, 3 (1991) 461–483.
11. Riis, S.K., Krogh, A.: Hidden Neural Networks: A Framework for HMM/NN Hybrids. In: Proceedings of ICASSP '97 (1997) 3233–3236.

State-Space Model Based Labeling of Speech Signals

Dušan Krokavec and Anna Filasová

Technical University of Košice,
Faculty of Electrical Engineering and Informatics,
Department of Cybernetics and Artificial Intelligence
Letná 9/B, SK 042 00 Košice, Slovak Republic
{krokavec, filasova}@ccsun.tuke.sk

Abstract. The paper present an application of one method for obtaining derivatives used in training a recurrent neural network that combine the technique of backpropagation through time and dual heuristic programming. The recurrent network was used in contextual causal phoneme recognition for complete (phoneme segmented) training set labeling.

1   Introduction

In general, optimization problem solving need methods that can capture high order complexity of the systems. Substantial interest has been generated regarding the use of methods which can be regarded as forms of approximate dynamic programming. Commonly used terms for such methods include adaptive critics design, reinforcement learning, heuristic dynamic programming and others, and systems that involve discrete states have dominant research in this idea.

Adaptive critic design (ACD) approximates dynamic programming for decision making in noisy, unstationary or nonlinear environments. This family consists of heuristic dynamic programming (HDP) and dual heuristic dynamic programming (DHDP), as well as their action-dependent forms (action dependent heuristic dynamic programming (ADHDP) and action dependent dual heuristic dynamic programming (ADDHDP). These methods can be considered as approximations of dynamic programming using neural networks.

A typical ACD includes action, critic and model modules, where the model module, in general, simulates an subject. The critic module estimates the function equation of dynamic programming and their derivatives with respect to states of the subject. The action module is supposed to find an optimal sequence of vectors of control by successively optimizing estimates from the critic module. Each module can be a neural network or, alternatively, any differentiable system.

The paper present an application of one method for obtaining derivatives used in training a recurrent neural network that combine the technique of backpropagation through time and dual heuristic programming. The recurrent neural network was used in contextual causal phoneme recognition for complete (phoneme segmented) training set labeling. This labeling can be regarded as one sort of state-space model based labeling of speech signals.

F. Jelinek, E. Nöth (Eds.): TSD'99, LNCS 1692, pp. 258–261, 1999.
© Springer-Verlag Berlin Heidelberg 1999

2   Recurrent Network Learning

Nonlinear adaptive models which have gained significant popularity are the feed-forward neural network and a substantial theoretical as well as practically oriented framework has been built up around this models. The framework includes usually methods for training, architecture optimization and model verification.

A more general model type is obtained if the connectivity of the feedforward network is extended to include feedback connections (recurrent network). The advantage of recurrent networks is due to an internal memory of past inputs, introduced by the feedback. This internal memory is adaptive, i.e. during training it may be adapted to encompass those previous inputs, which are relevant to the problem. This network at every time points stay at one of internal state and represents a non-linear state-space model.

Despite their advantages the recurrent networks have not gained popularity similar to that of feed-forward networks. A seeming result of this limited popularity is that training using widely preferred gradient descent method is not enough to train recurrent network, especially for fully recurrent network.

The basic algorithm to the cost function gradient for recurrent neural network is error back-propagation through time [5] and works by unfolding network in time. Unfolding in time is a procedure which turns a recurrent network into an equivalent deeply layered feed-forward network. The main drawback of this algorithm are the requirements for storage. If the total number of units in the network is N_u and a training sequence is of length L, it is necessary to store $N_u T$ activations and this can be impractical for long sequences. An advantage is that the number of operations is just as for the originally back-propagation.

Combination of backpropagation through time and dual heuristic programming technique gives a framework in which training of a network may be executed using derivatives obtained by truncated backpropagation through time.

Training of hybrid system for continuous speech recognition require the complete labeling (i.e. the phoneme segmentation) for a network training set. When dealing with incomplete labeling (i.e. the phonetic transcription) the label of each individual observation is unknown (time boundaries between the phonemes are unknown). If only the incomplete labeling is available, the iterative procedure using current network and current parameter set of hidden Markov models must be used to obtain the complete labeled training set.

The output of a recurrent network is based on the current and H previous inputs which may be expressed as

$$\mathbf{g}_j(\Theta, \mathbf{u}_l, \mathbf{u}_{l-1}, \ldots, \mathbf{u}_{l-H}) = P(C_j|\mathbf{x}_l), \quad \forall j \tag{1}$$

and this is equivalent to input providing with a causal acoustic context (a causal observation sequence $\mathbf{x}_l = \mathbf{u}_{l-H}^l = \mathbf{u}_l, \mathbf{u}_{l-1} \ldots, \mathbf{u}_{l-H}$) where Θ indicates the parameter estimate resulting from training. Scaled likelihoods

$$\frac{P(C_j|\mathbf{x}_l)}{P(C_j)} = \frac{P(\mathbf{x}_l|C_j)}{P(\mathbf{x}_l)} \tag{2}$$

are used as emission probabilities in hidden Markov models during decoding.

3   Experimental Setup

Parts of the recurrent network hybrid developed at Cambridge University [12] have been optimized through several years of research for phoneme recognition. For the standard structure of this system and their training based on the error back-propagation through time algorithm the account training times about of 4 CPU weeks on a fast workstation was uncommon for our experiments. Therefore in experiment were used a limited full connected recurrent network without hidden layers and adjoin internal state variables, and a constrained training set.

In the second evaluation stage there were use a sizeable collection of real data from TIMIT database. The TIMIT (Texas Instruments – MIT) database is a collection of hand-labeled speech samples compiled for the purpose of training speaker independent phonetic recognition systems. It contains acoustic data aligned with discrete labels from alphabet of 62 phones. In [8] was defined 39 phoneme class from full phone set which, in combination with recommended training and test set, was used in our experiment. The training set of 3416 utterances was accepted and recommended TIMIT test set of 1344 utterances was used for hybrid system evaluation.

In the first evaluation stage the goal was to construct a hybrid model for each word in the database, representing its phonetic structure as accurately as possible. Because the full TIMIT data set consists of 53355 phonetic samples for 6100 words to keep the task somewhat manageable, there were used a subset of this data consisting of word of intermediate frequency. This left to work with working data set of 206 word comprising total 7861 samples.

The causal context dependent phoneme recognition was realized in the configuration of the network for the 39 phoneme task and the network was fully connected recurrent neural network with sigmoid output functions and no hidden layers. The cost function for classification was the squared-error expectation between actual outputs and desired output of the network, the elements of the network weights was trained via the truncated error back-propagation in time algorithm, using a steepest descent procedure to a cost function space.

The results are reported in the form of the recommended TIMIT test set. The Table 1 shows how the total number of errors is distributed on phoneme substitutions, insertions and deletions.

Table 1. Class recognition accuracy

Substitutions	Insertions	Deletions	Total error	Accuracy
9.8 %	17.3 %	3.5 %	30.6 %	69.4 %

The neural network training phase for this limited complete labeled database was about 5 days for Sun SPARCstation 20 processor.

4   Concluding Remarks

The paper present some background material on applications of a fully connected recurrent neural network to the causal context dependent phoneme recognition.

A survey of this technique, considered from the point of hybrid system training is presented and a hybrid system for state-space based labeling of speech signals, is described in the context of methods which combine the technique of backpropagation through time and dual heuristic programming. The use of an independent validation test during the design process and official test set application at last step of model evaluation only seem to be a more objective method for speech recognizer evaluation.

References

1. Bridle, J.S.: ALPHA-NETS: A Recurrent Neural Network Architecture with a Hidden Markov Model Interpretation. Speech Communication, **9** (1992) 83–92.
2. Bridle, J.S., Dodd, S.: An Alphanet Approach to Optimizing Input Transformations for Continuos Speech Recognition. In: Proceedings ICASSP '90 (1990) 277–280.
3. Cichocki, A., Unbehauen, R.: Neural Networks for Optimization and Signal Processing. J.Wiley & Sons, New York (1993).
4. Filasová, A., Krokavec, D.: Heuristic Dynamic Programming for LQR Controllers and Kalman Estimators. In: Proceedings of the Conference on Artificial Intelligence in Industry AIII '98. High Tatras, Slovak Republic (1998) 110–119.
5. Herz, J., Krogh, A., Palmer, R.G.: Introduction to the Theory of Neural Computation. Addison-Wesley, Redwood City (1991).
6. Krokavec, D.: Neural Networks for Phoneme Probability Estimation. In: Proceedings of the Conference on Artificial Intelligence in Industry AIII '98. High Tatras, Slovak Republic (1998) 182–191.
7. Krokavec, D., Saxa, J.: Recursive Neural Network for Phoneme Recognition. In: Proceedings of the 3rd Conference on Digital Signal Processing DSP'97. Herlany, Slovak Republic (1997) 5–8.
8. Lee, K.F., Hon, H.W.: Speaker-Independent Phone Recognition Using Hidden Markov Models. IEEE Transactions on Acoustics, Speech and Signal Processing, **37** (1989) 1641–1648.
9. Matoušek, V.: Phonetic Segmentation Method for the Continuous Czech Speech Recognition. In: Proceedings EUROSPEECH '93 (1993) 713–717.
10. Richard, M.D., Lippmann, R.P.: Neural Network Classifiers Estimate Bayesian a Posteriori Probabilities. Neural Computation, **3** (1991) 461–483.
11. Riis, S.K., Krogh, A.: Hidden Neural Networks: A Framework for HMM/NN Hybrids. In: Proceedings of ICASSP '97 (1997) 3233–3236.
12. Robinson, T.: An Application of Recurrent Nets to Phone Probability Estimation. IEEE Transactions on Neural Networks, **5** (1994). 298–305
13. Werbos, P.J.: Backpropagation through Time: What It Does and How to Do It. Proceedings of the IEEE, **78** (1990) 1550–1560.
14. Werbos, P.J.: Consistency of HDP Applied to a Simple Reinforcement Learning Problem. Neural Networks, **3** (1990) 179–189.
15. Young, S.J.: Competitive Training in Hidden Markov Models. In: Proceedings ICASSP '90 (1990) 681–684.

Very Low Bit Rate Speech Coding: Comparison of Data-Driven Units with Syllable Segments[*]

Jan Černocký[1], Ivan Kopeček[2], Geneviève Baudoin[3], and Gérard Chollet[4]

[1] Brno Univ. of Technology, Inst. of Radioelectronics,
cernocky@urel.fee.vutbr.cz
[2] Masaryk University Brno, Faculty of Informatics,
kopecek@fi.muni.cz
[3] ESIEE Paris, Dpt. Signal et Télécommunications,
baudoing@esiee.fr
[4] ENST Paris, Dpt. Signal et Images,
chollet@sig.enst.fr

Abstract. Very low bit-rate (VLBR) coding of speech offers the opportunity to test methods of automatic generation of sub-word units. This paper describes two approaches to VLBR coding: the first based on AL-ISP (Automatic Language Independent Speech Processing) techniques, the second based on syllable segments. Experimental results are reported on a database of one Czech professional speaker. The obtained rates for unit encoding were approximately 135 bps for the former approach and 62 bps for the latter. The quality was evaluated by measuring the logarithmic spectral distortion (computed on LPC-spectra), and in informal listening tests. Possible mutual profits of each technique to the other are discussed.

1 Introduction

Current systems of automatic speech processing (ASP), including the recognition, synthesis, very low bit-rate (VLBR) coding and text-independent speaker verification, rely on sub-word units determined using phonetic knowledge (phonemes, diphones, etc). To make an automatic system use those units, one needs to know their position in the speech signal: in the recognition, phonetically labelled or at least transcribed databases are widely used, while in the synthesis, an important amount of human efforts must be devoted to the creation of (usually) diphone dictionary. The same holds for the use of such units in identification/verification or coding.

It is however possible to investigate the methods of *automatic determination* of such units based uniquely on raw speech data. In our previous works [10,11,12], ALISP (Automatic Language Independent Speech Processing) techniques were

---

[*] The research has been partially supported by the Ministry of Education, Youth and Sports of the Czech Republic, project Nbs. VS97060, VS97028, and by the Grant Agency of the Czech Republic under the Grant 201/99/1248.

F. Jelinek, E. Nöth (Eds.): TSD'99, LNCS 1692, pp. 262–267, 1999.
© Springer-Verlag Berlin Heidelberg 1999

used for unsupervised, data-driven determination of basic speech units. Temporal decomposition (TD), vector quantization (VQ) and hidden Markov models (HMM) are the main ALISP "tools". On the other hand, the use of syllable segments for speech processing was investigated in [4,5,3,6]. The syllable based approach is motivated by the fact that syllables create perceptually and acoustically coherent units and that they also represent the basic prosodic units. Automatic segmentation into syllable segments [4,9] is achieved using estimation of syllable time duration (by estimation of speech rate and articulation rate), followed by the estimation of the syllable boundary by means of functions of sonority decrease, acoustical intensity and acoustical changes.

Although the main target of proposed methods are speech recognition and synthesis, the *VLBR coding* offers an excellent possibility to test those methods without a passage to the lexical domain (a step, which is not straightforward for data-driven techniques). The efficiency of algorithms is evaluated by re-synthesizing the speech and by comparing it to the original. If the output is intelligible, one must admit, that this representation is capable of capturing acoustic-phonetic structure of the message and that it is appropriate also in other domains. Moreover (in contrast with classical approach, where the unit set is fixed a-priori and can not be altered), the coding rate in bps and the dictionary size carry information about the *efficiency* of the representation, while the output speech quality is related to its *precision*. This paper is organized as follows: section 2 describes the ALISP framework for VLBR coding while the following section 3 concentrates on syllable segments. Section 4 covers the experimental part of the work and section 5 details on the possible mutual enhancements of ALISP and syllable techniques. Section 6 concludes the paper.

2   Use of ALISP for VLBR Coding

The use of ALISP units for coding at very low bit rates was already described in [10,11,12]. Therefore, we include only a brief description of different "tools". In all experiments, the speech parameterization is done by classical LPC-cepstral (LPCC) coefficients calculated on fixed-length frames. *Temporal decomposition* [1] was used to initially segment the speech into quasi-stationary parts. A spectral parameter matrix is decomposed into a limited amount of *events*, each represented by a *target* and an *interpolation function* (IF). We used the td95 package[1] for this decomposition, where a short-time SVD with adaptive windowing takes place for the initial search of IFs, followed by post-processing of IFs (smoothing, decorrelation) and iterative refinement of targets and IFs. *Vector quantization* was used to initially cluster the events found by the TD. It was trained with one vector per segment (that in gravity center of TD interpolation function), but the VQ-coding worked with entire segments using cumulated distances. *Hidden Markov Models* are the standard framework for the speech recognition. Here, the HMMs are used to refine segments in the dictionary, to model them and to detect these segments in the input speech. The refinement consists of iterative steps of recognition with previously trained models, and of re-estimation of those

---

[1] Thanks to Frédéric Bimbot (IRISA Rennes, France) for the permission to use it.

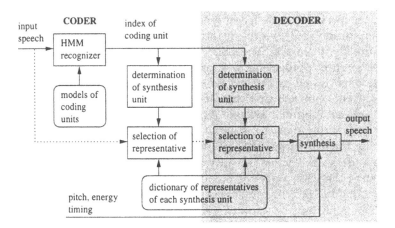

Fig. 1. Coding units, synthesis units and representatives in the coder and decoder. The information about chosen representative can be either transmitted (dotted line) or re-created in the decoder.

models using the new segmentations and labellings. Once the units are found (and the models trained), the *coding* of previously unseen signals can be done following Fig. 1.

3   Syllables and Syllable Segments

We use the notion *syllable segments* instead of syllables because we have no reliable specification of syllables. Although the feeling of syllables and syllable boundaries is usually very natural and strong, it is subjective and in many cases also not unique. A problem complicating the determination of an appropriate set of syllable segments is the necessity to respect the coarticulation between the adjacent syllables and to keep the number of segments reasonable. To solve this problem we use the fact that the coarticulation is in most cases reduced on the boundaries of the adjacent syllables in comparison to the inner parts of syllables. Also, syllabic onsets are often realized in a standard form [4,8] and acoustical intensity typically decreases at the syllable boundaries. For this purpose, some classes of syllable segments were proposed and analyzed (L-syllables, CVS-syllables, and other types [7,6]).

4   Experiments

4.1   Database

The experimental work has been conducted in speaker-dependent mode with a Czech database of one professional speaker. Two tapes with read texts of the well known actor Martin Růžek[2] were sampled at 11025 Hz (16 bit linear resolution), and split using energy minima detection into parts of 6 to 18 seconds. Parts with foreground and background music were discarded. The total amount of data is 126.5 minutes. This DB was split into training (7/8) and test (1/8) portions.

4.2   ALISP Experiments

The data were parameterized using 12 LPCC coefficients computed on 220-sample frames with 110-sample overlap (pre-distortion with $\alpha = 0.95$ and Hamming window were applied). The pitch was computed using FFT-cepstral method on larger frames (500 samples). The temporal decomposition was set to produce 15 events per second in average. The VQ codebook size for the initial clustering was 64. Prototype HMMs with 3 data-streams (LPCC, ΔLPCC, E and ΔE), each with 3 emitting states carrying a single Gaussian component, were initialized and trained using HTK tools HInit, HRest and HERest (5 iterations). The first HMM decoding on the training set was done using the obtained models, and then, 5 iterations of model re-estimation (5 × HERest) and Viterbi decoding were run (creating 1st to 5th "generation" of HMMs). The whole refinement took approximately 8 hours on a Pentium 233 MMX.

In the coding step, 8 longest representatives per coding unit were selected in the training corpus. The synthesis units from Fig. 1 were equal to coding ones. The original pitch and energy contours, as well as optimal DTW time-warps between the original segment and the coded one were used. The information about representative was transmitted (dotted line in Fig. 1) thus adding 3 bits per segment. The overall rates with 5-th generation models for coding unit transmission (assuming uniform encoding of their indices) are 115.46 bps and 116.96 bps for the training and test sets respectively. The average unit rate per second is 19.26 and 19.49, so that the total coding rates (not taking the prosody information into account) are **173.26 bps** and **175.44 bps**.

The *objective* measure of the quality was the mean of logarithmic spectral distortions:

$$D = \sqrt{\int_{-1/2}^{1/2} \left[10 \log \hat{S}(f) - 10 \log S(f)\right]^2 df} \quad \text{in dB}, \tag{1}$$

evaluated using the power frequency responses of prediction filters ($\hat{S}(f)$ stands for the response with coded coefficients and $S(f)$ for the same function with the original ones). For the part of test file df232000 (the same was tested with syllable segments), $D = 4.39$. *Subjectively*, the resulting speech is intelligible, but quite unnatural and with strongly audible artifacts. One must however take into account, that no smoothing was performed on the borders of segments, and that the synthesis (pure LPC) is quite primitive. Smoothing techniques and better synthesis (HNM, PSOLA) should improve the quality of the coder. Another issue is the encoding of prosody, which was not resolved in our work, and should be done as well on the segmental basis.

---

[2] Thanks to Czech Radio Brno for having granted us the access to those recordings for research and education purposes.

4.3 Syllable Segments Experiments

The procedure for segmentation to the syllable segments, used in the experiment, was based on the method which was used for semi-automatic segmentation procedure for syllable speech synthesizer *Demosthenes* [7,6]. It detects local maxima of the *function of sonority decrease*. Provided $v_i = (v_{i,1}, v_{i,2}, ..., v_{i,n})$ is the acoustic vector in the time i, the function of sonority decrease is based on the value

$$D(i) = \sum_{j=k}^{l} c_j(i),$$

where $c_j(i) = max\{v_{i,j} - v_{i-\delta,j}, 0\}$. The indices k, l correspond to the vowel formants frequency region, and δ is the length of the used difference. The coding consists in comparing a previously unseen syllable segment to all segments in the training corpus and in choosing the optimal one. For this comparison, 16-element acoustic vectors are obtained by modeling the Corti organ. Segments are compared using a metric based on a sum of differences of vector elements, weighted by ratios of those elements. The time is warped using linear transformation.

The resulting rate on the test corpus is 62.3 bps. The logarithmic spectral distortion for part of the test file df232000, evaluated using Eq. 1, was 8.90 dB. Subjectively, the quality of coded speech is poor and the utterances are hard to understand.

One of the goal of the experiment was to test whether the inaccuracy of the method will have substantial negative consequences for coding. Rather poor quality of the output shows, that such a simple type of the segmentation should be substituted by a more precise method. Also, use of a metrics developed specially for syllable segments could increase the quality of the output. On the other hand, low transmission rate and possibilities of enhancing the used method give a chance for further development in this direction. Statistical evidence of the syllable segments frequencies can be used to further decrease the bit rate (using entropy encoding).

5   ALISP and Syllables Helping each Other

It is obvious (though we have not yet managed to prove it experimentally), that ALISP techniques and syllable segments can be mutually profitable. So the later approach can take advantage from its ALISP counterpart in the following way:

- passage from semi-automatic to fully automatic determination of syllable segments in the corpus by application of clustering techniques.
- automatic segmentation of the corpus into syllable segments by training an HMM for each segment, and by Viterbi decoding.

while ALISP techniques can be enhanced by the works in syllable segments:

- studies of allowed transitions and coarticulation can be used as rules to drive the sequencing of units by multigrams [2], so that the intra-unit coarticulation is maximized, while the inter-unit one is minimized.
- prosody patterns determination techniques can be used for prosody encoding in an ALISP-based VLBR coder.

6 Conclusion

The coding using ALISP units gives fair and promising results; with the improvements mentioned in par. 4.2, this scheme should be suitable for very low bit rate transmission over the Internet, storage of huge amount of speech data, and other applications.

As for the possible use of the syllable segments, we would like to use more precise methods for segmentation, apply special metrics for comparison of the syllable segments and apply statistical method for evaluation of the optimal amount of the training data as well as for the optimization of the bit rate.

Our future works in the VLBR coding will aim at the mutual improvement of the approaches (as described in section 5), and at using those techniques in the continuous speech recognition.

References

1. F. Bimbot. An evaluation of temporal decomposition, Technical report, Acoustic research department AT&T Bell Labs, 1990.
2. S. Deligne. Modèles de séquences de longueurs variables: Application au traitement du langage écrit et de la parole, PhD thesis, École nationale supérieure des télécommunications (ENST), Paris, 1996.
3. G. Doddington. Syllable Based Speech Processing. Technical report, J. Hopkins University, 1997, WS97 Project Report, Research Notes No. 30.
4. S. Greenberg. Speaking in Shorthand – A Syllable-Centric Perspective for Understanding Pronunciation Variation. In *Proc. Workshop Modeling Pronunciation Variation for Automatic Speech Recognition*, pages 47–56, 1998.
5. L. Josifovski, D. Mihajlov, and D. Gorgevik. Speech Synthesizer Based on Time Domain Syllable Concatenation. In *Proc. SPECOM'97*, pages 165-170, Cluj-Napoca, 1997.
6. I. Kopeček. Syllable Based Speech Synthesis. In *Proc. 2nd International Workshop SPECOM'97*, pages 161–165, Cluj-Napoca, 1997.
7. I. Kopeček. Speech Synthesis Based on the Composed Syllable Segments. In *Proc. of Workshop on Text Speech and Dialogue (TSD'98)*, pages 259-262, Brno, Czech Republic, September 1998.
8. I. Kopeček. Syllable Segments in Czech. In *Proc. XXVII. Mezhvuzovskoy naucznoy konferencii*, pages 60–64, St. Petersburg, March 1998, Vypusk 10.
9. I. Kopeček and K. Pala. Prosody modeling for sylable-based speech synthesis. In *Proc. IASTED Conference on AI and Soft Computing*, pages 134–137, 1998.
10. J. Černocký, G. Baudoin, D. Petrovska-Delacrétaz, J. Hennebert and G. Chollet. Automatically derived speech units: applications to very low rate coding and speaker verification. In *Proc. of Workshop on Text Speech and Dialogue (TSD'98)*, pages 183–188, Brno, Czech Republic, September 1998.
11. J. Černocký, G. Baudoin and G. Chollet. Segmental vocoder – going beyond the phonetic approach. In *Proc. IEEE ICASSP 98*, pages 605–608, Seattle, WA, May 1998, http://www.fee.vutbr.cz/~{}cernocky/Icassp98.html.
12. J. Černocký. Speech Processing Using Automatically Derived Segmental Units: Applications to Very Low Rate Coding and Speaker Verification. PhD thesis, Université Paris XI Orsay, 1998.

Storing Prosody Attributes of Spontaneous Speech

Jana Klečková

Department of Computer Science
Faculty of Applied Sciences, University of West Bohemia
Univerzitní 22, CZ – 306 14 PLZEŇ, Czech Republic
kleckova@kiv.zcu.cz

Abstract. The paper deals with the problem of the storage of the prosody attributes which have been used in the Czech dialog system. In Czech language featured by a free-word-ordering the prosody serves a critical information for the recognition and understanding system. For some sentences the intonation is essential to determine the core of a communication, beeing used by a speaker who to emphasize a meaning of a sentence. The prosodic characteristics included in the sentence (features describing fundamental frequency F0, voice energy , the length of a pause behind and before the word, the speaking rate, flags indicating word finality and the lexical word accent) are stored in the database and consequently exploited by the linguistic module as an additional information used for recognizing and understanding the spontaneous speech. In case of storing digital speech signal in database we meet a problem of its high redundancy. Another problem is choosing cut points for segmenting speech. Suitable points are pauses, but they are not often present in the fluent speech. Instead a coarticulation between phonemes makes the placement of the cut points difficult. Their suitably for segmenting speech should be dependent on a context information. Having processed the characteristics by usual methods of statistics the database can also be used to generate answers in the dialog system. The module was implemented in the C language and supported by the ORACLE database.

1 Introduction

Spoken word recognition involves the classification and the segmentation of a variable and continuous speech input. Different proposal have been made to characterize the relation between these two operations and the nature of the resulting units. In Czech language intonation serves a critical information for the recognition and understanding system. For some sentences, the intonation is essential to determine the core of a communication, depending on a speaker who uses intonation to emphasive the meaning of a sentence. The design of the module for suprasegmental type processing is based on the partitioning the speech into sentences. In a such system prosodic attributes are determined by the acoustic–phonetic module. The time distribution of the voice energy and of the fundamental frequency is monitored within the period of a single sentence.

F. Jelinek, E. Nöth (Eds.): TSD'99, LNCS 1692, pp. 268–273, 1999.
© Springer-Verlag Berlin Heidelberg 1999

The length of a pause as well as flags indicating word finality and lexical word accent are determined. Consequently, this information is used to associate the sentence with a certain type. The attributes determined by this procedure are used as the second input to the linguistic module.

2 Prosodic Attributes

The ability of the listeners to identify correctly and almost instantly a word from amongst the tens of thousands of other words stored in their mental lexicon constitutes one of the most extraordinary human cognitive feats. The speech signal indeed presents a formidable challenge. Both the speech is variable (every word takes on a different phonetic shape each time it is produced - the existence of large numbers of a highly similar words in the lexicon makes this variability even more troublesome) and speech is continuous (unlike written text, it contains no systematic spaces or reliable markers to indicate where word or utterance ends and the next one begins. The intonation often serves an information of a broad meaning nature. For example, the falling pitch we hear at the end of a statement in Czech such as " Jan uz odesel. (John has already gone away.)" indicates that the utterance is complete. And on the contrary, the question " Jan uz odesel? (Has John already gone away?)" in Czech equivalent has rising intonation. For this reason, falling intonation at the end of an utterance is called a terminal intonation contour. On the other hand, rising or unvaried level intonation, often indicates incompleteness. However, Czech sentences that contain question words like *kdy, co, kdo, jak (when, what, who)* usually do not have any rising intonation. It is like that the words in the question suffice to indicate that the answer is expected. The fact that rising or level intonations are correlated with incompleteness and falling intonation with completeness admits other utilizations of the intonation. One of them helps to make clear the interpretation of potentially ambiguous utterances.

2.1 Pauses

The prosody is a very complex subject. Besides the intonation the hierachy of pauses is very important. Pauses of standard length in the places of punctuation marks between syntactic units are felt as bizzare in the spontaneous speech. After several experiments have been treid out, a three-tier pause hierarchy seems acceptable in Czech.
Examples:
...(P3) Prosim Vas, (P1) muzete mi rict, (P2) kdy jede vlak do Prahy? (P3) ...
        (...Please, can you tell me, when the train to Prague is going?...)
...(P3) Jak vidite, (P1) nezbyva nam mnoho casu. (P3) ...
        (...You see, there's not much time left....)
...(P3) Uz se prilis nezdrzuj, (P1) Vaclave, (P2) a pokus se vlak dobehnout. (P3) ...
        (...Don't get stuck, Vincent, and try to run out the train....)

    To make finer distinction of pauses would require to respect semantic relations of units in the dialog.

Table 1. Pause hierarchy

Pause	Duration of pause [ms] for speech rate	Classification of punctuation marks
P1	8 - 10	,
P2	80 - 100	- :
P3	200 - 240	; . ? !

3   Database of Prosody Attributes

One well-known proposal for Czech assumes that the syllable constitutes the basic perceptual unit of speech processing. The present oveview summarizes recent data from three different experimental paradigms that aim to clarify the processing role of the syllable. How listeners process the incoming speech signal so efficiently despite its variable and continuous nature? The some intermadiate representations serve as an interface between the speech signal and the phonological representations stored in the lexicon. A phonetic or phonological representation provides a more abstract and less variable format for representing the information in the input and for accesing the lexicon. Instead of trying to match each unique sensory input directly onto a huge number of lexical entries, listener first recode the variable input into a smaller number of abstract units like phonemes or syllables that in turn serve to contact the lexicon. The intermediate representation based upon units can potentially guide the segmentation process. For instance, the onset of prosodic units or strong syllables could be used as starting points for the lexical matching process. To the extent that these segmentation points are likely to correspond to word boundaries, such heuristics would be helpful in reducing wasteful attempts to match the input with misaligned lexical candidates. In the framework that attributes a central role to intermediate levels of representation, we are led to search for the nature of the units making up this representation.

The design of the prosody module is based on the two algorithms:

1. the partitioning the speech into sentences. The sentences are processed using the following method:
   - each sentence is divided into n-windows,
   - in each of the window is assigned with the voice energy, (the first feature) and the fundamental frequency (the second feature) .
2. the founding considerable variety in the types of perceptual units that have been proposed in the literature:
   - distinctive features,
   - phonemic segments,
   - syllables.

To classify sentences according to prosody, the prosodic characteristics must be computed and then considered as features. However, the features must be normalized and their number reduced to simplify recognition which then follows.

Taking into account properties of the neural network employed in the recognition the number associated with a simple sentence must be the same for each one. Thus, besides the intonation analysis the prosody module also involves subroutines which can evaluate the pitch, both in a sentence and in a word, and the pause. The sentence pitch indicates an expression which may be crucial for identifying the core of the statement. Position of the sentence pitch is a matter of the statement realization, thus, being subjected to a context of the communication, or to a standpoint of the speaker. Therefore, it cannot be estimated, or even defined in advance. The prosodic characteristics including the sentence (features describing F0, energy , the length of the pause after and before the word, the speaking rate, flags indicating word finality and lexical word accent) are stored in the database and consequently exploited by the linguistic module as an additional information used for recognition and understanding the spontaneous speech.

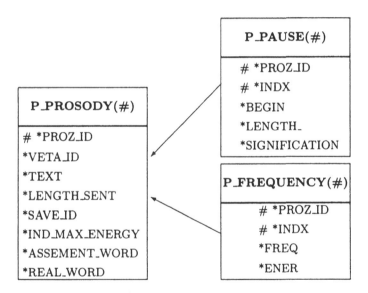

Fig. 1. ERA model of prosody characteristic database

All the program equipment including the code SNNS [5] has been used in a number of experiments focused on the sentence evaluation. The list now follows:

1. differentiating a sentence construction from a single sentence without pause classification;
2. differentiating a sentence construction from a single sentence with pause classification.

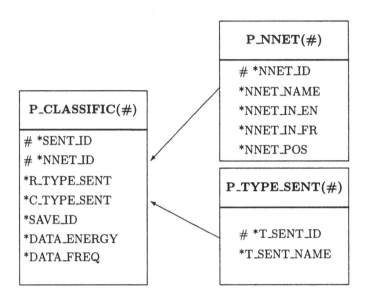

Fig. 2. ERA model of classification results

Table 2. Differentiating a complex sentence from a single sentence without pause classification.

Type of the sentence	Total	Correct		Incorrect	
		Number	%	Number	%
complex sentence - all sentences are significant	100	67	67	33	33
complex sentence - first clause is unsignificant	100	53	53	47	47
single sentence	100	51	51	49	49
TOTAL	300	171	57	129	43

Table 3. Differentiating a complex sentence from a single sentence with pause classification.

Type of the sentence	Total	Correct		Incorrect	
		Number	%	Number	%
complex sentence - all sentences are significant	100	92	92	8	8
complex sentence - first clause is unsignificant	100	85	85	15	15
single sentence	100	77	77	23	23
TOTAL	300	254	84	46	16

4   Experiments and Perspectives

One shortcoming of a syllabically-based segmentation strategy is that it does not easily handle cases of resyllabification of the input that result from phonological processes across word boundaries. The experiments reported in this paper were performed on a subset of Czech sentences - both read text and spontaneous speech. The sentences were generated using the ERBA templates [3]. To summarize the results is introduced in the tables Tab.2, Tab.3. In a recent study we showed that listeners can exploit certain durational cues in the signal to locate word boundaries. In the future experiment we shall use an algorithm with different sets of prosodic features like duration of words, syllables and syllables nucleus, etc.

5   Conclusion

It is currently that prosodic features have a very high significance for the dialog system. One important aspect of identifying elements of meaning (be it root morphemes, derivational or inflectional morphemes, words of various kinds or phrases) is their coding into segments, syllables and larger linguistic units like prosodic phrases. In the bottom-up process, acoustic cues in the speech signal are used by the listener in order to decode words. The aim of the experiments was twofold: First, the prosodic characteristics including the sentence (features describing F_0, energy, the length of the pause after and before the word, the speaking rate, flags indicating word finality and lexical word accent) are stored in the database and second consequently exploited by the linguistic module as an additional information used for recognition and understanding the spontaneous speech.

References

1. Klečková, J., Krutišová, J., Matoušek, V., Mautner, P., Netrvalová: *An Automatic Creation of the Language Model for the Spontaneous Czech Speech Recognizer*. In: Proceedings of the European Conference "EUROSPEECH '95", Madrid, September 1995, pp.1185 -1195.
2. Klečková, J., Matou"ek, V.: *Detection of the Sentence Types by the Integrated Prosody Module*. In: Proceedings of the First Workshop on Speech, Text, Dialogue - TSD'98, Brno, Czech Republic, September 1998, pp. 235 – 240.
3. Klečková J.: *Creation of the Language Model for the Spontaneous Czech Speech Recognizer*. In: Proceedings of the 4th International Workshop on Systems, Signals and Image Processing, Poznań, May 1997, pp. 93 – 96;
4. Kolinsky, R.: *Spoken word recognition: A stage-processing approach to language differences*. In: European Journal of Cognitive Psychology, 10, pp. 1 – 40.
5. Pal S.K., Dutta Majumder D.: *Fuzzy sets and decision making approaches in vowel and speaker recognition*. In: IEEE Trans. Syst., Man, Cybern., vol.7, pp. 625-629,1977.
6. *SNNS Stuttgart Neural Network Simulator*, User Manual, Version 4.1, červen 1995.

An Overview of the State of the Art of Coding Schemes for Dialogue Act Annotation

Marion Klein

German Research Center for Artificial Intelligence,
Stuhlsatzenhausweg 3,
D-66123 Saarbrücken, Germany
Marion.Klein@dfki.de
http://www.dfki.de/~mklein

Abstract. This paper investigates tag-sets of coding schemes used for two-agent, task-oriented, problem-solving dialogue act annotation. The observed schemes are analyzed using criteria like sufficient documentation, easy transferability to other domains, and reliability of annotation. Tag-sets of schemes are compared, and common phenomena are stated. The aim of the paper is to provide an overview of currently available dialogue act coding schemes and to give scheme developers support in their scheme design.

1 Introduction

Austin's 1962 posthumously published lecture notes [4] on speech act theory are commonly seen as the first fundamental work in this field. Searle [15] builds on and extends this work. Since then this theme was in the center of research using different notions like, dialogue moves [7], dialogue objects [12], communicative actions [2], communicative acts [3], and dialogue acts [5], according to the theoretical standpoint of the authors. A dialogue act (to adopt the most commonly used notion) represents a speaker's intention in an utterance and is defined in [5] as the functional unit used by the speaker to change the context.

Dialogue acts are used to annotate corpora. These annotated corpora are needed, for example, for training and testing purposes of NLP systems. One can derive statistical predictions about follow-up dialogue acts based on previous dialogue act annotations and on pattern recognition to improve performance of recognizers. Furthermore, dialogue act language models serve to study intonation, and to analyze referring expressions.

This paper focus on the state of the art of coding schemes for dialogue acts. World-wide available schemes are analyzed using criteria like sufficient documentation, easy transferability to other domains, and reliability of annotation. Tag-sets of schemes are compared, and common phenomena are stated.

This work is based on the review of schemes in [13], performed by the participating sites of the EU sponsored project MATE (Multi level Annotation Tools Engineering). The scheme review served as a starting point for research on best practice methods for scheme design, and was used to ascertain those coding schemes which should be supported by the MATE workbench [11].

F. Jelinek, E. Nöth (Eds.): TSD'99, LNCS 1692, pp. 274–279, 1999.

2   Comparison of Coding Schemes

Many schemes for different purposes exist. Not all of these schemes can be annotated reliably and are suitable for reuse. In the following we state criteria we have developed for selecting most appropriate schemes and represent the results of our scheme comparison according to these criteria (see Table 1).

scheme	criteria							
	1	2	3	4	5	6[1]	7[2]	8
ALPARON	+	3	experts	500 d.	77% agreem.	DES	D	+
CHAT	+	huge	experts	160MB	-	-	many	-
CHIBA	+	10	experts	22 d.	$0.57<\alpha<0.68$	DIR, BA, TR	J	?
COCONUT	+	2	experts	16 d.	+	FUR	E	+
Condon&Cech's	+	5	fairly exp.	88 d.	91% agreem.	TS	E	+
C-STAR	+	5	experts	230 d.	-	TR	E, J, K, I	+
DAMSL	+	4	experts	18 d.	$\kappa = 0.56$	-	E	+
Flammia's	+	7	trained	25 d.	$\kappa \geq 0.6$	DES	E	?
JANUS	+	4	experts	many	89% agreem.	BA	E	+
LINLIN	+	4	experts	140 d.	97% agreem.	TR, TS	S	+
MAP TASK	+	6	experts	128 d.	$\kappa = 0.83$	DIR	E	+
Nakatani's	+	6	naive	72 d.	-	INSTR	E	+
SLSA	+	7	experts	100 d.	+	COU	S	+
SWBD-DAMSL	+	9	experts	1155 d.	$0.8<\kappa<0.84$	-	E	+
Traum's	+	3	experts	36 d.	+	-	E	+
VERBMOBIL	+	3	naive	1172 d.	$\kappa = 0.84$	BA	E, J, G	+

Table 1. General Comparison of coding schemes.

Firstly, it is important for us that there is a coding book provided for a scheme (*criterion 1*). Without definition of a tag set, decision trees, and annotation examples, a scheme is hard to apply. Also the scheme has to show that it is easy to handle which means it should have been successfully used by a reasonable number of annotators (*criterion 2*) on different levels of expertise (*criterion 3*). For reusability reasons, language (*criterion 7*), task, and domain (*criterion 6*) independence is required. Additionally, it is crucial that a scheme has been applied to large corpora (*criterion 4*). The annotation of mass data is the best indicator for the usability of a scheme. Another indicator for scheme application is its integration in an NLP systems (*criterion 8*). Finally, it was judged positive

---

[1] BA = business appointments, COU = courtroom interaction, DES = directory enquiry services, DIR = giving directions, FUR = furnishing rooms interactivity, INSTR = giving instructions (e.g. about cooking), TS = transport, TR = travel;
[2] D = Dutch, E = English, G = German, I = Italian, J = Japanese, K = Korean, S = Swedish;

if schemes directly proved their reliability by providing a numerical evaluation of inter-coder agreement (*criterion 5*), e.g. the κ-value [6].

Information about schemes was collected from the world wide web, from recent proceedings and through personal contact. As shown in Table 1, we compared 16 different schemes, developed in the UK, Sweden, the US, Japan, the Netherlands, and Germany. Most of these schemes were applied to English language data. Only three of the reviewed schemes were annotated in corpora of more than one language, and thus, indicate some language independence.

A drawback in reusing schemes for different purposes is tailoring them to a certain domain or task. Nevertheless, most of the ongoing projects in corpus annotation look at two-agent, task-oriented dialogues, in which the participants collaborate to solve some problem. These facts are also reflected in the observed schemes which were all designed for a certain task and/or used in a specific domain.

With regard to the evaluation guidelines stated above we can positively mention that all schemes provide coding books. Also, all schemes were applied to corpora of reasonable size (10 K – 16 MB data). In 14 cases expertized annotators were employed to apply the schemes which leads to the assumption that these schemes are rather difficult to use. The inter-coder agreement, given by 10 of the schemes, shows intermediate to good results.

The comparison of tag sets was performed differently with regard to higher and lower order categories. The definition of higher order categories was mainly driven by linguistic and/or philosophical theories, e.g. [7], the schemes were based on. Whereas definitions and descriptions of lower order categories were influenced by the underlying task the scheme was designed for, e.g. information retrieval, and the domain of the corpus the scheme was applied to, e.g. conversation between children.

The only higher order aspect that was implicitly or explicitly covered in all schemes was forward and backward looking functionality. This means that a certain dialogue segment is related to a previous dialogue part, like a *response*, or to the following dialogue part, like a *claim*, that forces a reaction from the dialogue partner.

On the level of lower order tags we could see tags

- with nearly equivalent definitions, e.g.
  - the dialogue act *request* definition in D. Traum's scheme:
    "The speaker aims to get the hearer to perform some action."
    [16]
    compared to
  - the dialogue act *ra* definition in S. Condon & C. Cech's scheme:
    "Requests for action function to indicate that the speaker would like the hearer(s) to do something [...]" [9]
    compared to
  - the dialogue act *request* definition in the VERBMOBIL scheme:
    "If you realize that the speaker requests some action from the hearer [...] you use the dialogue act REQUEST" [1];

– which broadly seem to cover the same feature with slightly different description facettes, e.g.
  – the dialogue act *open-option* definition in the DAMSL scheme:
    "It suggests a course of action but puts no obligation to the listener." [2]
    compared with the examples above;
– which differ completely from the rest, e.g.
  – the dialogue act *update* definition in the LINLIN scheme:
    "where users provide information to the system" [10]
  — addressed to human-machine dialogues.

Especially the last group can be interpreted as highly task or domain dependent.

3   Common Phenomena Modeled by Dialogue Acts

Together with the comparison of tag sets dialogue acts were classified according to the phenomena they model. Approximately 20 classes were established of which the most common ones are summarized below.

The observed tag-sets are not necessarily appropriate for every-day conversations but were designed for problem-solving types of dialogues. As mentioned in [1] those dialogues are characterized by an opening, negotiation and closing phase.

Opening and closing of conversations usually contain conventional *greeting* phrases which are labeled in seven of the observed schemes. Other conventional phrases, like *thanking* and *apologizing*, which can appear at any place in a dialogue are also labeled by three and four schemes, respectively.

During the negotiation phase of a dialogue there are two different kinds of conversation – conversation about the current conversation, e.g. complaining about a bad connection during a telephone conversation, and conversation about a task that has to be performed. To the former kind of conversation also belong signals of understanding, like *acknowledging* or *correction of mis-speaking* by the listener, or *repetition* and *rephrasing* by the speaker. Meta-communication is marked by ten of the observed schemes.

The latter kind of conversation during a negotion phase, conversation about he actual task, is dominated by adjacency pairs like *question* (for information) – *answer*, *request* (of action) – *refusal/action performance*, *offer* (to do something) – *acception/rejection/indefinite reaction*, and *suggestion – acception/rejection/indefinite reaction* [8]. Adjacency pairs of the kind stated above appear in almost all observed schemes. Some schemes also mark simple information giving, i.e. *statements* which do not have to be enforced by a preceding question, like "Let me briefly explain how you can reach us ...".

A scientist with a certain phenomenon of interest to look at will notice that the general tag-set above does not fit his/her requirements. For an example, see the CHAT scheme designed for the CHILDES project [14] to label children's conversation. It consists of 67 different dialogue acts but still does not hit all dialogue acts identified in the general tag-set; on the other hand the general scheme is not fine grained enough to mark all distinctions made in CHAT.

4   Best Practice Method for Scheme Design

In order to reduce costs, reuse of schemes and annotated corpora in different projects is desired. As this cannot be achieved by the development of a general coding scheme a best practice method for dialogue act scheme design was explored in the MATE project based on ideas provided by members of the Discourse Resource Initiative (DRI):

1. Make a list of all phenomena you are interested in and assign a certain characteristic set of tags to each phenomena.
2. Use this multi-dimensional scheme[3] by yourself, apply it to some training corpora and test reliability.
3. Flatten the multi-dimensional scheme to a single-dimensional one, by merging tags which always occur together and deleting those which were never used. Remember not to have any extremely small categories as coders tend to overuse them and use natural, easy observable distinctions of tags which are easy to remember.
4. Provide a coding book for the single-dimensional scheme, including tag set definition, decision tree and example annotations, like in [1], for example.
5. Develop a mapping mechanism to convert multi-dimensional annotation to single-dimensional annotation and the other way around.
6. Randomly check the coding by reliability tests and improve your scheme(s), if necessary.
7. Document all steps!

The advantage of this method is that the multi-dimensional scheme supports reusability because it models the different phenomena best, and thus, is easier to understand by foreign scheme developers. The single-dimensional scheme, on the other hand, is easier to apply which speeds-up the annotation process and hence, makes annotation less expensive. For these reasons the flattened scheme is much more appropriate for mass data annotation. Of course, a flattened scheme is not necessarily required and if coders are happy with multi-dimensional coding, the time-consuming process of flattening (3.) can be omitted.

5   Conclusion and Future Work

Having reviewed a large amount of currently available coding schemes for dialogue acts we presented a general comparison to show the reliability of schemes and their inappropriateness for reusability in other domains. Furthermore the tag-sets are compared and common features spotted are pointed out. Also a best practice method for dialogue act scheme design is proposed.

---

[3] In contrast to single-dimensional schemes, multi-dimensional schemes, like DAMSL [2], are based on the assumption that an utterance covers several different orthogonal aspects, called dimensions. Each dimension can be labeled.

This work was performed to ascertain which coding schemes should be supported by the MATE workbench. At the time being this decision is fairly established but we are still open to look at new schemes which came up recently, if information is provided.

Acknowledgement. I would like to thank Claudia Soria for her support on reviewing coding schemes, and Norbert Reithinger who gave valuable feedback on earlier versions of this paper.

The work described here is part of the European Union funded MATE LE Telematics Project LE4-8370.

References

1. J. Alexandersson, B. Buschbeck-Wolf, T. Fujinami, M. Kipp, S. Koch, E. Maier, N. Reithinger, B. Schmitz, M. Siegel. Dialogue Acts in VERBMOBIL-2, Second Edition. Verbmobil Report 226, 1998.
2. J. Allen, M. Core. Draft of DAMSL: Dialogue Act Markup in Several Layers. http://www.cs.rochester.edu:80/research/trains/annotation, 1997.
3. J. Allwood. An activity based approach to pragmatics. Gothenburg Papers in Theoretical Linguistics 76, 1995.
4. J. L. Austin. How to do things with words. Oxford: Clarendon Press, 1962.
5. H.C. Bunt. Dynamic interpretation of dialogue theory. In: M.M. Taylor, D.G. Bouwhuis and F. Neel. The Structure of Multimodal Dialogue, volume 2. John Benjamins, 1995. Amsterdam.
6. J. Carletta. Assessing agreement on classification tasks: the kappa statistic. In: Computational Linguistics, volume 22(2), pages 249–254, 1996.
7. L. Carlson. Dialogue Games: An Approach to Discourse Analysis. Reidel, 1983.
8. H. Clark. Arenas of language use. The University of Chicago Press, Chicago, 1992.
9. S. Condon, C. Cech. Manual for Coding Decision-Making Interactions. ftp://sls-ftp.lcs.mit.edu/pub/multiparty/coding_schemes/condon, 1995.
10. N. Dahlbäck, A. Jönsson. A coding manual for the Linköping dialogue model. http://www.cs.umd.edu/users/traum/DSD/arne2.ps, 1998.
11. A. Isard, D. McKelvie, B. Cappelli, L. Dybkjær, S. Evert, A. Fitschen, U. Heid, M. Kipp, M. Klein, A. Mengel, M. Baun Møller, N. Reithinger. Specification of Coding Workbench. MATE Deliverable D3.1. http://www.cogsci.ed.ac.uk/$\sim$amyi/mate/report.html, 1998.
12. A. Jönsson. Dialogue actions for Natural Language Interfaces. In: Proceedings of IJCAI-95, 1995. Montreal, Canada.
13. M. Klein, N. O. Bernsen, S. Davies, L. Dybkjær, J. Garrido, H. Kasch, A. Mengel, V. Pirrelli, M. Poesio, S. Quazza, C. Soria. Supported Coding Schemes. http://www.dfki.de/mate/d11, 1998.
14. B. MacWhinney. The CHILDES Project: Tools for Analysing Talk. http://poppy.psy.cmu.edu/childes/index.html.
15. J. R. Searle. Speech Acts. Cambridge University Press, 1969.
16. D. Traum. Coding Schemes for Spoken Dialogue Structure. ftp://sls-ftp.Lcs.mit.edu/pub/multiparty/coding_schemes/traum, 1996.

Structural and Semantic Dialogue Filters

Zdeněk Mikovec, Martin Klíma, and Pavel Slavík

Czech Technical University, Dept. of Computer Science and Engineering
Karlovo namesti 13, 121 35 Praha 2, Czech Republic
xmikovec@fel.cvut.cz, xklima@hwlab.felk.cvut.cz, slavik@cs.felk.cvut.cz

Abstract. The problem of the access of blind people to the graphical information (in a form of picture) is discussed. It is necessary to reduce amount of complex information of the picture to keep understandability of the picture description for blind user. This is solved by special semantic and structural dialog filters, which filter required information. The research has been conducted in the framework of several projects. The result of this research is an implementation of tools for creating and browsing the picture description. Because of the similarity between picture description and hypertext document description the filtering methods developed could be also used on any hypertext document, which follows our document structure [3] definition made in XML [6]. The method developed allows the user to transform two-dimensional information into textual one. In such a way complex manipulations can be performed in textual form.

1 Introduction

The graphical information is more complex than information commonly exchanged between user and information system (IS). The graphical information consist of three parts:

- geometrical (e.g. position, shape)
- structural (e.g. one object is a part of other one)
- semantic (e.g. behavior, semantic relation - photons are exciting electrons).

The data structure used should store information in very general form to keep complexity of graphical information. Therefore we choose the object oriented approach. That means the entire world consists of objects with their behavior and relations among them (see Figure 2, 3). The behavioral and relational information should be also included into data structure. The information about any object usually consists of a structural and semantic information.

This methodology for describing the picture information could also be used in other applications, where some structured information is used – e.g. for hypertext. Objects in hypertext are documents, sections, paragraphs, etc. The structural information is: section, sub-section, paragraph. The semantic information is: type of hyperlinks in document (examples, detailed specification), category of text (example, for advanced users, for novice users, related topic – sport, culture, price offer).

F. Jelinek, E. Nöth (Eds.): TSD'99, LNCS 1692, pp. 280–285, 1999.

2   Methodology of Filtering

The complexity and special structure of graphical information needs a specific form of dialogue between blind users and IS to keep efficiency of communication. For blind users it is very difficult to understand the meaning of graphical information because of large volume of complex information that could be browsed. To help blind users we must simplify the access to information required. This could be reached by separate handling of different types of information (e.g. structural and semantic).

Therefore in the dialogue there are used special filters which allow blind users to reduce amount of information dependent mostly on structural and semantic information. It means, that blind users communicate with IS in three steps:

- select information type (structural or semantic)
- reduce amount of information – filter out inessential information (objects, descriptions of objects) by defining structural and semantic constraints
- browse in reduced information using queries

When browsing it is possible to switch between structural (see Figure 2) and semantic (see Figure 3) view of picture.

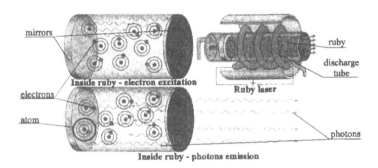

Fig. 1. Ruby laser

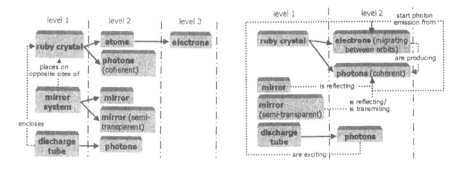

Fig. 2. Ruby laser – Structural view   **Fig. 3.** Ruby laser – Semantic view

2.1   Selecting Information Type (Structural and Semantic Filter)

When analyzing creation and browsing of the picture description we found out that we could perceive the picture from very different points of view, which means that final description could consist of different objects, different object descriptions and different relations. Two types of view were defined:

A. Structural view (see Figure 2) – focused on structural relations

- Goal of this view is to give user standard for picture description.
- Major relation is relation "is in" (full arrow head lines). The objects are ordered into a hierarchical tree. The hierarchy means that the child object is a part of the parent object (e.g. child object "electron" is a part of parent object "atom" – see Figure 2)
- There is a minimum of other relations – dotted arrowhead lines (e.g. Mirror system is placed on opposite sides of ruby).

B. Semantic view (see Figure 3) – focused on semantic relations

- The goal of this view is to describe only significant objects and relations.
- It is not necessary to order objects into a hierarchical tree as in the structural view.
- This view is useful if the picture should be described from several totally different views (these views consist of different objects and relations). In our example (Figure 1) the structural view (Figure 2) describes structure of ruby laser and the semantic view (Figure 3) describes functionality.

2.2   Object and Relation Filter

First method how to filter out information is to choose the type of object description we want to see. While there are defined categories of object descriptions:

- geometrical (position, shape)
- material, color, weight
- category, action (exciting, reflecting on), etc.

the browsing user can choose which description categories he wants to "see" (e.g. color of objects). Applying such a filter will cut out descriptions and objects that don't match chosen categories.

Second method how to filter out information is to define relations between objects we want to see. That means we can define query such as "Find all object which are reflecting photons". For concrete examples see Section 4 "Example and testing of filters".

These filters can be combined in a logical expression (using logical operations AND, OR) or applied in a sequence (see Section 4).

3   Implementation

The conceptual scheme of Blind Information System – BIS (see Figure 4) consists of three modules:

- Grammar&Creating/reading description file (module for creating/reading description files in XML format and executing filter queries)
- Description creation (module for picture description creation)
- Description browser (module for browsing in picture description by blind users)

For detailed information about implementation of BIS see [3,4,5].

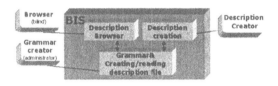

Fig. 4. Blind Information System (BIS) – Conceptual scheme

Implementation of filters is based on two structures. One structure corresponds with structural information and the second with semantic structure of graphical information (see section 2.1 "Selecting information type").

The system has been implemented in JAVA programming language and the data structure is implemented in XML. The implementation has been intensively tested with group of blind and non-blind users.

The modification of creator and browser module for processing hypertext resides in different manipulating with hypertext document. Instead of working with graphical objects we will manipulate with textual objects. Thereby we transform graphical operation to textual. The rest of functionality of the modules will be unchanged.

4   Example and Testing of Filters

The example shown below illustrates the possibility of filtering. The filter conditions may be combined as shown in example C. Using the filters we can build queries, which specify objets, object descriptions and relation between objects. The result of searching is sub-graph (tree) of the description tree (see Figure 2, 3). The leaves of sub-graph (tree) represent filtered objects and edges represents filtered relations between them.

For representation of objects and relations (structural and semantic) we will use following notation:

```
object in the picture: ruby crystal
filtered object: PHOTONS
semantic relation: [is reflecting PHOTONS]
```

```
description of object: (semi-transparent)
hierarchical relation: ruby crystal
                              photons
```
As a testing picture we used the ruby laser (see Figure 1). The following example describes the way of obtaining information about interaction of laser photons. It consists of a sequence of three queries. First need is to locate photons that come from a laser (query A, B) and then to find which objects interact with these photons and how (query C):

A. **Find photons in structural view.**
   Result (sub-graph of graph in Figure 2):
   ruby crystal
      PHOTONS (coherent)
   discharge tube
      PHOTONS
   In Figure 5 we can see result of filtering.

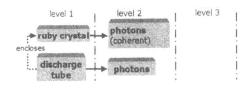

Fig. 5. Result of query A: Sub-graph of graph in Figure 2

B. **Find coherent photons in structural view.**
   Result (sub-graph of graph in Figure 2):
   ruby crystal
      PHOTONS (coherent)
C. **Find any object in any relation with coherent photons in semantic view. Result (sub-graph of graph in Figure 3):**
   ruby crystal
      ELECTRONS [are producing PHOTONS]
      MIRROR [is reflecting PHOTONS]
      MIRROR (semi-transparent)[is reflecting PHOTONS]
         [is transmitting PHOTONS]

4.1   Hypertext Filtering Examples

In this section we will show possible use of our methodology in real applications working with hypertext document. Pictures and hypertext documents (e.g. Web structure) have similar structure. They both consist of types of objects (graphical/textual) with description, structural and semantic relations between objects. When we describe hypertext documents using our methodology we will be able to use structural and semantic filters for automating intelligent search in applications working with hypertext structure.

One application is "Analysis of access on Web structure" [9]. The problem is how to define logical structure of the Web for better access to information. The user has to manually define logical structure of the Web. Using our methodology we can automate creation of logical Web structure by defining structural and semantic information for each Web page. The application will automatically analyze the logical Web structure defined by structural and semantic filter. For example: The pages are divided into several categories – for top managers or for information technology managers. We can define which topics are Web pages related to – prices, technical information, manager information etc. The application can now analyze the access to each category and prepare profile of common visitor of Web pages.

Next application is "Downloading the Web structure". Our methodology can help to filter required information to be downloaded. For each Web page we will define structural and semantic information. Than using semantic and structural filters we can define which Web pages we want to download. For example: "Download pages with technical information related to laser devices.", "Download pages with prices.", etc.

5   Conclusion and Future Work

We have defined methodology and also have implemented modules for the picture description creation and browsing the description. While this description contains structural and semantic information we can use special structural and semantic filters. These filters allow us to automate "intelligent" searching both in pictures and hypertext trees (see Section 4 "Example and testing filters"). The use of XML [6] as the definition language for description file structure simplifies an implementation of our methodology in any application.

The future work will deal (beside the IS improvement) with development of formal model of the type of communication described. We are preparing implementation of our description methodology for intelligent filtering in picture database systems and hypertext structures.

References

1. Melichar B.: Languages and Compilers (in Czech), 1996, CTU Prague.
2. Mark C. Chan, Steven W. Griffith, Anthony F. Iasi: 1001 "Java Programmer's Tips", 1997, UNIS Publishing.
3. Mikovec Z.: BSc thesis – Picture Description for Blind – Object Library, 1998, CTU Prague.
4. Klima M.: BSc thesis – Picture browser for Blind, 1998, CTU Prague.
5. Pavlica D.: BSc thesis – Creating Picture Description for Blind, 1998, CTU Prague.
6. SGML/XML: http://www.w3.org/XML/.
7. Slavik P.: Syntactic Methods in Computer Graphics. Proc. of Eurographics 83, North Holland, 1983, pp. 134–144.
8. Slavik P.: Grammars and Rewriting Systems as Models for Graphical User Interfaces. In: Cognitive Systems. No.4-3, 1997, pp.381-399. ISSN0256-663X.
9. Fuchs M.: Visualization of access in Web structure (in Czech), 1999, Student project, CTU Prague.

A Retrieval System of Broadcast News Speech Documents through Keyboard and Voice

Hiromitsu Nishizaki and Seiichi Nakagawa

Toyohashi University of Technology, Toyohashi, 441-8580, Japan

Abstract. To retrieve interesting broadcast news documents out of an enormous number of TV news programs, if no indexing is done on the news and word-based retrieval is required, it is inevitably necessary to transcribe all the broadcast news documents automatically[1,2]. And this task can be done only by using a Large Vocabulary Continuous Speech Recognition (LVCSR) system.

In this study, first the retrieval performance was experimentally compared between the system using automatically transcribed database and the one using manually transcribed database. Next, to input keywords via voice, we proposed a retrieval technique using the association degree between keywords.

1   System

1.1   An Outline

First, news speech was passed the Large Vocabulary Continuous Speech Recognition system(LVCRS), then, the "database(automatic)" was created. A database that was manually transcribed is called as "database(manual)". Keywords(typing or speech) are inputted to this database, and some related documents are retrieved by a retrieval module. The retrieval module executes full-text retrieval using an indexing method. The documents in which most of keywords are included are retrieved.

1.2   Grouping Keywords

When input keywords via voice are recognized, we should consider the following three cases:

1. the keyword is correctly recognized.
2. the keyword is correctly recognized as a syllable sequence, but incorrectly recognized as a word, i.e. homophone word.
3. the keyword is incorrectly recognized as an incorrect syllable sequence.

In any of the three cases, the system cannot judge whether the recognized result is correct or not. This means that the system has to execute the retrieval processing using the recognized results. And if the recognized result has its homophone, all the homophones of the result must be considered in parallel. Moreover

F. Jelinek, E. Nöth (Eds.): TSD'99, LNCS 1692, pp. 286–289, 1999.

even if it has not any homophone, several words listed in the N-best candidates of the speech recognizer have also to be used as the input keywords to the retrieval system. In any case, since a larger number of keyword candidates than one uttered by a user should be regarded as the input, irrelevant documents might be retrieved by the system. So, to solve this problem, we propose a novel method, where, before the retrieval processing, unproductive candidates are discarded by a grouping processing using the association degree between words.

The association degree is the measure which shows how degree any two keywords are related each other. We used mutual information as the measure. This value is the scale which objectively shows the collocation and relation of the words, and is calculated as:

$$I(W_1; W_2) = \log \frac{\frac{f(W_1, W_2)}{N}}{\frac{f(W_1)}{N} \frac{f(W_2)}{N}}$$

$f(W_i)$: the number of documents in which W_i existed.$(i = 1, 2)$
$f(W_1, W_2)$: the number of documents in which both W_1 and W_2 existed.
N: the total number of documents.

The association degree was trained using the database(automatic). As shown in Figure 1, keyword candidates having high association degree were grouped together. Figure 1 illustrates an example of keyword candidates in the case of three words uttered in isolation. Some keyword candidates connected in the arrow have high association degree to one another and they form a group. In Figure 1, three groups are formed, then the group G_1 that has the most number of keyword candidates is used to retrieve documents.

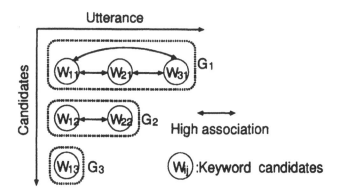

Fig. 1. Grouping of keyword candidates

1.3   Retrieval Method

By the difference between the input manners of keywords, the experiment was carried out by the following five-ways methods.

1. text input
2. voice input (1-best, with grouping)
3. voice input (1-best, without grouping)
4. voice input (3-best, with grouping)
5. voice input (3-best, without grouping)

The "1-best" is the case which the highest likelihood in keyword recognition is used. And the "3-best" is the case which the results of keyword recognition hypotheses up to rank 3 are used.

In the case of voice input, it is very dangerous if only documents that agree with all of the keyword candidates are retrieved, because the constraint is too strong. Therefore, a threshold by which we can retrieve plausible documents with the accuracy more than 95 % was set. In short, it requires the maximum N which satisfies the following inequality:

$$\sum_{i=N}^{M} {}_M C_i p^i (1-p)^{M-i} \geq 0.95$$

where M is the number of keywords input, N is the number of the keywords which exist in a document, and p is the speech recognition rate of keywords.

2   Experiments

2.1   Databases

The speech database which is Japanese NHK broadcast news (1st/June/96 $\sim$ 14th/July/96) were used for this experiment. The number of total documents was 976 (7099 sentences). Half of them were noisy utterances including background music and so on.

The speaker-independent LVCSR system to transcribe the speech database was used a frame-synchronous one pass Viterbi algorithm, while using HMMs as syllable units. The Language models for the first pass was word bigram that has a vocabulary size of 20000. And the second pass used the word trigram. The covering rate of news speech was 96.7 %, the word recognition rate was 54.3 %, and the word accuracy was 38.0 %.

2.2   Results

When the text-input keywords were used, the result for database (automatic) is shown in Table 1, where the recall and precision rate are defined as follows:

$$recall = \frac{the\ number\ of\ retrieved\ ``correct''\ documents}{the\ number\ of\ ``correct''\ document}$$

$$precision = \frac{the\ number\ of\ retrieved\ ``correct''\ documents}{the\ number\ of\ retrieved\ documents}$$

Note that the retrieved documents from the database(manual) using a set of text-input keywords were assumed as "correct".

In the result, the comparatively high recall rate in spite of lower recognition rate was obtained. This is because the recognition rate of the keywords in the transcribed database is better than the whole recognition rate(93%). Using the database(automatic), the performance is lowered at about 20%, but it can be said that the whole recognition rate dose not so influence the retrieval performance.

Table 1. Text_input keyword for database(automatic)

recall[%]	precision[%]
76.2	50.1

Table 2 shows the result when keywords were inputted via voice uttered by two males. From Table 2, it was proven that using up to 3-best hypotheses improves the performance further than using only 1-best. Because the recognition rate of keywords with including up to 3-best(90.0%) was higher than that with only 1-best(88.0%).

Moreover, comparing the reults using the grouping technique with no grouping, the precision in the former case was improved drastically, although the recall in both cases was lowered a little. This means that the many unproductive documents were retrieved without grouping, and the grouping technique reduced the candidates.

Table 2. Voice_input keyword

	for database(manual)		for database(automatic)	
	recall[%]	precision[%]	recall[%]	precision[%]
1-best (with grouping)	82.1	34.6	70.9	23.9
1-best (without grouping)	84.3	22.5	71.7	18.5
3-best (with grouping)	86.7	63.3	70.2	42.5
3-best (without grouping)	91.1	16.3	82.1	6.6

3 Conclusion

We developed the broadcast news speech retrieval system, and it was shown that the retrieval performance was preserved with a high accuracy for the automatically transcribed database. In the experiments by voice input keyword, the keyword candidates that are unnecessary were removed using the proposed grouping technique before the retrieval process, and we obtained the high retrieval ability.

References

1. A.G. Hauptmann,H.D. Wactlar: Indexing and Search of Multimodal Information, Proc. ICASSP, pp. 195–198 (1997).
2. Kenney NG: Towards Robust Methods for Spoken Document Retrieval, Proc. IC-SLP, pp. 939–942 (1998.12).

Situations in Dialogs

Petr Hejda

Czech Technical University, Faculty of Electrical Engineering,
Department of Computer Science and Engineering,
Karlovo nám. 13, 121 35 Praha 2, Czech Republic
hejda@fel.cvut.cz
http://www.cgg.cvut.cz/~hejda

Abstract. A net-based application is a software technology using a
server, clients (browsers), and data describing what the server and the
clients should do. Languages used for communication between the client
and the server define objects presented to the user. They are descriptive
and media dependent. This results into unsatisfying complexity when
developing a net-based applications with browsers running on computers with various input/output capabilities. Our approach is based on
functional description of a dialog. Instead of defining what should be
presented, the browser is instructed about the structure and functional
parts of dialogs. Dialogs are described as collections of standardized parts
called situations. Each situation induces a conversation with the user.
This approach has following advantages. The description of the dialog
is media independent, and its length is not directly proportional to the
number of semantic tokens exchanged with the user. The resulting dialog
can be either sequential or multi-threaded.

1   Introduction

This article has the following structure. Section 2 defines basic terms and introduces a notion of a net-based application. Languages used in common net-based
applications are discussed in Section 3. Section 4 lists problems which inspired
our research. The intrinsic of our solution are introduced in Section 5. The following section argues some issues resulting from our research.. Section 7 lists
some of our experimental results. Summary and future directions are contained
in the final section.

2   Net-Based Applications

Although the boom of the World Wide Web was adherent to browsing static text
with hyper-links, the Web has soon expanded into distributed transaction systems and interactive applications. The architecture of these applications usually
comprises four main parts: a **server** running on a server computer, and a client
(called **browser**) running on a client computer. Both server and client are usually common, general programs running on computers connected via network.

F. Jelinek, E. Nöth (Eds.): TSD'99, LNCS 1692, pp. 290–295, 1999.
© Springer-Verlag Berlin Heidelberg 1999

The functionality of the application is defined by server scripts, and by the data transmitted between the server and the browser.

Figure 1 depicts common structure of a net-based application. When the user demands some information, the browser connects to a server using a specialized protocol (for example HTTP[1]) and it requests some information. In a successful case, the server loads the data from the data storage. Using a specialized language (for example HTML[2]), the data are transmitted to the browser. The browser parses the data and uses the parser results for calling device drivers (or window managers) to present objects described by the data.

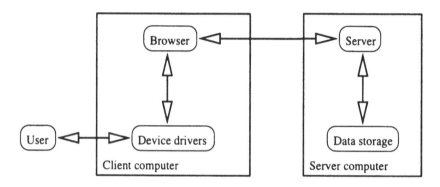

Fig. 1. Architecture of a net-based application

The browsers provide net-based applications with a user interface. The expressive power of the user interface is defined (or rather limited) by the language used for communication between the browser and the server. We can again use HTML as an example. It is capable of describing user interfaces consisting of text, hyper-links, images, image maps, text input fields, buttons, radio buttons, choice selectors, and frames[3].

3   Markup Languages

The existing languages for data transmission between the server and the client are always centered on the capabilities of the client computer. The elements of the language describe presentation objects, their properties, and optionally their behavior.

For example, HTML is oriented towards visual media. Its elements define paragraphs, text formating, tables, layouts, images, and hyper-links. It is capable of presenting a few interaction objects [4]. Lately, it has incorporated basic audio

---

[1] HTTP – Hyper-Text Transfer Protocol
[2] HTML – Hyper-Text Markup Language
[3] There is a possibility to embed an application (applet) into HTML formated data. This might further enhance the expressiveness of the user interface.

features, like sound replay on the background. The side effects of extending the language originally intended for document presentation are incompatibility of clients, and difficult application development.

Other languages, like SpeechML [5] and VoxML [6] focus on sound, voice synthesis and recognition. Their elements include sound data for replay, texts which are to be used by a synthesizer, and grammars used by voice recognition.

Yet another languages are tailor-made to special client computers. HDML[4] is intended to be used by devices with limited display capabilities [1]. The elements of this language are visual, though the structure is different from HTML. WAP WML[5] is oriented to small narrow-band wireless devices [3] like mobile phones or PDAs[6].

All the languages listed above have many common features. They are markup languages easy readable by humans as well as machines. They mark tree-like structures of documents. Each element can have sub-elements and a set of parameters. In spite of these similarities, it is usually very hard to develop a system with several types of client computers which would behave consistently. Another challenge might be a browser which is adapting to a changing environment or changing user's needs.

4   Problem Definition

If we rephrase the previous paragraph, there is a special language for each class of client computers. Each language requires a special server or at least server setup. These are the questions which inspired our research:

- Is it possible to design a universal language for at least a class of applications?
- How to deal with different natures of visual and spoken dialogs?

5   Dialog as a Set of Situations

As mentioned above, the browser provides the net-based application with a user interface. From this point of view, the main task of the browser is dialog management. The data transmitted between the browser and the server should deal with dialog issues. The language has to describe dialogs, their structure and parameters.

Our solution is based on the notion of situations. A **situation** is the smallest part of a conversation, which is instrumental for obtaining a piece of data from the user. For example conversation held with a person while enquiring about the current time can be considered as a situation. The description of the situation may be as simple as:

<TIME> What time is it now? </TIME>

---

[4] HDML – Handheld Device Markup Language
[5] WAP WML – Wireless Application Protocol's Wireless Markup Language
[6] PDA – Personal Digital Assistant

The conversation, on the other hand, may vary. One browser may display a watch-like widget and process the events while the user manipulates the widget. Other browser may voice the question and listen to a user until it is sure about the answer.

For each type of sought after information there is a special class of situations. This has two benefits. The browser can limit the user to input only acceptable values. For instance, if the result is supposed to be a boolean value (answer to a simple question), the browser may display only two radio buttons. In case of a voice-based browser, the recognizer might be sensitive only to words "yes, no, not, yap", etc... The other benefit is that the browser can check if the result is suitable or not. This principle can be further enhanced by constraining the possible set of answers. An example can be setting a minimum and a maximum value while expecting a number.

Each situation has to result into an outcome in order to be able to proceed with the execution of the application. We call this process **solving** the situation. Which means of communication are used to solve a situation is fully under the control of the browser. The result is what counts.

A **dialog** is a set of situations which can be processed at one time. They should be independent in the sense that none of the situation's parameters depend on the results of some other situation from the same dialog. The intention is to generate and transmit a dialog at once. A dialog is an analogy to the page in hypertext systems. One session usually involves solving of several dialogs.

6   Dialog System Based on Situations

The interaction with user is described as a sequence of dialogs. Each dialog consist of one or more independent situations. This section discusses some aspects of a system based on this approach.

6.1   Browser Issues

The browser receives from the server one dialog at a time. There are two possible ways how the situations are presented to the user. The first way is to solve the situations in sequential manner. This is more suitable for example for speech-based user interfaces.

The second option is to display some presentation for all the situations in the dialog at once. This results in a multi-threaded interface style, where user can work with any interaction object at any time. This style is more suitable for instance for client computers equipped with large screens. In any case, the browser can decide which style to use according to the capabilities of the client computer.

The sizes of description of a situation and of the resulting conversation are not directly proportional. The browser can for example demand multiple confirmations of user's input while using voice recognition in a noisy environment.

The mapping between classes of situations and interaction styles need not be static. The same situation can result in a different kind of conversation according to user abilities, user preferences, or environmental issues.

6.2 Server Issues

The server's task is to generate situations, collect them into dialogs, and send dialogs to a browser. Using our approach, there is one server for all types of client computers. This reduces significantly the effort made while developing a widely accessible application.

7 Experimental Results

In order to prove our ideas workable, we have implemented Dialog Markup Language [2]. It is an XML1.0 application which ensures platform independence and network transparency. The latest version (June 1999) has seven classes of situations. Table 1. summarizes their names, types of result, and additional parameters.

Table 1. Situation classes in DML

Class	Result	Parameters
Message	(none)	id
Question	boolean	id, default
Choice	option(s)	id, default, multi
Text	string	id, default, grammar
Number	double	id, default, min, max, step
Date	Date	id, default, from, to
Time	Time	id, default, from, to

We have implemented several experimental browsers simulating browsers running on computers with various capabilities. The list and brief characterizations follows:

– Desktop computer. Large screen, keyboard, mouse.
– PDA. Small screen, pointing device, gesture recognition, three buttons.
– Phone. Voice synthesis and recognition.
– Fax. Form printing and OCR[7].
– HTML bridge.
– Text terminal.
– Applet.

To be able to test the clients, we have developed several test servers. They mainly allow for browsing a database, or for changing it. The results are available for public at http://www.cgg.cvut.cz/~hejda/dml/.

---

[7] OCR – Optical Character Recognition

8   Summary

A novel approach for describing a man-machine conversation was presented. It is suitable for net-based applications. The conversation is described as a sequence of dialogs, where dialogs consist of one ore more independent situations. The situation is defined as the smallest part of a conversation instrumental for obtaining a piece of data from the user. Situations are standardized and parametrized.

Our approach is useful when developing net-based applications and using computers with various capabilities as clients. The description of a dialog is media independent. The resulting dialog can be either sequential or multi-threaded. We can see its benefits in easy adaptability and customization.

So far, DML does not incorporate any interaction between situations. They are completely independent. Some situations may operate on common data structures. In the future, we would like to extend DML to incorporate context as a mediator between related parts of a dialog.

References

1. Handheld Device Markup Language
   http://www.w3.org/TR/NOTE-Submission-HDML-FAQ.html.
2. Hejda, P.: Dialog Markup Language. In: Poster 99, page. Faculty of Electrical Engineering, Czech Technical University in Prague, 1999.
3. Herstad, J. and Van Thanh, D. and Kristoffersen, S.: Wireless Markup Language as a Framework for Interaction with Mobile Computing and Communication Devices. Proceedings of the First Workshop on Human Computer Interaction with Mobile Devices. Scotland, 1998.
4. Putz, S.: Using World-Wide Web Hypertext as a Generic User Interface. Proceedings of ACM Hypertext 1993.
5. Speech Markup Language
   http://alphaworks.ibm.com/tech/sml.
6. VoxML http://www.voxml.com/voxml.html.

Components for Building an Automatic Voice-Dialogue Telephone System

Miroslav Holada

SpeechLab, Department of Electronics and Signal Processing,
Technical University of Liberec, 460 17 Liberec, Czech Republic
tel. +420 48 53 53 182, miroslav_holada@vslib.cz

Abstract. This paper is focused on components for building and providing an automatic voice – dialogue system. The components are represented primarily by the software obtaining the speech data for the speech recogniser and the software for user-easy designing dialogue applications.

1   Introduction

During the last 3 years we have developed our own platform for creating voice dialogue applications intended mainly for an automated information system operating over telephone. The system uses speech synthesis at the output side and a HMM-based recogniser of single words and short phrases at the input side.

The communication between the computer and a user is based on a controlled dialogue in which the system provides prompts and offers corresponding entries according to the dialogue scenario. Nowadays, the system platform has been used and publicly tested in a concrete application called InfoCity. The InfoCity provides callers by information from the area of culture, sport and transport in the town Liberec [1].

The aim of this contribution is to describe the components of the dialogue system that are necessary for its operation and, namely, for its preparation and adaptation for new tasks. These components include programs and program modules for creating new application vocabularies, recording telephone data, checking and visualising recorded data and trained Markov models. We have also developed our own dialogue description language for easy creating dialogue application and its graphic interactive environment.

All described programs and components work on MS Windows 95 or NT platform. Our application Infocity runs on Pentium MM 200Hz.

2   The Recording and Checking Tools

In this part are described components, that allow recording, detecting and preparing data for HMM. Figure 1. shows connections among components and data flows. The first part is focused to recording data from speakers by RECTEL program. The second part converts recorded data from wave file format to our

F. Jelinek, E. Nöth (Eds.): TSD'99, LNCS 1692, pp. 296–301, 1999.
© Springer-Verlag Berlin Heidelberg 1999

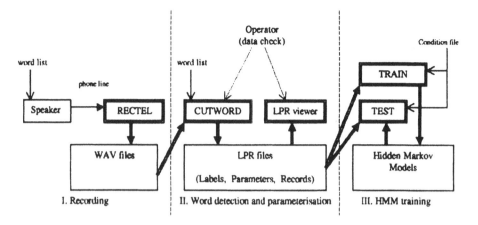

Fig. 1. Block scheme of location recording and training components in system.

internal LPR file format (Labels, Parameters and Records) and detects words in the stream and computes recognition features. In the third part is included software that trains HMM from saved LPR data and also software that off-line tests obtained HMM. Incoming condition file allows exactly define HMM conditions for training single- or multi-mixture word models.

2.1 RECTEL

The recording tool RECTEL was developed for recording and working with larger vocabularies (for example, the InfoCity application database consists of 220 words that had to be recorded from 40 native speakers). This program automatically records the data from a telephone line or from a microphone and saves them in the 'wav' format. The speaker can do that either with or without prompting. The latter is, of course, much faster. The program can work fully automatically. It is necessary when speakers record data by telephone line from various places (other cities) at various time.

First experience has shown that the best way of recording large vocabularies is continuos vocabulary reading by speaker. It takes 8 – 12 minutes per 200 words and short phrases width short pauses between words. When speaker makes a mistake he can say wrong word again.

2.2 CUTWORD

After that the recorded data are divided into speech and non-speech parts. This is done by the program CUTWORD which finds the beginnings and ends of single words in the wave files.

The program operates automatically but the found beginning of a word or the end of word are not exact, it allows the user for hand correcting. The words are detected by same 'endpoint' algorithm that is used in the real recognizer.

At the same time it does the signal parameterisation and provides acoustic and visual check of the data.

Because LPR file format is in binary form we must develop the LPR data viewer. This viewer is used to browsing and checking recorded data. At first sight we can compare different quality of recorded data.

3   Description Language

To build up and test new dialogue system applications we have developed a language (RDL – readable description language) that allows the easy description of the dialogue process. In our case, it is a simple BASIC like language. We use it to compile easy readable text files that are interpreter listed by the dialogue manager during the operation of the system. At this time the dialogue description language includes about 25 commands and can be used in a wide range of applications. We may split the RDL commands in some few partitions:
– declaration commands – they allow declare variables, used vocabulary, directory with HMM models and prompt waves.
– system commands – they describe start and end of program and subroutines, remarks, jumps, cycle, if clause.
– input and output – these commands operate with recogniser and synthesiser or saved prompts.
– database commands – access to application databases. This database contains for example in Infocity application departures and arrivals of trains.
– other – mat functions, string functions . . .

The RDL supports only one type of variables, it is string. When user needs to make basic mathematical operation, he can insert numerical values in string variables and makes mat operation by standard signs (plus, minus, . . .). The

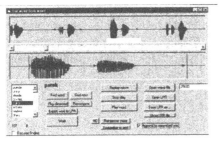

Fig. 2. CUTWORD program

```
REM "Test program: yes or no"
SRC_FILE Test_YN\Test_YN.SRC
WAVE_DIR waves\
DIM       yes_no REM "declare variable"
INTRO:    SAY "Say yes or no" intro.wav
RECOG_YES_NO: RECOG yes      OK
                    no
                    breath noise
              END
              ANSWER yes_no
SAY_RESULT:   SAY "You said" answ.wav
              SAY WORD        yes_no
END_DIALOG:   STOP
```

Example 1. Short dialogue application described by RDL (textual form of description language)

dominant values in variables are words from recogniser, results from database and system constants.

In the last version of RDL we must implement subroutines supporting. It is necessary for large applications and subroutines makes source code well-arranged.

4 Graphic Environment

Recently we made a prototype of graphic interactive environment. This tool allows to display a text description file as visual diagram. The user can easy browse dialogue application in the visual diagram, make changes and update, edit variables. Graphic environment also supports a creating of new application.

The operator can save the visual diagram in it's own format (GDL) or convert it to the text description file. The GDL (Graphic Description Language) is used for storage current diagram to disk. It is similar as RDL but file contain information of location boxes in graph, colours and breakpoints. But when we can use application in real interpreter, we must convert it in the RDL format.

Graphic environment has a three general areas of use:

1. Easy to create new application
User can create new spoken dialog. He must fill 'General properties' form, where is declared name of application, used variables, vocabulary, working and other directories.

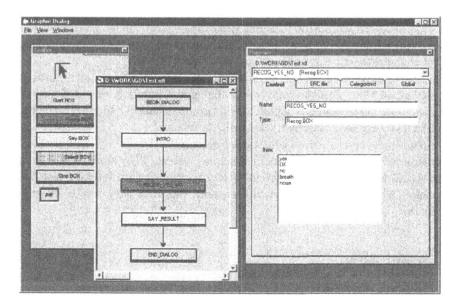

Fig. 3. Graphical environment with simple TEST dialog application.

After it user must create the graphical diagram of application. The diagram is composed of boxes and connection lines with arrows. All drawn objects are moveable by common drag and draw process. User can dislocate boxes in workspace at pleasure. Each box represents one command from RDL and has particular properties. For example recognizer command is represented by next properties: name, list of words, time out for response, alias and noise models. We must fill properties form or copy it from other box and make necessary changes.

User also can create new subroutines, that are presented as new window in workspace. The subroutines can have arguments that are passing by reference in subroutine.

However, when we begin to design new application, we probably don't known all words that we will use. At this time we can use option 'Automatic generate new vocabulary'. System has in its basic setting declared all since existing databases and automatically offers stored words. If required word doesn't exist in databases, we can add it as new word. After finishing new application system generate list of used words with numbers of speakers in existing databases. The list of words is used in the Rectel and Cutword components to obtain training data for HMM.

2. Debug existing applications

When our system runs on phone line, it makes log file, where is described dialog path with all prompts, recognised word and response time. The graphic environment allows to load this log file and display current way across dialogue. It is useful when we want to debug and optimise application.

Saved log session could by off-line replayed and stopped at position where we insert breakpoint. At this time we can debug critical part of our application. Application should by traced step by step too.

In special watch window are displayed variables with values. Current value we can change and observe what happen in the next step. In debug mode is impossible change structure of application.

The system allows replay all prompts and we can change it and immediately replay again. The most optional sets of dialogue prompts is very difficulty task. It must be apposite and intelligible to wide group of callers. On other side prompt mustn't be very long, because caller calls by phone.

3. Graphical representation of designed system

Some people ask us how works our dialogue system. They would know branching and kind of offered information. When we can explain function and design principle of our spoken dialogue application, the most easy way is by graphic scenario.

5   Conclusion

All described components are useful for designing and debugging dialogue applications in various national languages. The components also allow to check quality of recorded data and thereby increase quality of speech recogniser. We have created large system that allows to user easy create new spoken application in user friendly interactive environment. This is a way of designing dialogue systems for widely connection users.

Acknowledgement. The work was supported by the COST 249 European Action and GACR grant no. 102/96/K087. Basic part of graphic studio was made during short time mission in CPK Aalbork, Denmark. Also I would like to thank my colleagues from Speechlab for they remarks.

References

1. Holada M., Nouza J.: A City Information System Operating on Telephone. To appear on IVVTA Workshop, Torino, Italy, Sept. 29–30,1998, pp. 141–144.
2. Larsen, L.B.: Voice Controlled Home Banking – Objectives and Experiences of the ESPRIT OVID Project. Proc. Of IVTTA'96, Basking Ridge, 1996.
3. Nouza J., Hajek D.: A Bus Time-Table Information System with Voice Input and Output. In Proc. Of ECMS'95 workshop, Liberec 1995, pp. 62–65.

Modeling of the Information Retrieval Dialogue Systems

Ivan Kopeček

Faculty of Informatics, Masaryk University
Botanická 68a, 602 00 Brno, Czech Republic
kopecek@fi.muni.cz
http://www.fi.muni.cz/~kopecek/

Abstract. Formal models of dialogue systems and related notions (dialogue, dialogue strategy, evaluation function, dialogue communication, goal of dialogue) are introduced in the paper. A model of the information retrieval dialogue systems based on the dialogue strategies described by the Mealy-type automata and information systems (in the sense of Pawlak) is introduced and discussed.

1 Introduction

As a consequence of their applicability, information retrieval dialogue systems are nowadays in the centre of attention of researchers as well as software companies. Their structure and properties are becoming more and more sophisticated and the focus approaches subtle and complicated problems of syntax, semantics and pragmatics of the dialogue act as well as the field of user and system modeling.

Abstract modeling of the information retrieval dialogue systems, i.e. the modeling of the real systems on the base of some formal structures and terms of discrete mathematics, is motivated by the following assets that the formalization brings:

1. Systemization; this is without doubt an important aspect especially in such a young and dynamically developing field as the dialogue systems theory is.
2. Generalization unifying points of view and approaches.
3. Standardization of terms and notions.
4. Last but not least, applicability of some general methods (or methods developed for related formal fields) and the related results for our formal models.

2 Notation

If M is a set then M^* denotes the free monoid over the set M. By 2^M we denote the set of all subsets of the set M and by $\emptyset$ the empty set. An alphabet is a finite nonempty set.

F. Jelinek, E. Nöth (Eds.): TSD'99, LNCS 1692, pp. 302–307, 1999.

3 Dialogue and Dialogue Strategy

Let Σ be an alphabet, $L \subseteq \Sigma^*$ and $u \in L^* \cup \{\emptyset\}$. Then u will be called *dialogue*.

Consider a dialogue $u = (u_1, u_2, ..., u_n)$. The odd indexed elements of u correspond to the utterances of first participant of the dialogue and the even indexed elements correspond to the utterances of the second participant of the dialogue. The utterances u_i will be also called *steps* of the dialogue u. The case $u = \emptyset$ corresponds to the empty dialogue.

Let $\#$ be a special symbol which does not belong to the alphabet Σ. A function

$$s : \{L^* \cup \{\#\}\} \to \{L \cup \{\#\}\}$$

will be called *dialogue strategy*. The set of all dialogue strategies will be denoted by $S(L)$.

Let s be a dialogue strategy, $u \neq \#$ and $s(u) = v$. Based on the present dialogue u, the dialogue strategy s determines the continuation of the dialogue, which is uv, provided $v \neq \#$, or terminates the dialogue, if $v = \#$.

Suppose now that $u = \#$ and $s(u) = v$. Then the dialogue strategy s determines the beginning element of the dialogue, which is v, provided $v \neq \#$, or terminates the dialogue, if $v = \#$ (resulting in the empty dialogue).

Let $s_1, s_2 \in S(L)$. We shall say that the ordered pair (s_1, s_2) *generates* the dialogue $u = (u_1, u_2, ..., u_n) \in L^*$ (or that the dialogue u is *generated* by (s_1, s_2)), if $s_1(\#) = u_1, s_2(u_1) = u_2, ..., s_1(u_n) = \#$ provided n is even or $s_2(u_n) = \#$ provided n is odd. We will use the notation $u = [s_1, s_2]$.

4 Dialogue Communication and its Goal

A function E from L^* into the set of real numbers will be called *evaluation* function. The set of all evaluation functions will de denoted by $E(L)$.

Let $S_1, S_2 \subset S(L)$ and $E_1, E_2 \in E(L)$. Then the ordered quadruple

$$M = (S_1, S_2, E_1, E_2)$$

will be called *dialogue communication*. Let α be an equivalence on the set L^* satisfying $(u, v) \in \alpha \Rightarrow E_1(u) = E_1(v)$ and $E_2(u) = E_2(v)$. Such equivalence will be called *E-equivalence*. The greatest E-equivalence in the dialogue communication M will be denoted by $\sigma(M)$.

Further, let β be an equivalence on the set L. Let us denote by $[\beta]$ the induced equivalence on the set L^* (i.e., $(u, v) \in [\beta]$ if the length of u and v is the same and corresponding elements are β-equivalent). If $[\beta]$ is an E-equivalence, the equivalence β will be called E_L-*equivalence*. The greatest E_L-equivalence on the set L will be denoted by $\sigma_L(M)$. The set of all classes of the equivalence $\sigma_L(M)$ will be denoted by $\sigma_L[M]$. In other words, the elements of a class of $\sigma_L[M]$ are in any step of the dialogue interchangeable with respect to the evaluation of the dialogue.

Goal of the dialogue communication for the first (second) participant of the dialogue is to minimize the function $E_1([s_1, s_2])$ $(E_2([s_1, s_2]))$, i.e. to choose such a strategy $s_1 \in S_1$ $(s_2 \in S_2)$ that $E_1([s_1, s_2]) = min$ $(E_2([s_1, s_2]) = min)$. Typically, the participant wants to minimize the number of steps that are needed to obtain desired information.

Let us remark, that dialogue communication can be considered from the game theoretical point of view as a two-person game with the sets of strategies S_1, S_2, and with the payoff functions E_1, E_2 (see e.g. [1]).

From another point of view, dialogue communication can be also considered as an algebra with unary operations that are represented by the natural extension of the range of the correspondent strategies to the set $L^* \cup \{\#\}$. Both viewpoints induct possibilities for applying their techniques and results.

5   Classification

A dialogue communication $M = (S_1, S_2, E_1, E_2)$ will be called:

a) *cooperative*, if $E_1 = E_2$.
b) *non-cooperative*, if $E_1 \neq E_2$.
c) *zero-sum*, if $E_1 = -E_2$.

Further, M will be called *finite*, if the semantic equivalence $\sigma^*(M)$ has a finite number of classes.

Let $f, g \in S(L)$. We shall say, that a dialogue strategy f is *finite*, if the set $\{f(u); u \in L^*\}$ is finite. If f is not finite, it will be called *infinite*. Finite dialogue strategies can be expected by computer dialogue systems. On the contrary, infinite dialogue strategies can be typically used by humans.

6   Dialogue Strategies Represented by the Mealy-Type Automata

Let us briefly recall that a Mealy-type automaton is an ordered quintuple $\mathbf{A} = (A, X, Y, \lambda, \delta)$ where A, X, Y are finite non-empty sets, $\delta : A \times X \rightarrow A$ and $\lambda : A \times X \rightarrow Y$ are functions called *transition function* and *output function*, respectively (see e.g. [2]).

Sets A, X and Y are the sets of *states*, *input symbols* and *output symbols*, respectively. Being in the state $a \in A$, the automaton detects the actual input symbol, changes its state according to the function δ and outputs a symbol according to the function λ.

Hence, the Mealy-type automaton can convert an input string (consisting of input symbols) into an output string (consisting of output symbols). To define this conversion uniquely, one element of A must be distinguished as the initial state. In what follows we will always suppose that such an element is determined and we will indicate it by zero index.

Let us consider a finite dialogue communication $M = (S_1, S_2, E_1, E_2)$. To each strategy s from S_1 or S_2 we assign the Mealy automaton

$$\mathbf{A} = (A, \sigma_L[M] \cup \{\#\}, Y \cup \{\#\}, \lambda, \delta)$$

as follows:

1. $s(\#) = \lambda(s_0, \#)$
2. $s((u_1, ..., u_n)) = \lambda(\delta([u_1], ..., [u_{n-1}]), [u_n])$,

where $[u_i]$ is the class of $\sigma_L[M]$ containing u_i. (Here we suppose the function δ to be naturally extended to $A \times X^*$.)

We put $\sigma_L[M] \cup \{\#\}$ to be the set of input symbols because we do not expect a finite strategy from the user. On the other hand, we expect that M is finite, and hence the set of input symbols for the Mealy automaton is also finite. Further, we suppose that the strategy of the system will be finite, i.e. that Y is a finite subset of L.

7   Information Retrieval Dialogue Systems

First, let us recall the notion of *information system* (see e.g. [8]): Let U, T, V be finite nonempty sets and let f be a mapping $U \times T \to V$. Then the ordered quadruple $S = (U, T, V, f)$ is said to be *information system*. The elements of U are called objects, the elements of T attributes, and the elements of V values of attributes. Two objects u, v are called interchangeable if $f(u, t) = f(v, t)$ for all $t \in T$. In what follows, we will assume that no objects are interchangeable.

Let us consider the following model of information retrieval dialogue system. The goal of the user is to get full information about an object (which can be described by means of attributes) in minimal number of steps. This goal is shared also by the system. We will consider a cooperative dialogue communication model

$$M = (S_1, S_2, E)$$

for which E is defined as the weighted sum of steps needed to get the information (i.e. to identify the object by means of attribute values). Let us consider the following types of utterances that can be used in the investigated dialogue:

u_1. Give me more information ($f(x, ?) = ?$);
u_2. Tell me the value of an attribute of the demanded object ($f(x, t) = ?$);
u_3. Tell me whether the information $f(x, t) = v$ is true ($f(x, t) = v?$);
r_1. Response to u_1 in the form ($f(x, ?) = ?$) $= (f(x, t) = v)$;
r_2. Response to u_2. in the form ($f(x, t) = ?$) $= (f(x, t) = v)$;
r_3. Response to u_3 in the form ($f(x, t) = v$) $= true$ or ($f(x, t) = v$) $= false$.

The evaluation function E can assign weights to the utterances according to the type of utterance. Suppose that the utterance space L contains natural language representations of these types of utterances and that $\sigma_L[M]$ may be represented using the following representatives:

$$u_1, u_2(t), u_3(t,v), r_1(t,v), r_2(t,v), r_3(true/false)$$

where $t \in T$ and $v \in V$. For the sake of brevity we restrict ourselves to the utterance space

$$\sigma^*[M] = \{u_1, u_2(t), r_1(t,v), r_2(t,v)\}.$$

Let us first define a simple dialogue strategy of the user that will be represented by the Mealy automaton $\mathbf{A}_U = (A_U, X_U, Y_U, \lambda_U, \delta_U)$ as follows:

$A_U = \{s_0\}$,
$X_U = Y_U = \sigma_L[M] \cup \{\#\}$,
$\lambda_U(s_0, \#) = r_1(t,v)$,
$\lambda_U(s_0, u_1) = r_1$,
$\lambda_U(s_0, u_2) = r_2$.

(We have omitted not essential data in the description of the transition function). Now, let us define the system dialogue strategies that will be modeled by the following Mealy automaton $\mathbf{A} = (A, X, Y, \lambda, \delta)$:

$A = 2^T$,
$s_0 = \emptyset$,
$X = Y = \sigma_L[M]$,
$\delta(\{t_0, ..., t_n\}, r_1(t,v)) = \{t_0, ..., t_n\} \cup \{t\}$,
$\delta(\{t_0, ..., t_n\}, r_2(t,v)) = \{t_0, ..., t_n\} \cup \{t\}$.

We will distinguish the following strategies with respect to the definition of the function λ:

Strategy s_1:
$\lambda(\emptyset, r_1(t_1, v)) = u_2(t)$ $(t_1 \neq t)$,
$\lambda(\{t_0, ..., t_n\}, r_2(t_{n+1}, v)) = u_2(t)$, $t \notin \{t_0, ..., t_n, t_{n+1}\}$,
$\delta(\{t_0, ..., t_n\}, r_2(t_{n+1}, v)) = \#$ if $\{t_0, ..., t_n, t_{n+1}\}$ determines an object.
For this simple strategy $max\ E[s, s_1]$ is dependent on $card(T)$.

Strategy s_2:
$\lambda(\emptyset, r_1(t_1, v)) = u_2(t)$ $(t_1 \neq t)$, $\lambda(\{t_0, ..., t_n\}, r_2(t_{n+1}, v)) = u_2(t), t \in R, t \notin \{t_0, ..., t_n, t_{n+1}\}$, $\delta(\{t_0, ..., t_n\}, r_2(t_{n+1}, v)) = \#$ if $\{t_0, ..., t_n, t_{n+1}\}$ determines an object,

where R is reduct of the information system (see [6,7,9]).

For this strategy $max\ E[s, s_2]$ depends on $card(R) \leq card(T)$ (provided the weight coefficients equal 1).

By finding a minimal reduct and by improving the strategy so that only the values of the corresponding attributes are required, we can obtain optimal results for this type of strategies. Unfortunately, the problem of finding a minimal reduct is NP-hard. However, heuristics or approximative algorithms can be used. An algorithms for finding reducts is also described in [6].

Strategy s_3: Strategy s_2 can be improved by determining of an optimal order of inquires. We do not know, however, any polynomial algorithms which would solve this problem. Generally, the problem lies in finding optimal decision tree based on the given information system.

Strategy s_4: This strategy uses both u_1 and u_2 and combines them suitably to obtain optimal results. Provided the user answers u_1 inquiries reasonable (the response must not be the worst characteristics of the object; this also means that the user applies more suitable strategy than we suppose in this discussion), we can obtain better results than for strategy s_3. These strategies are nowadays supported by some heuristics and we are trying to obtain also more general results.

8   Conclusions

The presented model of the information retrieval dialogue systems is in a general form being implemented the dialogue programming system Dialog (see [4,5]) and also to the developed dialogue system for support of programming of visually impaired people (see [3]). The future work in this field will be aimed at the analysis of the decision trees based on the information systems and their applications for the optimization of the information retrieval dialogue systems, and also at the possible applications of the game theory and automata theory.

Acknowledgement. The author is grateful to K. Pala and M. Novotný for reading the draft of the paper and valuable comments. The research has been partially supported by the Czech Ministry of Education under the Grant VS97028 and by the Grant Agency of the Czech Republic under the Grant 201/99/1248.

References

1. Blackwell, D., Girshick, M.A.: Theory of Games and Statistical Decisions, J. Willey and Sons, New York, 1954.
2. Gécseg F., Peák I.: Algebraic Theory of Automata, Akademiai Kiado, Budapest, 1972.
3. Kopeček, I., Jergová, A.: Programming and Visually Impaired People; in Proceedings of the XV. World Computer Congress, ICCHP'98, Wien-Budapest, September 1998, pp. 365–372.
4. Kopeček, I.: Dialog Based Programming; Proceedings of the First Workshop on Text, Speech and Dialogue – TSD'98, 1998, pp. 407–412.
5. Kopeček, I.: Basic Principles of the Programming System Dialog, FI MU Report Series, FIMU-RS-98, 1998.
6. Novotný, M., Pawlak, Z.: Algebraic Theory of Independence in Information Systems, Fundamenta Informaticae XIV (1991), pp. 454–476.
7. Novotný, M., Pawlak, Z.: On a Problem Concerning Dependence Spaces, Fundamenta Informaticae 16(1992), pp. 275–287.
8. Pawlak, Z.: Information Systems, ICS PAS Reports, 338, Warszawa 1978.
9. Rauszer, C.M.: Reducts in Information Systems, Fundamenta Informaticae, 15 (1991), pp. 1–12.

Improvement of the Recognition Rate of Spoken Queries to the Dialogue System[*]

Václav Matoušek and Jana Ocelíková

Department of Computer Science and Engineering
Faculty of Applied Sciences, University of West Bohemia
Univerzitní 22, CZ – 306 14 PLZEŇ, Czech Republic
matousek@kiv.zcu.cz, jnetrval@kiv.zcu.cz

Abstract. This paper describes an approach to the significant improvement of the sentence accuracy before the linguistic analysis of spontaneous pronounced queries to the "natural speaking" dialogue information retrieval system by using language models of several kinds. The results of an application of language models of three different types (n-gram, polygram and permugram ones) to the output of a standard word recognizer are presented after the description of the theoretical background of all used language models.

1 Introduction

A robust continuous and spontaneous speech recognition and understanding system has been implemented within the framework of the development of a "natural speaking" information retrieval dialogue system at the Department of Computer Science and Engineering of the University of West Bohemia in Pilsen. The main problem to be solved was the significant improvement of the sentence accuracy before the linguistic analysis of processed utterances. The achieved sentence recognition rate using a classical word recognizer [7] (see the left part of Fig. 1), which was fully described e. g. in [6], didn't exceed 46 %, although the recognizer trained on large amounts of domain specific speech data yielded a word error rate of only 8 – 16 %. A majority of wrong recognized words was composed of prepositions; their use in Czech is unfortunately indefinite and – twice unfortunately – very important. E. g., the sentence *V kolik hodin jede vlak do Prahy? (At what time is leaving the train to Prague?)* is semantically absolutely different from the sentence *Kolik hodin jede vlak do Prahy? (How many hours goes the train to Prague?).* Therefore a special procedure for the sentence structure correction by using several kinds of language models had to be developed to improve the sentence accuracy. The classical bigram to pentagram language models [3], as

---

[*] The work presented in this paper was partly supported by the Copernicus Joint Research Project COP-94/1634 SQEL and by the Czech Ministry of Education under contract number MSM 2352 00005. The authors would also like to sincerely thank the colleagues from the University of Erlangen for valuable suggestions and discussions.

F. Jelinek, E. Nöth (Eds.): TSD'99, LNCS 1692, pp. 308–314, 1999.
© Springer-Verlag Berlin Heidelberg 1999

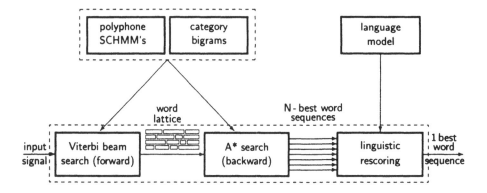

Fig. 1. A principal word recognizer structure

well as the polygram [5] and permugram [8] language models with perplexities
5.2 to 11.9 were created and tested. The model perplexity values seem rather
low, but Czech is a very inflective language with e. g. noun declension into seven
cases and the words occur in a large number of word forms. For example, the
word "město" *(the town)* can occur altogether in twelve different word forms.
The number of processed word forms is then very high and it essentially reduces
the perplexity of the language model.

Sentence hypotheses recognized by the word recognition module are corrected with the help of a special program module, that examines step by step
the obtained best word sequence, searches for all possible n-grams, polygrams,
or permugrams, and tries to correct the sentence hypothesis according to the
occurrence probabilities of corresponding n-grams, etc. E. g., the above mentioned sentence can be corrected at least by the trigram, more frequently by
the tetragram, polygram, or permugram language model. The results of corrections of several word sequences (sentence structures of several kinds) by above
mentioned language models will be evaluated in the third paragraph.

2   Theoretical Background

The classical n-gram language model determines the apriori probability of a word
sequence. This probability can be determined from a product of conditional
probabilities, where we need for each word from the lexicon a probability that
it follows any possible word sequence. The joint distribution $P(w_1, \ldots, w_n)$ for
a given sequence $\mathbf{w} = w_1, \ldots, w_n$ of words from a vocabulary $\mathcal{V}$ of size L may
be written as a conditional decomposition

$$P(w_1, \ldots, w_n) = \prod_{i=1}^{n} P(w_i|w_1, \ldots, w_{i-1}) =$$

$$P(w_1) * P(w_2|w_1) * \ldots * P(w_n|w_1, \ldots, w_{n-1}).$$

The conditional n-gram probabilities on the right hand side of the above presented equation are replaced by the maximum likelihood estimators

$$\hat{P}(w_i|w_1, \ldots, w_{i-1}) = \frac{\Phi(w_1, \ldots, w_i)}{\Phi(w_1, \ldots, w_{i-1})}$$

where the function $\Phi(...)$ counts the frequency of a given word sequence in the training corpus [3]. Since it is impossible to determine conditional probabilities of sequences of various lengths, the word sequence probability has to be approximated with conditional probabilities, where each word depends only on some previous words. The polygram language model [7], in which each word depends only on some previously occuring words (it is impossible to determine conditional probabilities of sequences of various lengths), was used as the more appropriate language model. Conditional word probabilities were estimated by relative frequencies of different word sequences from the training corpus. Even for small i the above frequency ratios are far from being reliable estimates. Smoothing of these statistics can be achieved by pruning the word histories or by partitioning the vocabulary into word categories. The *polygram language model* does without history pruning and evaluates the conditional n–gram probabilities by the linear interpolation formula

$$\hat{P}(w_i|w_1,\ldots,w_{i-1}) = \rho_i \hat{P}(w_i|w_1,\ldots,w_{i-1}) + \ldots + \rho_1 \hat{P}(w_1) + \rho_0 \hat{P}_{uniform}$$

where $\hat{P}(...)$ denotes the ML estimate, $\hat{P}_{uniform}$ is the zerogram or uniform word probability $1/L$, and the weights $\rho_1, ..., \rho_i$ are iteratively optimized by the estimation–maximization algorithm using a cross-validation data set.

An another approach to the problem of missing word sequences used in our case is the introduction of word categories. Words from the recognition vocabulary are assigned to word categories (the categories group words with same grammatical and semantical characteristics), and so the number of parameters of the language model can be drastically reduced. But, the estimation of word probabilities in category polygrams is not simple procedure for languages that are distinguished for the free sentence word order (free sentence structure). The created polygrams have to be sorted by the occurence probabilities to minimize the processing time of the linguistic rescoring of N-best word sequences. Altogether 212 word categories to classify the words were defined from the recognition vocabulary. Unfortunately, there are some words showing indefinite classification to the word categories – e. g. the words "dlouhá", "nové" occur in sentences both as the noun describing the city name (Dlouhá, Dlouhá Třebová, Nové Město) and as the adjective in "dlouhá cesta" (long way), "nové spojení" (new connection). In a similar way, the word "Pátek" denotes the village name, or the weekday (Friday). In opposite to this example, the word "česká" has the same meaning in the word connections "Česká Kamenice", "Česká Skalice" as the adjective in "Czech language". These category assignments have to be carried out by hand or – more appropriate – automatically from the sentence context. However, the second approach significantly complicates the creation of the polygram language model which have to be determined in several iterations. Furthermore, the free word order of Czech sentences considerably influences on the estimation of conditional probabilities of the polygram language model. The polygram language model must be trained only with recordings of the most probably occuring sentence structures. But, the occurence probability of different sentence structures is strongly regional and semantic dependent. Unfortunately, the limited page number of this paper disables to describe this problem more detailed.

The last alternative to be used for the successful correction of the recognized word sequence is the application of the *permugram language models*. These models are very appropriate for the correction of Czech sentences, which are distinguished for a very variable structure. A small example of possible word ordering in the sentence "At what time will be leaving the fast train from Carlsbad to Prague early morning?" see please below:

V kolik hodin	*pojede*	*brzy ráno*	*z Karlových Varuů*	*rychlík*	*do Prahy*	?
V kolik hodin	*pojede*	*brzy ráno*	*rychlík*	*z Karlových Varuů*	*do Prahy*	?
V kolik hodin	*brzy ráno*	*pojede*	*rychlík*	*z Karlových Varuů*	*do Prahy*	?
V kolik hodin	*brzy ráno*	*pojede*	*z Karlových Varuů*	*rychlík*	*do Prahy*	?
V kolik hodin	*pojede*	*z Karlových Varuů do Prahy*	*brzy ráno*	*rychlík*		?

The permugram language models were described in detail in [8]. The principle of these language models is based on the analysis of positions of possible word permutations in the structure of analyzed sentence. Assume we are going to rearrange the word order in $w_1, ..., w_K$ according to a permutation

$$\pi : \{1, \ldots, K\} \xrightarrow{1-1} \{1, \ldots, K\}$$

The conditional decomposition of $P(\underline{w})$ remains valid after reordering of the input sequence:

$$P(\underline{w}) = P_\pi(\underline{w}) = P(w_{\pi(1)}, \ldots, w_{\pi(K)}) = \prod_{k=1}^{K} P(w_{\pi(k)} | w_{\pi(1)}, \ldots, w_{\pi(k-1)})$$

As a matter of fact, the identity $P(\underline{w}) = P_\pi(\underline{w})$ becomes inapplicable as soon as the permuted n-gram probabilities are replaced by the polygram-like interpolation rule. Consequently, it is tempting to formulate a permugram model by the linear combination

$$\hat{P}(\underline{w}) = \sum_{\pi \in \mathcal{P}} \lambda_\pi \hat{P}_\pi(\underline{w})$$

of permutation-dependent joint probability estimates, ranging over the set $\mathcal{P}$ of all possible permutations of $\{1, \ldots, K\}$. Observe that our permugram formula in fact incorporates conditional bigram probabilities $P(w_j|w_i)$ for each possible pair i, j of relative word positions; the same is true for trigrams, and tetragrams, and so forth. Consequently, stochastic dependences between word pairs or triples lying widely separated in the input text can already be modelled without the need of higher-order statistics of the word generation process.

The essential challenge in permugram modelling is to check the combinatorial explosion caused by the vast amount of theoretically possible sentence permutations. By means of configurations, which represent the local probability contribution of word permutations, we are able to define a doubly stochastic process that generates sentences according to a convex combination of permuted polygram models. Of course, only a restricted class of subsets of all possible permutations can be realized in a finite state process of reasonable size.

A *hidden permutation model* (HΠM) consists of a set $\{S_1, \ldots, S_M\}$ of states S_i with associated configurations $\underline{c}(S_i)$. The non-deterministic sentence production process starts in S_1 and moves from S_i to S_j with probability a_{ij}. The state identity at time t is hiddent to the observer, but using the configuration attached to the state and its corresponding permutation, an open word position $w_{\pi(t)}$ is filled according to the selected conditional (non-standard) n-gram distribution.

3   Experimental Results

In order to compare all three approaches (with traditional n-gram models, polygrams, and permugrams), a corpus of 500 test sentences was especially generated with the aim to involve there all irregularities of the Czech language occuring in the application domain. A lot of several experiments were realized and the obtained results are summarized in Tab. 1 for classical n-gram models and in Tab. 2 for polygram and permugram language models.

Table 1. Numbers of word corrections by using the classical n-gram models:

Number of \ Model	bigram	trigram	tetragram	pentagram
sentences shorter then n words	0	8	35	48
words in sentences	5912	6013	6041	6051
vocabulary size	549	549	549	549
all n-grams	5412	5013	4549	4049
"right" n-grams	3152 (58,2%)	3638 (72,6%)	3036 (66,7%)	2291 (56,0%)
inserted words	648 (11,0%)	101 (1,7%)	28 (0,5%)	10 (0,2%)
replaced words	406 (6,9%)	296 (4,5%)	94 (1,6%)	58 (1,0%)
wrong sentences before the correction	437 (87,4%)	294 (58,8%)	289 (57,8%)	295 (59,0%)
wrong sentences after the correction	145 (29,0%)	242 (48,4%)	275 (55,0%)	285 (57,0%)

Table 2. Numbers of word corrections by using the polygram and permugram language models:

Number of	polygram model	permugram model
words in sentences	6051	6051
vocabulary size	549	549
all poly- | permugrams	19023	27886
"right" poly- | permugrams	3269 (17,2%)	4318 (15,5%)
inserted words	224 (3,7%)	431 (7,1%)
replaced words	365 (6,0%)	641 (10,6%)
wrong sentences before the correction	436 (87,2%)	436 (87,2%)
wrong sentences after the correction	274 (54,8%)	137 (27,4%)

Obviously, the best results were achieved by using the "classical" bigram and permugram language models. The first one is not surprising and only confirms the expectations. The second result – the success in the application of the permugram model – depends on the variability of used sentence structures. The Czech language, in contrary to e. g. German or English, does not preserve the fixed sentence word ordering and the variability of Czech sentence structures is, in addition, strongly regional dependent. If we want to respect all regional irregularities and to process the sentences of all possible kinds, we have to use very variable language model involving a lot of word permutations. But, it requires to do a detailed analysis of all language structures based on the analysis of large language corpora, which unfortunately were not completely created to this day.

4 Concluding Remarks

The above language models for the Czech language recognizer were created by training with the read speech – the especially generated corpus of 10,000 domain dependent sentences was recorded by 50 male and 50 female speakers. The creation of a large corpus of unique sentences read by many speakers guarantees the maximal phonetic and syntactic variation of recorded sentences for a given corpus size. The use of permugram language model obtained by linear interpolation of a large number of conventional bigram, trigram, or polygram language models which operate on different permutations of input word sequences, increases the word accuracy for more then 20 %, but depending on the application.

In spite of this success, we feel that more dramatic improvements were possible if sentence scores could be computed in a decision-directed fashion, i.e., by evaluating the HΠM using the Viterbi algorithm. Under this condition, the permugram model would decide on the best-fitting word permutation for each input sentence rather than averaging over the entire subspace spanned by the HΠM.

References

1. Aretoulaki M., Gallwitz F., Harbeck S., Ipsic I., Ivanecky J., Matousek V., Niemann H., Noeth E., Pavesic N.: SQEL: A Multilingual and Multifunctional Dialogue System. In: Proceedings of the ICSLP, Sydney, Australia, 1998
2. Chelba C., Jelinek F.: Exploiting Syntactic Structure for Language Modeling. In: Proc. of the IEEE, No. 12, 1998
3. Jelinek F.: *Self-Organized Language Modeling for Speech Recognition*. In: Waibel A. and Lee K. F. (eds.): *Readings in Speech Recognition*, Morgan-Kaufmann Publ., San Mateo, CA, 1990, pp. 450 - 506.
4. Klečková J., Matoušek V., Netrvalová J.: *An Automatic Creation of the Language Model for the Spontaneous Czech Speech Recognizer*. In: Proceedings of the EUROSPEECH, Madrid, 1995, pp. 1185 - 1188.
5. Kuhn T., Niemann H., Schukat-Talamazzini E. G.: Ergodic Hidden Markov Models and Polygrams for Language Modeling. In: Proc. of the ICASSP, Adelaide, Australia, 1994

6. Matoušek V., Mautner P.: *Word Recognition in Spontaneously Pronounced Czech Sentences*. In: Proc. of the 4th Int. Workshop on Systems, Signals and Image Processing, Poznaň, 1997, pp. 97 – 110.

7. Schukat-Talamazzini E. G., Niemann H.: *Speech Recognition for Spoken Dialog Systems*. In: Niemann H., De Mori R., Hannrieder G. (eds.) *Progress and Prospects of Speech Research and Technology*, CRIM/FORWISS Workshop, Infix, Munich, 1994, pp. 110 – 120.

8. Schukat-Talamazzini E.G., Hendrych R., Kompe R., Niemann H.: Permugram Language Models. In: Proc. of the EUROSPEECH, Madrid, 1995, pp. 1773–1776.

Analysis of Different Dialog Strategies in the Slovenian Spoken Dialog System*

Ivo Ipšić[1], France Mihelič[2], and Nikola Pavešić[2]

[1] Faculty of Philosophy, University of Rijeka, Croatia,
ivoi@mapef.pefri.hr
[2] Faculty of Electrical Engineering, Laboratory of Artificial Perception,
Tržaška 25, 1000 Ljubljana, Slovenia,
name@fe.uni-lj.si

Abstract. In the paper we first present the architecture of the Slovenian spoken dialog system. The Slovenian spoken dialog system was developed within the joint project in multilingual speech recognition and understanding Spoken Queries in European Languages (SQEL- Copernicus -1634). Such a system can handle spontaneous speech, and provide the user with correct information in the domain of air flight information retrieval. The architecture of the Slovenian system is based on the Erlangen Train Time Table Inquiry System, developed within the European Esprit project Sundial. The major modules perform word recognition, linguistic analysis, dialog management and speech synthesis.

We further describe experiments we have performed using different dialog strategies in the system - user conversations over the microphone and telephone. In the paper we describe the different dialog strategies and show some examples.

We present the modules of the Slovenian spoken information system in more detail and show some results with respect to word accuracy, semantic accuracy and dialog success rate for different dialog strategies.

1 Introduction

The Slovenian spoken dialog system is being developed within the joint project in multilingual speech recognition and understanding *Spoken Queries in European Languages* (SQEL-Copernicus -1634). The final aim of the project is to built a multilingual and multifunctional system, capable of having a dialog with the user in one of the four European languages (German, Slovenian, Czech and Slovak) about a task oriented topic. Such a system has to be able to handle spontaneous speech, and to provide the user with correct information. The information system being developed for Slovenian speech is used for flight inquiries.

The development of a spoken dialog system concerns solutions to speech recognition problems as well as to speech understanding and human machine interaction problems. One approach to spoken dialog systems design is to solve the

---

* This work was funded by the Commision of the European Communities and the Slovenian Ministry of Science and Technology.

F. Jelinek, E. Nöth (Eds.): TSD'99, LNCS 1692, pp. 315–320, 1999.
© Springer-Verlag Berlin Heidelberg 1999

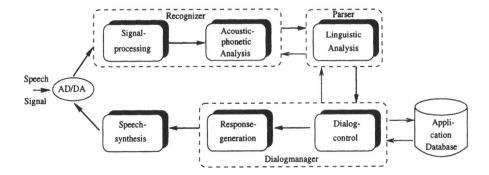

Fig. 1. Architecture of the Slovenian spoken dialog information retrieval system.

problems in different communicating modules. The architecture of the Slovenian system is based on the Erlangen Train Time Table Inquiry System [2], developed within the European Esprit project Sundial [1]. The German system can perform human machine continuously spoken dialogs about train connections via telephone. Figure 1 shows the architecture of the spoken dialog information retrieval system. The major modules perform word recognition, linguistic analysis, dialog management and speech synthesis.

The information system being developed for Slovenian speech is used for air flight information retrieval. The system has to answer questions about air flight connections and their time and date. The main differences between the German system and the Slovenian are: a word recognition module for the Slovenian language, a Slovenian parser and Slovenian text-to-speech synthesis.

The first step in spoken dialog system design is the collection of speech material, which is used for dialog modeling as well as for building statistical models for the word recognizer. We have collected recordings of dialogs between anonymous clients and telephone operators at the Adria Airways information service. From these dialogs we created the Slovenian speech database GOPOLIS of 5000 sentences, which was read by 50 speakers [5]. The read speech database was used for the creation of acoustic and language models, as well as for testing of the parser and the dialog manager. We have tested the separate modules with microphone and telephone quality speech. The overall system was tested with cooperative users and dialog determination and success rate was estimated.

In the following subsections we describe the main modules, the adaptation steps and show some results.

2 Word Recognition

The speech signal is sampled at 16 kHz with 16 bit and mel–cepstrum features and their derivatives were computed every 10 ms. The speech signal vectors were transformed into symbols using the soft vector quantization technique. The spoken words recognizer is based on hidden Markov models (HMM) of context dependent phone units - polyphones[1]. To define the context dependent units we have proposed a set of 33 Slovenian phones and their HMM models, which give the highest recognition accuracy. To train the polyphone models the ISADORA system [6] was used. The 824 words were modeled with 2086 polyphone models of different length.

The recognizer is built by parallel concatenation of word models, which were constructed from context dependent phone models. Best word sequences were generated using the Viterbi beam search and a categorical bigram language model with perplexity 10. 127 word categories[2] to classify the words from the Slovenian recognition vocabulary have been defined. The bigrams were trained on 10 000 sentences comprising 824 different vocabulary words [7].

3 Linguistic Analysis

Input to the linguistic module is the recognized word sequence; output of the linguistic module is its semantic interpretation in the semantic interface language (SIL) [4]. SIL was defined in the Sundial project for interfacing different parsers with the dialog manager and for knowledge representation within the dialog manager. Since the Slovenian dialog system uses the Sundial dialog manager, the Slovenian parser produces a SIL structure from the recognized utterance.

Besides the known problems which occur during the design of a linguistic analysis system additional problems arose due to the nature of the Slovenian language: a large number of inflected word forms and a rather free word order. To overcome these problems, we developed a robust semantic-driven parser [8], which extracts the most important information from the recognized sentences in three steps:

- parsing of temporal expressions,
- parsing of simple noun words. Their meaning is often deduced from a simple semantic category, which they belong to, and from their position in the sentence. This method is used for departure/arrival city determination.
- The rest of the sentence is parsed using a very simple parser that tries to locate as many keywords as possible. These keywords are used in further processing to determine the sentence type.

The semantic-driven parser is based on the DCG formalism, and is implemented in Prolog. It can handle also ungrammatical and colloquial expressions. Phrases, which the linguistic analysis module cannot parse are:

- sentences where the time phrase is divided throughout the sentence,

---

[1] A polyphone consists of a phone with arbitrary length of left and right context of phones, and it is determined from the training database. The criterion to form a polyphone was the minimal number of its occurrences in the words of the training sentences.

[2] The categories group words with same grammatical and semantical characteristics.

- determination of arrival or departure town can fail in those rare examples where there is no difference between the first and fourth case of a noun and
- some complex time expressions.

The output of the Slovenian parser is the SIL structure of the recognized user sentence, and it contains only application relevant information. The example shows the SIL structure of the user sentence *Tomorrow in the morning I would like to fly to London.*:

```
O = [[id:1456,semantics:
                    [id:1457,type:dbflight,
                     date:21049,
                     goalcity:london,
                     sourcetime:[300,1000]]]]
```

The user intention spoken in the sentence is described with the *dbflight* SIL structure, which contains information about travel date and time, and the goal-city.

4   Dialog Management and Message Generation

The dialog manager takes the semantic representation of the user utterance and performs the interpretation within the current dialog context. The dialog manager from the German demonstrator is language and application independent, so it can be used for the Slovenian dialog system. The dialog manager consists of a number of communicating sub-components [3]:

- the linguistic interface, which enables the input from different parsers,
- the belief module, responsible for interpreting user utterances in the current dialog context,
- the task module, responsible for handling the task and performing a database query,
- the dialog module, responsible for the pragmatic interpretation and planning of system utterances and
- the message planer, responsible for message generation.

The adaptations for a new language and a new application have to be done within the task module and the message planer. For the Slovenian dialog system the flight timetable, provided by the Slovenian airline company Adria Airways is used. The timetable, which is available over the WWW, is transformed into Prolog facts representing the timetable data.

In the Message Generation Module the SIL representations of the system answers were converted to (Slovenian) text via sentence tabloids.

5 Dialog Strategies

Different dialog strategies are defined in the dialog module. They define the initiative in the dialog acts and different confirmation strategies of the recognized parameters, which are needed for a database query. These strategies are: separate confirmations of SIL parameters, confirmation alone, confirmation of SIL parameters and dialog initiative and no confirmation [4]. On the example of the recognized user sentence *Tomorrow in the morning I would like to Fly to London.* different dialog strategies are demonstrated:

- Separate confirmations:

  system: You want to fly to London?
  user: Yes.
  system: You want to leave from Ljubljana?
  user: Yes.
  system: When do you want to travel?

- Alone confirmations:

  system: You want to fly from Ljubljana to London?
  user: Yes.
  system: When do you want to travel?

- Confirmation of parameters and dialog initiative:

  system: When do you want to travel from Ljubljana to London?

- No confirmation of parameters:

  system: When do you want to travel?

Depending on the dialog strategies the dialogs can be long and boring and they can be short but unfortunatly unsuccessful. The experiments with the Slovenian spoken dialog system show that the no confirmation strategy leads to unsuccessful dialog ends. The reason lies in the word recognition rate, which is about 80 %. Best dialog success rates are achieved with separate confirmation and alone confirmation dialog strategies. Confirmation of recognized parameters or rejection of wrong recognized parts of sentences leads to more successful dialogs. The ideal solution would be a dynamical dialog strategy, which would change depending on the recognition and linguistic analysis success rates.

6 Speech–Synthesis

The generated sytem answers are transformed to speech signals. This can be done with a simple concatenation technique of prerecorded time signals for isolated words. Another possibility is the use of synthesized speech. A system for

Slovenian synthesis was developed [9]. The Slovenian synthesizer uses the concatenative diphone synthesis technique TD-PSOLA (Time Domain Pitch Synchronous Overlap and Add) [10]. The TD-PSOLA scheme enables pitch and duration transformations directly on the waveform, at least for moderate ranges of prosodic modifications without considerably affecting the quality of synthesised speech.

7   Conclusion

In the paper we have presented the architecture of the Slovenian spoken dialog system. We have described the modules of the system and the tasks they perform. Besides word recognition rates and linguistic analysis success rates a major role in the dialog success rate plays the dialog strategy. We have presented different dialog strategies in the dialog manager, and shown which strategy leads to best dialog success rate.

References

1. J. Peckham: *Speech Understanding and Dialogue over the Telephone: an Overview of Progress in the SUNDIAL Project*. In *Proc. European Conf. on Speech Technology*, volume 3, pages 1469–1472, 1991.
2. W. Eckert, T. Kuhn, H. Niemann, S. Rieck, A. Scheuer, E.G. Schukat–Talamazzini: *A Spoken Dialogue System for German Intercity Train Timetable Inquiries*, Proceedings of the European Conference on Speech Communication and Technology, Berlin, September 1993, pp. 1871–1874.
3. W. Eckert: *Customizing the Erlangen Dialog Manager*. Technical report, Lehrstuhl für Mustererkennung, FAU Erlangen–Nürnberg, 1996.
4. W. Eckert: *Gesprochener Mensch–Maschine-Dialog*, Ph.D.Thesis, Universität Erlangen–Nürnberg, 1996.
5. S. Dobrišek, J. Gros, F. Mihelič, N. Pavešić: *Recording and Labelling of the GOPOLIS Slovenian Speech Database*, First International Conference on Language Resources and Evaluation, edts.: A. Rubio, N. Gallardo, R. Castro, A. Tejada, May 1998, Granada, Spain, Vol. II, pp. 1089–1096.
6. E. G. Schukat–Talamazzini: *Automatische Spracherkennung*, Vieweg, Braunschweig, 1995.
7. I. Ipšić, F. Mihelič, N. Pavešić, E. Nöth: *Slovenian Word Recognition*, Proceedings of IAPR Workshop on "Speech and Image Understanding", N. Pavešić, H. Niemann, S. Kovačič and F. Mihelič eds., Ljubljana, Slovenia, pp. 87–96, 1996.
8. K. Pepelnjak, F. Mihelič, N. Pavešić: *Semantic Decomposition of Sentences in the System Supporting Flight Services*, Journal of Computing and Information Technology, Vol. 4, No. 1, 1996, pp. 17–24.
9. J. Gros, N. Pavešić, F. Mihelič: *Text-to-Speech Synthesis: A complete system for the Slovenian Language*, Journal of Computing and Information Technology, Vol. 5, No. 1, 1997, pp. 11–19.
10. E. Moulines, F. Charpentier: *Pitch-Synchronous Waveform Processing Techniques for Text-to-Speech Synthesis Using Diphones*, Speech Communications 9, pp. 453–467, 1990.

Dispersion of Words in a Language Corpus*

Jaroslava Hlaváčová[1] and Pavel Rychlý[2]

[1] Institute of the Czech National Corpus, Faculty of Arts, Charles University,
nám. J. Palacha 2, 116 38 Prague 1, Czech Republic
`hlava@mathesius.ff.cuni.cz`
[2] Faculty of Informatics, Masaryk University, Botanická 68a, 602 00 Brno,
Czech Republic
`pary@informatics.muni.cz`

Abstract. This paper proposes new measures for dealing with word dispersion in a language corpus – *reduced frequency* and *rarity*. Their calculation is described and some results from the Czech National Corpus (CNC) presented. Some previous approaches are briefly mentioned.

1 Preface

One of the most popular statistical characteristics of a language is frequency of its words. It is often the main criterion for lexicographers for the decision whether a word should be included into a dictionary or not.

The frequency of a word is calculated as number of occurrences of the word in a text. Let us call this frequency *pure frequency*. If pure frequencies of all the words occurring in the text are compared, we can see easily which words are current and which are rare (in the text). Of course pure frequencies depend on the text. Not only on its length, but also on the subject of the text, on its author, its style, and so on. But these are difficult to measure.

We can solve the problem by joining all the texts into one. If we could include into the whole all the existing written texts together with all spoken texts, then we could calculate the real frequency of all the words in the language. Because of the impossibility of such a task, we have to do with creating only a sample of all the existing texts. Such a sample is called language corpus.

As a sample it has all the drawbacks that all samples have. One of the most important is uneven distribution of words.

An example will explain this. Into the CNC (at the time of the calculations it contained 100 million words) we have incorporated a chronicle of a small town Toušeň. We guess that apart from citizens of the town only very few people of the Czech Republic have ever heard the word. The pure frequency of the lemma *toušeňský* (adjective created from the name *Toušeň*) is 384, which places it within the 14 603 most frequent lemmas of the CNC. It has the same pure frequency as lemmas *vidlička* (fork), *sršeň* (hornet) or *pirátský* (piratical).

---

* This research was supported by the GACR, Grant Nr. 405/96/K214.

F. Jelinek, E. Nöth (Eds.): TSD'99, LNCS 1692, pp. 321–324, 1999.

1.1   Previous Approaches

The problem is not new. Several authors have already tried to invent a dispersion measure, that would allow to decide automatically which words are really frequent and which are not. A brief comparison of three of them is given in [2]. A larger summary of the existing methods described Králík [1].

The proposed measures have quite complicated formulae. Their calculation from a corpus having a hundred million words would be too long. Another disadvantage is that they assume pre-division of the corpus into several parts. If the parts have equal lengths, the question arises, how many parts it should be. If the corpus is divided according to another criterion (usually the type of texts – i.e. fiction, newspapers, professional books), the division is always subjective.

Another approach to the problem is *logarithmic frequency*, which was described at this conference last year by Rychlý [3]. This method is based on division of the corpus into *documents*, that are more natural than previous divisions, but not very reliable either, because the division depends largely on the original arrangement of texts, as we get them from publishers. For instance, we get one title of newspapers divided into as many files (documents) as is the number of its articles, an other title we get as one file (document) for every copy. It is not possible to unify divisions into documents throughout the whole corpus.

2   Reduced Frequency

Let us try to propose another measure – *reduced frequency* or *r-frequency* – which would reflect the (un)evenness of word distributions within the whole corpus (the word dispersion) and which would not have the above mentioned drawbacks. We have 2 main requirements:

1. Words that occur only in one or several limited portions of the corpus should have lower r-frequency than words with the same pure frequency but occurring more evenly. In other words, the more even the distribution of the word within the corpus, the higher value of its r-frequency.
2. If a word occurs entirely evenly within the whole corpus, its r-frequency should be the same as its pure frequency. It implies that the pure frequency is the maximum that the r-frequency could reach.

2.1   Notation and Definitions

Let x be a word. It can be a word form or a lemma (or a grammatical or another tag). Let $f(x)$ designate the number of occurrences of the word x in the corpus (its pure frequency). We index all the words of the corpus according to their position in the corpus, starting with 1. In a corpus of N words, the last word has the position number N. For every word x, let us divide positions of the whole corpus into $f(x)$ intervals, each of them having (approximately) the same length. Thus, i-th interval for the word x is: $\left\langle \left[\frac{(i-1)N}{f(x)} \right] + 1, \left[\frac{iN}{f(x)} \right] \right\rangle$, for $i = 1,...,f(x)$. ([ ] denotes the integer part of a real number.)

Reduced frequency is the number $r(x) = \sum_{i=1}^{f(x)} f_x(i)$,
where $f_x(i) = 1$,   if the word x occurs at least once in the i-th interval,
     $f_x(i) = 0$,   if the word x does not occur in the i-th interval.
Rarity of the word x is the quotient f(x)/r(x).

2.2   Basic Properties of the Reduced Frequency and the Rarity

- If a word is distributed entirely evenly in the whole corpus, its reduced frequency is equal to its pure frequency and its rarity is 1.
- The same is valid for all the words with the pure frequency 1.
- A word which occurs only in one small portion of the corpus falls very probably into not more than 2 following intervals. Therefore its reduced frequency will be 1 or 2, even if its pure frequency was quite high. The rarity of such a word is high.

The main drawback of the reduced frequency is its dependence on the arrangement of texts within the corpus. If we rearrange the texts, results could be different. We have not made any calculations so far to find out, how much it changes results for different words. Evidently, the rearrangement wouldn't influence the r-frequency of words with pure frequency 1. Evenly distributed words would not be influenced neither. But it can happen for a word, that in one arrangement all its occurrences fall into one sect ion, while in another one they are scattered in several sections, especially if they are included in short texts.

However, really rare words, e.g. special terms or names, occur usually in one text only. If they occur in more texts, they were very probably published by one publishing house and according to our experience with creation of a corpus, they were incorporated to the corpus together, so they do not disperse. And if they occur in production of more publishers (are more dispersed), they are probably not so rare. Of course, it would require more calculations and comparisons.

3   Results

We calculated r–frequencies for all the lemmas of the lemmatised CNC.
     At first let us have a look at our example – the adjective *toušeňský*. The comparison with the 3 mentioned words of the same frequency shows the table:

lemma	in English	pure freq. $f(x)$	r-frequency $r(x)$	rarity $f(x)/r(x)$
toušeňský	Toušeň's	384	2	192.00
vidlička	fork	384	130	2.95
sršeň	hornet	384	122	3.15
pirátský	piratical	384	170	2.26

This small example gives an impression how the r–frequency could be useful in recognising the rare words which have only incidentally high pure frequency.
     There are 366 891 lemmas with the pure frequency greater than 1. Of these, 26 577 (7.24 %) is entirely distributed within the whole corpus, having the pure

frequency and reduced frequency equal. However, all these words are very rare. The highest frequency within this group is 8 and it is reached only by two lemmas. One of them is *předimenzovanost* (something like outsized or overblown dimension), the second is probably an error. There is no really common word with the rarity equal to 1. The values for the 10 most frequent words of the CNK are in the following table.

Czech word	in English	pure freq. $f(x)$	r.-frequency $r(x)$	rarity $f(x)/r(x)$
a	and	2 809 239	1 767 491	1.59
v	in	2 665 425	1 601 007	1.66
být	to be	1 689 681	937 477	1.80
na	at	1 632 182	983 832	1.66
z	from	862 104	503 969	1.71
že	that	849 759	455 581	1.87
o	about	770 509	425 573	1.81
který	which/that	765 873	456 381	1.68
s	with	747 057	438 487	1.70
o	into	596 828	340 884	1.75

There are 5018 words having the rarity equal or higher than 10. We can assume, that the majority are proper names, because only one third (1646) of them begin with a small letter. Among these words approximately one third are terms, another one third are errors and the rest are foreign words (parts of quotations), special abbreviations or unique words discovered by individual authors or invented for a special purpose.

From this brief summary we can see other possible uses of the reduced frequency. It can be used for searching errors or terms in the corpus.

4   Conclusion

We have proposed the measures – *reduced frequency* and *rarity* – that allow to compare usage of words with the same pure frequencies in a language corpus. Their calculation is very simple. There is no need to pre-divide the corpus into artificial sections and can be calculated entirely automatically. It could help lexicographers as an objective tool to decide whether a word is really common or if its high frequency is only a chance. It can be used also for searching unusual sequences of letters (errors, terms, special abbreviations) in the corpus.

References

1. Králík, J.: On the dispersion and its computation. Prague Studies in Mathematical Linguistics, Prague, Academia 1978, pp. 149–158.
2. Oakes, M.P.: Statistics for Corpus Linguistics. Edinburgh University Press, 1998.
3. Rychlý, P.: The Improvement of Common Statistical Measure. Proc. TSD '98, Brno 1998, pp. 109–112.

Corpus-Based Rules for Czech Verb Discontinuous Constituents*

Eva Žáčková and Karel Pala

Faculty of Informatics, Masaryk University Brno
Botanická 68a, 602 00 Brno, Czech Republic
{glum,pala}@fi.muni.cz

Abstract. In this paper we present a method for extracting general structures of the verb groups from a tagged and fully disambiguated corpus and consecutive exploitation of these structures for the building a formal grammar in the Prolog DCG fashion. Our goal is to apply them as a rules for the analysis of the Czech verb groups in the non-disambiguated grammatically tagged Czech corpus texts. The problem of the recognition of verb discontinuous constituents in Czech is also approached and obtained statistical data are presented.

1 Introduction

For various applications in the field of natural language processing it is necessary to have rules capable to recognize and analyze reliably the verb groups in texts. The previous attempts to build the rules describing Czech verb constituents have been based rather on the introspective (Chomskyan) approach, which can hardly cover all the relevant cases occurring in the corpus texts. Thus we try to arrive at a more complete algorithmic description of the Czech verb groups that could serve as a ground on which such rules can be built.

In the presented research we use corpus based techniques to find all the verb groups in DESAM corpus (Czech, general, tagged and fully disambiguated corpus containing at present about one million positions [1,2]). The obtained results have been classified into several clusters according to their structure. Then the representative Prolog rule for each cluster has been created.

We are also touching the problem of the recognition of verb discontinuous constituents in Czech: without having it solved hardly any reasonable syntactic parser can be built. This means that it is inevitable to have a usable algorithm that would enable us to recognize such verb groups since DCG formalism in Prolog does not offer a reasonable solution of this problem. There are gapping grammars [3] that address this issue but we have not met any implementation that could yield realistic and acceptable results for Czech.

---

* This research has been partially supported by the Czech Ministry of Education under the grant VS97028

F. Jelinek, E. Nöth (Eds.): TSD'99, LNCS 1692, pp. 325–328, 1999.

2   The Extraction of the Structures of Verb Groups

The occurrences of the verb groups (from which the general structures have been extracted afterwards) were searched out in the above mentioned DESAM corpus using CQP (Corpus Query Processor [4]) queries satisfying the following conditions:

- the particular occurrence of the verb group is found exactly once,
- the whole group is found (not only a part of it),
- two independent groups are not merged together,
- the improper words (not belonging to the verb group) are not included into the group,
- groups consisting of discontinuous constituents are found.

The queries constructed take advantage of the properties of the verb groups [5]: their components are either verbs or the pronoun *se* (*si*)[1] and a verb group cannot cross the boundary of a sentence but between its two components there can be a gap consisting of an arbitrary number of non-verb words or even a clause.

The queries were divided into several categories according to the number of "clusters" creating a searched verb group. A cluster consists of an arbitrary number of continuous components forming a verb group (see e.g. the verb group in the sentence *To by měl tento zákon umožnit.* (*It should be enabled by this law.*) consists of two clusters. Every continuous verb group consists of one cluster.).

3   Results and Statistics

From the found verb groups we extracted about 1 500 different structures which after the necessary generalizations produced about 150 skeletons of rules with the following format:

```
mít/k5e?p?n?tMmPa? k5e?p?n?tPmCa? sebe/k3xXn?c? gap k5e?mFa?
```

The notation similar to that produced by morphological tagger LEMMA [6] is used. The base form of the word (can be omitted) is followed by a slash and tag which specifies required POS (part of speech) and grammatical categories. The small letters in the tag denote attributes (grammatical categories) and capital letters (or numbers) their values (e.g. k5 denotes verb; question mark stands for any value). The string gap can be substituted by an arbitrary number of words which do not belong to the particular verb group. The following example shows a possible instance of the above mentioned rule (in the format: word/base-form/tag).

```
nemělo/mít/k5eNpNnStMmPaI by/by/k5eAp3nPtPmCaI se/sebe/k3xXnSc4
vedení/vést/k1gNnSc1 spíše/spíše/k6xMeA zaměřit/zaměřit/k5eAmFaP
```
(literary translation: *wouldn't the management better focus...*)

---

[1] In this research we do not consider the complete verb groups consisting of the copula (mostly with *být* (*to be*)) and the noun, adjective or adverbial group. This problem will be solved in the course of the further and more complex analysis.

Verb groups found in the corpus have been also exploited for extracting some statistical data. Especially we focused on the discontinuous verb groups and the examination of gaps. Table 1 shows the representation of verb groups classified by number of its components. The second column contains the ratio of occurrences of each category in the corpus, the third column shows the percentage of the discontinuous groups of each particular category. From the table can be observed that the groups consisting of at most five components were found (we estimate that this is probably an upper limit for Czech language).

The discontinuous verb groups found in the corpus represent about 50 % of all verb groups consisting of two or more components.

Table 1. Representation of the verb groups classified by the number of its components.

# of comp.	% of all	% with gaps	example
1	61.7	–	*pršelo*
2	30.0	47.5	*se brzy rozhodne*
3	7.2	57.3	*chtěl jsem u toho být*
4	1.0	60.6	*měli bychom si to uvědomit*
5	0.1	66.7	*mohl by se oxid uhličitý začít hromadit*

The frequencies of the different word types which occur in the gaps can be found in Table 2. When a sentence occurred in a gap, it was counted only once (category "sentence"), its components were not evaluated separately. The average size of a gap is two words.

Table 2. Components of gaps: percentage.

category	per cent
noun	35.5
adjective	10.5
pronoun	12.5
numeral	1.8
adverb	16.7
preposition	14.7
conjunction	0.7
particle	6.9
abbreviation	0.2
sentence	0.4

The largest gap found consists of the ten components (gaps including sentences are not taken into account here): *je pověra o léčebných schopnostech tohoto preparátu v čínské medicíně hluboce zakořeněna (the superstition of therapeutic abilities of this stuff is widely spread in the Chinese medicine).*

4 Converting the Rules into Prolog

Before converting the skeletons of the rules created from corpus data to DCG rules the problem of parsing of discontinuous constituents in Prolog has to be faced. Since the DCG formalism does not help in this respect our solution of the problem uses a special D(iscontinuous)-predicate which enables us to collect words occurring in the gaps during parsing the verb group and then move them to the beginning of the rest of (not yet parsed) sentence (the idea of gapping grammars is employed here). The D-predicate can be extended by the specification of the words that can or cannot be skipped. It is also possible to use D-predicate to solve some free word order problems in Czech but its performance in this kind of analysis is not efficient enough yet.

The above mentioned skeleton of the rule (see page 326, section 3) converted into Prolog will have the following form (pass_gap is the D-predicate):

```
verb_group(vg(Verb1,Verb2,Se,Verb3),Rest_to_Analyze) -->
    word(Verb1,'mít',k5,_,_,_,tM,mP,_),
    word(Verb2,_,k5,_,_,_,tP,mC,_),
    word(Se,'sebe',k3,xX,_,_),
    pass_gap(Rest_to_Analyze),
    word(Verb3,_,k5,_,mF,_).
```

5 Conclusions

A method for extracting skeletons of the rules capable to recognize verb groups in Czech texts from tagged and fully disambiguated corpus has been presented. We have described practical usage of these skeletons for building a formal grammar in Prolog and we also suggested and implemented the mechanism for the analysis of the discontinuous constituents. Further exploitation of created rules in various linguistic applications (e.g. together with the idea of verb valencies [7,8]) is to be expected.

References

1. Karel Pala, Pavel Rychlý and Pavel Smrž. DESAM — approaches to disambiguation. Technical Report FIMU-RS-97-09, Brno, 1997.
2. Karel Pala, Pavel Rychlý and Pavel Smrž. DESAM — annotated corpus for Czech. In *Proceedings of SOFSEM'97*. Springer-Verlag, 1997.
3. Veronica Dahl. More on Gapping Grammars. In *Proceedings of the International Conference on Fifth Generation Computer Systems*. Tokyo, 1984.
4. Bruno Maxmilian Schulze and Oliver Christ. The CQP User's Manual. Universität Stuttgart, Stuttgart, 1996.
5. Jan Petr et al. The Grammar of Czech III. Academia, Praha, 1987.
6. Pavel Ševeček. *LEMMA — a Lemmatizer for Czech*. Brno, 1996. (manuscript).
7. Karel Pala and Pavel Ševeček. Valencies of Czech Verbs. Studia Minora Facultatis Philosophicae Universitatis Brunensis, A45, 1997.
8. Pavel Smrž and Eva Žáčková. New Tools for Disambiguation of Czech Texts. In *Proceedings of TSD'98*. Masaryk University, Brno, 1998.

Automatic Modelling of Regional Pronunciation Variation for Russian

Kseniya B. Shalonova

Department of Phonetics,
St.Petersburg State University,
Universitetskaya emb. 11, Saint-Petersburg, Russia
paul@phonet.lang.pu.ru

Abstract. The present paper describes an Automatic modelling system of regional pronunciation variation for Russian. The system contains two models: Automatic Adaptive Transcriber and a corpus of dialectal pronunciation configurations containing divergences from the pronunciation standard both on the phonological and phonetic transcription level in accordance with their distribution on the national territory.

1 Introduction

One of the main problems facing speech recognition is robustness to pronunciation variability. Working out of ASR's for independent speaker stimulates generating possible word pronunciation variants due to different speaking styles, speech rate and dialectical phenomena. The aim of the present work is to get the possibility of generating all pronunciation variants that may occur in Russian dialectal pronunciation. There exists a number of flexible transcribers for representing dialectal variants of pronunciation, e.g., for exhibiting Italian regional pronunciation variants [1]. All these flexible transcribers generate possible variants on the level of phonetic transcription without phonological interpretation of sound modifications. The proposed automatic modelling of Russian regional pronunciation variants is carried out by means of an Adaptive Automatic Transcriber for Russian connected speech and a corpus of dialect pronunciation configurations containing both phonological and phonetic rules that can occur in dialects.

2 Design of AAT

Adaptive Automatic Transcriber for Russian connected speech (AAT) has being worked out as a mechanism for generating pronunciation variation [1]. Besides literary pronunciation standard, AAT can represent different divergences from the normative canonical model of a word. The process of pronunciation variant generation is carried out by converting any orthographic text (containing stress and intonation marks inserted automatically or manually) into the transcription sequence including/excluding and modifying normative phonological and phonetic rules in accordance with the following phonetic conditions:

F. Jelinek, E. Nöth (Eds.): TSD'99, LNCS 1692, pp. 329–332, 1999.

1. context influence;

2. word and syntagmatic stress position;

3. the quality of the stressed, pre-accented and post-stressed vowels in the word.

AAT can also generate pronunciation variants in accordance with morphological and lexical conditions which are formalized in special dictionaries.

3   Corpus of Dialect Pronunciation Configurations

The creating of dialect pronunciation configurations (the sets of phonological and phonetic rules re-realized in different conditions) was based on the systematized phonetic knowledge obtained in experimental study of the great amount of dialect speech corpora. These systematized dialectal data, being adapted to AAT, include hierarchically organized possible dialect pronunciation features presented as disparities from the literary pronunciation standard.

Those the dialectal phenomena in Russian are very numerous it is possible to subdivide them in into 3 main categories in accordance with the level of their generation.

The following 3 levels are presented in the reference table.

1. Forming dialectal consonant and vowel phoneme inventory

There are 6 vowel phonemes and 36 consonant phonemes in Russian literary speech. The number of phonemes in dialects may differ from that in normative speech. The reference table includes possible dialectal vowel and consonant phonemes-which are not included into the phonological system of Russian literary speech. These new dialectal phonemes can substitute normative phonemes by changing the set of their distinctive features-In order to model the dialectal phonological system it seems necessary to carry out the process of forming dialectical phoneme inventory on the first level.

2. Modelling of dialectal orthoepic rules

The reference table contains possible dialect orthoepic rules for vowels which can be modelled in accordance with the following phonetic conditions:

- word stress position;

- character of the neighboring sound (including initial and terminal word position);

- character of the stressed vowel.

The fragment of this reference table is shown in Fig. 1.

Normative orthoepic rule	Dialect orthoepic rule	Phonetic conditions of dialect orthoepic rule realization		
		Word stress position	Stressed vowel	Phonetic context
a→/a/				
	└→/o/	1-st pre-accented syllable	any vowel	any context

Fig. 1. Fragment of the reference table for vowel dialectal pronunciation generating.

As can be seen in Fig. 1., the normative orthoepic rule for vowel is represented by transforming the letter into the phoneme; the not normative dialectal
orthoepic rule is represented as a new phoneme the letter is transformed into.

Be means of adapting this reference table to AAT it becomes possible to
model the following disparities from the pronunciation standard:

– modification of the normative vowel interchanges in the unstressed syllables
 forming different types of dialectal vocalic systems, e.g.
 poca'
 rosa→rasa (see the rule in Fig.1.)
– obtaining new orthoepic rules for representing various replacements of normative vowel phonemes by not normative vowels both in the stressed and
 unstressed syllables, e.g., realization of the unstressed /i/ instead of normative /a/ in the word initial position: ostrova → istrova.

The reference table contains the following possible dialect orthoepic rules
for consonants which can be modelled in accordance with the character of the
phonetic context:

1. switching off 2 types of normative consonant assimilation:
 – switching off normative assimilation in accordance with consonant palatal-
  ization, e.g., realization of hard bilabial consonants instead of palatalized
  ones before the vowel /e/: b'eg→ beg;
 – switching of normative assimilation in accordance with voiced-voiceless
  character of consonant phonemes, e.g., realization of the voiced consonant
  instead of the voiceless one in the syntagmatic final position: kot→kod;
2. switching on new rules for representing dialectal elisions, consonant insertions
 and assimilations, e.g., insertion of the consonant /d/ after initial /n/ before
 /r/: nraf→ndraf.

It is important to note that certain orthoepic rules can be modelled in the
limits of a concrete morpheme or a concrete lexem. In this case a special vocabulary containing about 100000 lexems with morphological characteristics is
adapted to AAT.

3. Modelling of dialectal phonetic rules

The reference table includes new phonetic rules for modelling disparities from normative combinatory-positional phoneme modifications, e.g., realization of bilabial /w/ instead of labiodental /v/ before vowels /u/, /o/: vot→wot.

On the basis of this reference table there was elaborated a corpus of dialect pronunciation configurations. These configurations containing divergences from the literary pronunciation standard are subdivided into 3 main groups accordingly to their distribution on the national territory: northern, southern and central part of Russia. In one's tern each of these configurations is subdivided into several groups in correspondence with the concrete region.

4   Conclusion

For testing the reliability of the reference data a number of tests for comparing "ideal" and "real" dialect pronunciation realizations was carried out. Certain discrepancies between "ideal" and "real" pronunciations revealed some new phonetic regularities which should be verified by further dialect researches. It is important to note that the process of dialect pronunciation formalization itself has formed a set of criteria that should be used while investigating regional variants of pronunciation.

The working out of AAT was stimulated mostly by applied tasks. The corpus of dialect pronunciation rules being adapted to AAT can be used as a module in robust ASRs. In this case the correlation of each dialectal rule with it's frequency characteristics may increase the reliability of ASR's.

References

1. Bonaventura P., Fissore L., Leprieur H., Giorgoi Micca.: A Flexible Vocabulary Recognition System for Italian. Proc. ICPhS, Stockholm, 1995, Vol 4, pp. 252–255.
2. Shalonova K.: Flexible Transcriber for Russian Continuous Speech. Proc. SPECOM, Cluj-Napoca, 1997, pp. 171–175.

Experiments Regarding the Superposition of Emotional Features on Neutral Korean Speech

Cheol-Woo Jo[1], Attila Ferencz[2], and Dae-Hyun Kim[1]

[1] Dept. of Control & Inst. Eng., Changwon National University,
#9 Sarim-dong, ChangWon, Kyeongnam, 641-773 South Korea
cwjo@sarim.changwon.ac.kr

[2] Software ITC, 109 Republicii street, 3400 Cluj-Napoca, Romania
Dept. of Control & Inst. Eng., Changwon National University,
#9 Sarim-dong, ChangWon, Kyeongnam, 641-773 South Korea
Attila.Ferencz@sitc1.dntcj.ro, aFerencz@sarim.changwon.ac.kr

Abstract. This paper contains a brief presentation about the analysis, re-synthesis and evaluation of emotional Korean speech. Four different emotional styles were taken into consideration that were induced from the actors. Prosodic features of these sentences were extracted and analyzed. Based on these statistical results, one neutral uttered sentence was re-synthesized into these four different emotional styles. The generated artificial emotional speech was evaluated by judges based on a MOS test.

1 Introduction

During the last three to four years several new approaches were started worldwide in order to improve the naturalness of synthesized speech. It seems that among the many other factors, emotion is one of the key features which has to be implemented.

Research regarding emotion-speech (prosody) interaction for the Korean language started at Changwon National University in the framework of the project "Implementation of Emotional Speech in Multimedia Environment", which has been supported by the Korea Science and Engineering Foundation (KOSEF) since 1997. This paper emphasizes, among other aspects, the implementation of a re-synthesis system, which should allow the generation of a given emotional style, based on an initial neutral utterance.

2 Assessment of the Collected Emotional Speech

Kinds of emotions which were considered in the project were: Happy, Angry, Sad and Afraid. The list of analyzed utterances is presented also in [5]. The collected emotional speech was assessed by judges, using a Mean Opinion Score (MOS) test to evaluate the initial emotional state. A self-estimation for each speaker was also achieved, and Table 1. contains the evaluated results. Another assessment method was based on a judgement using auditory and visual images.

F. Jelinek, E. Nöth (Eds.): TSD'99, LNCS 1692, pp. 333–336, 1999.
© Springer-Verlag Berlin Heidelberg 1999

Table 1. Self-estimation (Actor-initiated)

Emotion Current state	Happy	Angry	Sad	Afraid
Happy	3.08	1.75	2.00	1.33
Angry	1.58	4.00	2.11	3.00
Sad	1.38	3.13	4.13	2.83
Afraid	2.13	3.88	3.63	4.00

(scale 1 – 7)

3   Analysis and Synthesis

The collected emotional speech was analyzed, focusing on the variations of pitch frequency, variation of average pitch, and duration of the utterances. Although there are several parameters related to emotional speech, we used in our experiments only these parameters to achieve emotional effects.

3.1   Syntactic Labeling

Before analyzing emotional speech, we made the assumption that all sentences consist of the following five syntactic components: subjective part (SUB); descriptive part (DES); objective part (OBJ); the final syllable of each part (DEE); as well as the pause between all these (PAU). This classification is based only on observations performed upon collected emotional speech signals. Based on this categorization, the four sample sentences were segmented and labeled.

Table 2. shows an example of a sample sentence, which was segmented and encoded using this method. The meaning of this sentence (in English) is: I said "Don't go away", and I closed the door.

Table 2. Example of syntactic segmentation

Nan		Gazimalagohamyon	seo	Munul	Datat	da
SUB	PAU	DES	DEE	OBJ	DES	DEE

3.2   Signal Analysis

In order to extract prosodic features, emotional speech from only a single speaker was analyzed, because analyzing speech from more speakers can decrease the emotional characteristics of each speaker, and can weaken the emotional effect. The characteristic pitch tracks for each emotional style are presented in [5].

For this single speaker, duration, pitch range and mean pitch was evaluated. Table 3 shows the measured duration for three sentences. Average pitch and pitch variation intervals were also evaluated and statistically analyzed.

Table 3. Measured durations for three sentences (in ms)

	Sent 1	Sent 2	Sent 3
Angry	2131	677	874
Happy	1768	566	2145
Sad	2595	821	1753
Afraid	1959	683	1972
Neutral	2970	1050	1760

3.3 Re-synthesis of Emotion

Based mainly on the results of the previous analysis step, we extracted a rule table (Table 4) containing the variations of pitch and duration for each emotional style. These values are also based on some previous experimental results and assumptions.

As can be seen, in the case of synthesizing the sad emotion, a random alteration is applied to the normal pitch track, which was extracted from the neutral uttered sentence.

Based also on the labeling information, the rules appearing in Table 4. were encoded for each case of emotional style into a tabular format containing specific commands related to the synthesizer module.

Table 4. Rules for varying pitch and duration

	Pitch	Duration
Angry	Modify +40 %	Modify −30 %
Happy	Range: +10 % +20 % to −20 %	Modify +20 %
Sad	Random effect	Modify +30 %
Afraid	Modify +100 %	Modify −10 %

Pitch and duration alteration, in order to transform a neutral sentence into an emotional one, is performed by using the Pitch Synchronous Overlap and Add (PSOLA) method, and for this a dedicated analysis and synthesis tool was developed.

4   Evaluation Procedures

The transformed emotional speech was evaluated with the MOS test. All the synthesized emotional speech was played for 20 different listeners. Each listener was asked to specify the perceived emotional state and the correspondent confusion matrixes were filled in, including one for the originally neutral uttered sentence (Sent(4)) and also one for another new emotional one.

Table 5 shows the correct identification rates for each sentence per each emotion. Table 5 shows that the ratio of correct identification is not so high, but in the case of sadness the identification rate is higher (65 % – 90 %) than for other kinds of emotions. This is quite a noticeable aspect compared to other cases of emotion and this means that the initial assumptions for the parameters (in case of sadness) were correct.

Table 5. Identification Rate (%)

	Happy	Sad	Angry	Afraid	Neutral
Sent(1)	5	80	50	40	70
Sent(2)	10	65	40	30	40
Sent(3)	0	85	5	65	60
Sent(4)	40	90	40	20	80

In the case of happiness, even the original emotional speech is hard to distinguish, so it is difficult to derive a proper rule for this case of emotion. In the cases of anger and fear, the correct identification ratio is around 50 %. This is due to the fact that these two emotions have similar characteristics, i.e., pitch level increases and duration increases.

5   Conclusions

In this paper some experiments regarding emotional speech analysis, re-synthesis and evaluation procedures were briefly introduced.

Emotional speech data was collected from actors. The collected speech data was segmented and labeled with a syntactic description method based on observations performed upon emotional speech signals. Finally a neutral utterance was transformed into a specific emotional style.

In concordance with the evaluation of the synthesized emotional speech, sadness can be synthesized most accurately. In the case of the other emotions, further improvements are requested.

References

1. Marray, I.R. and Arnott, J.L.: Toward the Simulation of Emotion in Synthetic Speech: A Review of the Literature on Human Vocal Emotion, JASA, 93(2) (1993) 1097–1108.
2. Klasmeyer, G. Sendmeier, W.F.: Objective Voice Parameters to Characterize the Emotional Speech, Proceedings of ICPhS'95, Stockholm, Sweden (1995) Vol.1, 182–185.
3. Mozziconaxxi, S.: Pitch Variations and Emotions in Speech, Proceedings of ICPhS'95, Stockholm, Sweden (1995) Vol.1, 178–181.
4. Cheol-Woo Jo et.al: Analysis of Emotional Speech According to Various Emotional Contents, Trans. On KASA, Vol.16, No.3 (1997) 33–37.
5. Cheol-Woo Jo: Collection and Analysis of Korean Emotional Speech, Proceedings of Oriental COCOSDA Workshop'99, Taipei, Taiwan (1998) 27–31.

Modeling Cue Phrases in Turkish: A Case Study

Bilge Say

Middle East Technical University,
Informatics Institute,
06531, Ankara, Turkey
bsay@ii.metu.edu.tr
http://www.ii.metu.edu.tr/~bsay/

Abstract. Cue phrases are lexical units that carry various signals within discourse to phenomena such as discourse relation detection. Their characterization have been found to be helpful in practical Natural Language Processing systems including dialogue planning agents and text summarizers.

A corpus based study of cue phrases in Turkish is to be presented. The results show that there is no significant tendency in sentences with cue phrases that have a noun full noun phrase subject to be followed by sentences with null subject sentences. Variations of this result is explained in the article.

1 Motivation

Cue phrases (also known as discourse markers, discourse particles or connectives) have been used as heuristics by various researchers in computational linguistics (Stede and others, [5]). In particular, they have been used to contribute to discourse segmentation, discourse relation detection and judging the coherence of essays. The motivation of this study is to characterize the interaction between nominal reference and occurrence of cue phrases. Especially, part of the inspiration stems from the memory-based paradigm to text comprehension, in the sense of research done by Myers and Brien ([4]). As an extension to such research, cue phrases can be seen as a factor increasing the elaboration of possible antecedents, thus actively facilitating the resolution of anaphora. The question that whether such a hypothesis is plausible from a generation point of view on existing test is a starting point for this research.

2 A Corpus Analysis of Turkish Cue Phrases

As an initial step, all the sentences with cue phrases (78 sentences in all) in ten expository articles (two articles on academic ethics, the rest newspaper or magazine editorials), each ranging from 600 to 1500 words were studied. Clause-level usage of cue phrases were not counted.

The percentage of sentences with cue phrases with a noun phrase subject (NP) followed by another NP subject is slightly larger than those followed by

F. Jelinek, E. Nöth (Eds.): TSD'99, LNCS 1692, pp. 337–340, 1999.

zero pronoun (Turkish being a pro-drop language) subject. Additionally, in only half of the 11 cases, the zero pronoun refers to the same NP (see Table 1). The rate of the following sentence having a overt pronoun referring to the previous subject NP is significantly lower.

Table 1. Number of S_{CPNP} (*Sentences with cue phrases that have NPs as subjects*) followed by S_{ACPNP}, S_{ACPO}, S_{ACPZ} (*a sentence with an NP, overt and zero pronoun subject respectively*)

	Number and % of S_{CPNP}	
Followed by S_{ACPNP}	14	51.9 %
Followed by S_{ACPO}	2	7.4 %
Followed by S_{ACPZ}	11	40.7 %

On the other hand, when the cue phrased sentence is the last sentence of a paragraph, which can be taken as an orthographic indicator of discourse segmentation, the rate of following sentence having another NP as a subject greatly increases (upto 80 %) as seen in Table 2.

Table 2. Number of S'_{CPNP} (*Paragraph-last sentences with cue phrases that have NPs as subjects*) followed by S_{ACPNP}, S_{ACPO}, S_{ACPZ} (*a sentence with an NP, overt and zero pronoun subject respectively*)

	Number and % of S'_{CPNP}	
Followed by S_{ACPNP}	12	80.0 %
Followed by S_{ACPO}	1	6.7 %
Followed by S_{ACPZ}	2	13.3 %

When an NP comes as the subject of the sentence before the sentence containing the cue phrase, the number of the cue phrased sentences continueing the same subject with a zero pronoun is lower (see Table 3) than the case depicted in Table 1. The use of an overt pronoun remains low, too.

Table 3. Number of S_{BCPNP} (*Sentences that have NPs as subjects preceding a cue-phrased sentence*) followed by S_{CPNP}, S_{CPO}, S_{CPZ} (*a cue-phrased sentence with an NP, overt and zero pronoun subject respectively*)

	Number and % of S_{BCPNP}	
Followed by S_{CPNP}	24	63.2 %
Followed by S_{CPO}	3	7.9 %
Followed by S_{CPZ}	11	28.9 %

If the cue phrased sentence is starting a new paragraph (thus presumably a new discourse segment), it contains a noun phrase as the new subject in 71 % of the cases and never a zero pronoun (see Table 4).

Table 4. Number of S_{BCPNP} (*Sentences that have NPs as subjects preceding a cue-phrased sentence*) followed by S'_{CPNP}, S'_{CPO}, S'_{CPZ} (*a paragraph-initial cue-phrased sentence with an NP, overt and zero pronoun subject respectively*)

	Number and % of S_{BCPNP}	
Followed by S'_{CPNP}	5	71.4 %
Followed by S'_{CPO}	2	28.6 %
Followed by S'_{CPZ}	0	0 %

Related data with annotations, which is not presented here due to lack of space, is available from the author.

3 Related Work and Discussion

The primary indication of this corpus study is that cue phrases by themselves from a text production point of view do not show adequate tendency for facilitating usage of overt or zero pronouns in subject positions instead of noun phrases. It has been found out in Turkish that if the information structure is retained then zero pronouns are used. Usage of overt pronouns or full noun phrases signal the introduction of a new focus-topic or beginning of a new discourse segment (Turan, [6]). Cue phrases can be taken as a means to make this kind of transitions smooth or in other words coherent though they do not by themselves facilitate the usage of zero pronouns or overt pronouns in the sentences they occur in or the following sentence.

(1) a. Murat yolculuktan   hoşlanmadı.
       Murat journey+ABL like+NEG+PAST+3sg
       Murat did not like the journey.

   b. Zaten o gitmek istememişti.
      In fact he go+INF want+NEG+PPAST+3sg
      In fact, he had desired not to go.

   c. Gitmek   istememişti.
      Go+INF want+NEG+PPAST+3sg.
      ⊘ had desired not to go.

   d. O gitmek   istememişti.
      He Go+INF want+NEG+PPAST+3sg.
      He had desired not to go.

Sentences (1a) followed by either (1b) or (1c) form a coherent discourse segment whereas (1d) is less felicitous because of the overt pronoun without an otherwise signalled information structure.

Previous studies (Litman and Passonneau, [3]) have found out that cue phrases fall at discourse segment boundaries more than 40 % of the times but when combined with other factors such as prosody can be an indicator of new discourse segments. The effect of prosody in conjunction with cue phrases is yet to be clarified for Turkish. Another further point can be clarified by means of a classification study like that of Knott and Sanders ([2]) that classifies cue phrases according to various discourse relations they are related to and their specificity. Some cue phrases via their discourse relation signalling characteristic will be more likely to start discourse segments and thus more likely to affect nominal structure.

4 Conclusion and Future Work

This initial work shows that there are no explicit strong links between occurrence of cue phrases and pronominal expressions as subjects in Turkish. Further corpus analysis using a theoretical framework such as Centering Theory (Grosz and others, [1]) could be beneficial as well as psycholinguistic experimental settings to relate to psychological plausibility of the role of cue phrases' interaction with nominal reference from a comprehension and production point of view.

References

1. Grosz, B., Joshi, A., Weinstein, S.: Centering: A Framework for Modeling the Local Coherence of Discourse. Computational Linguistics. 21(2) (1995) 203–225.
2. Knott,D., Sanders, T.: The Classification of Coherence Relations and their Linguistic Markers: An Exploration of Two Languages. Journal of Pragmatics, 30 (1998) 135–175.
3. Litman, D., Passonneau, R.: Combining Multiple Knowledge Sources for Discourse Segmentation. 33rd Annual Meeting of the Association for Computational Linguistics, (1995) 108–115.
4. Myers, J.L., O'Brien, E.J.: Accessing the Discourse Representation During Reading. Discourse Processes. 26(2&3) (1998) 67–86.
5. Stede, M., Wanner, L., Hovy, E. (eds.): Proceedings of Discourse Relations and Discourse Markers COLING-ACL'98 Workshop (1998).
6. Turan, Ü., Ranking Forward-Looking Centers in Turkish: Universal and Language-Specific Properties. In Walker, M., Joshi, A. K., Prince, E. F. (eds.) Centering Theory in Discourse, 8, Oxford University Press, (1997) 139–160.

Speaker Identification Based on Vector Quantization*

Vlasta Radová and Zdeněk Švenda

University of West Bohemia, Department of Cybernetics,
Univerzitní 22, 306 14 Plzeň, Czech Republic
{radova, svendaz}@kky.zcu.cz

Abstract. In this paper a method of text-independent speaker recognition using discrete vector quantization is presented. The identification experiments were performed in a closed set of 599 speakers and two various types of features were tested: cepstral mean subtraction coefficients and mel-frequency cepstral coefficients. The effect of the various codebook size on the speaker identification performance was investigated.

1 Introduction

There are many various methods that can be used for the design of speaker recognition systems. At present, a majority of systems is based on vector quantization, hidden Markov models or artificial neural networks [1,3,4]. The method described in this paper is based on discrete vector quantization and can be used for text-independent speaker identification. The principle of the method is explained in Section 2. In Section 3 the structure of the speech database and the feature analysis techniques used in our experiments are described. Achieved experimental results are presented in Section 4.

2 Principle of the Method

The principle of the speaker identification using vector quantization consists of 2 phases – training and identification.

During the *training phase* a codebook for each reference speaker is formed from the training data of that speaker. To do it, we used so-called uneven binary decision and the modified MacQueen (k-means) algorithm described in [5]. Achieved codebooks have a tree structure, where each leaf node represents a codebook vector. Due to this structure, a very simple algorithm can be used to find the nearest codebook vector to an input vector.

During the *identification phase*, the speech signal of an unknown speaker is first analyzed and converted to a sequence of N feature vectors $x(1), \ldots, x(N)$. Then each feature vector of the unknown speaker is quantized using codebooks

---

* This work was supported by the Ministry of Education of the Czech Republic, project no. VS 97159.

F. Jelinek, E. Nöth (Eds.): TSD'99, LNCS 1692, pp. 341–344, 1999.

of all reference speakers and the average quantization distortion Q_k is computed for each of K reference speakers according to the formula

$$Q_k = \frac{1}{N} \sum_{n=1}^{N} d(x(n), v_{*k}),$$ (1)

where $d(x(n), v_{*k})$ is a distortion measure between the input vector $x(n)$ and its nearest codebook vector v_{*k} of the k-th reference speaker. The unknown speaker is then identified as the reference speaker with the minimum average distortion, i.e.

$$k^* = \operatorname*{argmin}_{k=1,...,K} Q_k,$$ (2)

where k^* represents the identified speaker.

3   Speech Database and Feature Analysis

The method described in Section 2 was used to identify "unknown" speakers in a group of 599 speakers (317 male, 282 female). Every speaker spoke a different set of several short Czech sentences and isolated words in the total duration of about 70s. Approximately 40s of speech of each speaker were regarded as training data and were used to form the codebook of that speaker. The remaining about 30s of speech were used for identification tests.

The speech signal of each speaker was recorded during one session through a telephone channel, sampled at the rate of 8kHz and stored in a μ-law 8bit format. Before further processing, however, the 8bit μ-law samples were converted to linear 16bit PCM samples. Finally, the digitalized utterances were converted to sequences of segments using a 16ms rectangular window.

In our experiments various types of features were tested – cepstral mean subtraction coefficients (CMSCs) and mel-frequency cepstral coefficients (MFCCs). To obtain the vectors of CMSCs the utterances were first converted to sequences of vectors of 12 LPC coefficients. These vectors were then converted to vectors of 12 LPC cepstral coefficients (LPCCs) and finally to the vectors of 12 CMSCs according to the formula (see [2])

$$c_{cmsc}(n) = c_{lpcc}(n) - \bar{c}_{lpcc},$$ (3)

where $c_{lpcc}(n)$ is the n-th vector of LPC cepstral coefficients and $\bar{c}_{lpcc}$ is the mean vector of LPC cepstral coefficients over the whole speech data of a speaker, i.e

$$\bar{c}_{lpcc} = \frac{1}{N} \sum_{n=1}^{N} c_{lpcc}(n),$$ (4)

where N is the number of segments in the speech of the speaker.

In the experiments with MFCCs the speech signal was processed first by a set of 26 triangual filters spaced uniformly on a mel-scale. The output energy of the filters were then transformed into vectors of 16 MFCCs using the discrete cosine transform [2].

The feature vectors were formed in several various ways. First, either vectors of only 12 CMSCs or vectors of only 16 MFFCs were used as feature vectors. Next the feature vectors were extended and to the CMSCs or MFCCs their delta coefficients or delta and acceleration coefficients were added. The delta coefficients were computed using the formula (see [6])

$$d(n) = \frac{\sum_{i=1}^{I} i[c(n + i) - c(n - i)]}{2 \sum_{i=1}^{I} i^2},$$ (5)

where $d(n)$ is a vector of delta coefficients for the n-th segment of speech and $c(n)$ is a vector of 12 CMSCs or 16 MFCCs for the n-th segment of speech. The value of I was set equal to 2 in our experiments. Similarly the vectors of acceleration coefficients $a(n)$ were computed from the delta coefficients $d(n)$ according to the formula

$$a(n) = \frac{\sum_{i=1}^{I} i[d(n + i) - d(n - i)]}{2 \sum_{i=1}^{I} i^2}.$$ (6)

Since equations (5) and (6) rely on past and future speech segments some modification is needed at the beginning and end of the speech. One of the possible modifications is simply to replicate the first or last vector as needed [6].

4   Experimental Results

All types of feature vectors described in Section 3 were used to form a codebook of either 80 or 320 codebook vectors. To compute the distortion measure $d(.,.)$ in (1) the cepstral measure [5] was used. Achieved results are presented in Table 1 and Table 2. In these Tables, CMSCΔs means feature vectors composed of CMSCs and delta coefficients, and CMSC$\Delta\Delta$s means feature vectors composed of CMSCs, delta coefficients and acceleration coefficients. Similar denotation is valid for the MFCCs as well.

Table 1. Identification results using the codebook of 80 vectors

Coefficients	# correct	Correct [%]	Coefficients	# correct	Correct [%]
CMSCs	564	94.16	MFCCs	595	99.33
CMSCΔs	569	94.99	MFCCΔs	595	99.33
CMSC$\Delta\Delta$s	567	94.66	MFCC$\Delta\Delta$s	586	97.83

Table **2.** Identification results using the codebook of 320 vectors

Coefficients	# correct	Correct [%]	Coefficients	# correct	Correct [%]
CMSCs	570	95.16	MFCCs	597	99.67
CMSCΔs	578	96.49	MFCCΔs	597	99.67
CMSC$\Delta\Delta$s	572	95.49	MFCC$\Delta\Delta$s	594	99.17

As the Tables show, using the feature vectors with the MFCCs better results were achieved than using feature vectors with the CMSCs. The size of the codebooks has a certain effect on the speaker identification performance as well – when the codebook with 320 vectors was used more speakers were identified correctly than with the codebook of 80 vectors.

5   Conclusion

A method of speaker identification based on vector quantization was presented in this paper. Various types of feature vectors were tested and as many as 99.67% of speakers in a group of 599 speakers were identified correctly. Such a result may be regarded as a promising way to a high-performance speaker identification system. However, it has to be taken into account that the speech data used in the experiments were recorded during one session. When there is a time interval between the recording of training and test data the achieved results may not be so excellent. Our experiments on data where the time interval between the recording of test and training data are from several days to several weeks are now in progress.

References

1. Cheng, Y., Leung, H. C.: Speaker Verification Using Fundamental Frequency. In: Proc. of the ICSLP (1998) 161–164.
2. Mammone, R. J., Zhang, X., Ramachandran, R. P.: Robust Speaker Recognition. A Feature-based Approach. IEEE Signal Processing Magazine **5** (1996) 58–71.
3. Monte, E., Adolf, A., Miró, X., Hernando, J.: Text Independent Speaker Identification on Noisy Environments by Means of Self Organising Maps. In: Proc. of the ICSLP (1996) 1804–1087.
4. Monte, E., Arqué, R., Miró, X.: A VQ Based Speaker Recognition System Based in Histogram Distances. Text Independent and for Noisy Environments. In: Proc. of the ICSLP (1998) 185–188.
5. Psutka, J.: Communication with Computer by Speech. Academia, Praha (1995) (in Czech).
6. Young, S., Jansen, J., Odell, J., Ollason, D., Woodland, P.: The HTK Book (for HTK V2.0). Cambridge University (1996).

Robustness in Tabular Deduction for Multimodal Logical Grammar - Part 1

Geert-Jan M. Kruijff*

Institute of Formal and Applied Linguistics (ÚFAL)
Faculty of Mathematics and Physics, Charles University
Prague, Czech Republic
gj@ufal.mff.cuni.cz
http://kwetal.ms.mff.cuni.cz/~gj/

1 Introduction

Although opinions may differ on what to understand by the notion of "robustness", it is clear that -whatever we believe robustness to entail- we need it for parsers (and grammars!) which are to work on 'real-life' text or speech. In this paper we present several extensions to a deduction method for parsing with *multimodal logical grammars*, aiming to provide more robustness. The deduction method was originally developed by Hepple [1] for linear logic and has been extended by Kruijff in [2] to cover multimodal logical grammar (MMLG). MMLG is a grammar framework in the spirit of categorial grammar and type-logical grammar [3]. Within the last couple of years, various authors have used MMLG to formulate grammar-fragments dealing not only with 'pure' syntax but also with prosody and information structure (notably Hendriks, Steedman). MMLG thus appears interesting as a grammar framework on which both discourse analysis and dialogue analysis can build.

2 Tabular Deduction for MMLG

Tabular deduction for MMLG avoids redundant computation like chart-based parsing does for CFGs (i.e. the method is *memoetic*). Earlier work by König, Hepple on chart-based methods for parsing with categorial grammars based on natural deduction (ND) suffered from administrative overhead due to the possibility (in ND) to *introduce assumptions*. So-called "mini-charts" had to be used to keep track of the introduction (and proper discharge) of assumptions. Morrill proposed to overcome this problem by using proofnets and charts. Hepple discusses in [1] a method for tabular deduction that provides a solution by means of *compilation*. Hepple's compilation procedure extracts from higher-order formulas the hypotheses that are implicitly in there. The result is that a higher-order

* The research reported of in this paper was supported by grants from the Grant Agency of the Czech Republic (GAČR), 405/96/K214, and from the Ministry of Education, Project No. VS96151. Thanks to Mark Hepple for many illuminating discussions. Needless to say, any mistakes are my own.

F. Jelinek, E. Nöth (Eds.): TSD'99, LNCS 1692, pp. 345–348, 1999.
© Springer-Verlag Berlin Heidelberg 1999

formula can be compiled into a set of first-order formulas that are linked by means of *index-lists* showing their interdependency, and where the formula expressing the result/conclusion has the introduction explicitly compiled into its semantics (by means of a λ-abstraction - to which ND-introduction corresponds under the Curry-Howard correspondence between formulas and λ-terms). Due to the possibility to have introduction explicitly in the λ-semantics of a term, there is no longer a need to have introduction rules in the deduction system. Hepple proposes a single elimination rule, which operates on formulas and which ensures *linearity* and *proper use of hypotheses* by means of constraints on the index-lists of the formulas to be combined.

Hepple formulates his proposal for linear logic, which is the strongest logic in the hierarchy of resource-sensitive logics. The logics used in MMLG-fragments are usually weaker (more constrained). We elaborate in [2] Hepple's method in order to include such weaker logics, by expounding the ND system to a labelled deductive system, and making use of the labelling to (further) constrain inferences. A labelled deductive system ($\mathcal{LDS}$) deals with terms of the form $\langle label, formula \rangle$. The label *annotates* the formula with information that can be used to control the inference over the formula. In the context of MMLG, we use terms of the form $R \vdash C : S$, with R the label or resources, and C a category with semantics S.

For our tabular deduction method we also specify a compilation procedure and an elimination rule, like Hepple, but we operate on slightly different terms. The basic form into which we compile linguistic resources of the form $R \vdash C : S$ is $\langle \rho, \varphi, \kappa, \sigma \rangle$ with ρ the label (or, technically, the "resource"), ϕ an index list, and κ and σ (compilations of) the category and semantics, respectively. The compilation procedure and the elimination rule are defined as follows - see [2] for discussion.

Definition 1 (Compilation procedure). *We define the compilation τ in a ecursive fashion. We distinguish rules for compiling directional implications (τ1-τ4), one rule for compiling atomic categories without the unary modal $\Box^{\downarrow}$ (τ5), and rules for compiling categories that are boxed (τ6-τ7).*

Compilation of Directional Implications

(τ1) $\tau(\langle \rho, \varphi, X/_i Y, \sigma \rangle) = \tau(\langle \mathbf{right}(i, \rho), \varphi, X \multimap (Y : \emptyset), \sigma \rangle)$,
   where Y has no (inclusion) index set.

(τ2) $\tau(\langle \rho, \varphi, X \backslash_i Y, \sigma \rangle) = \tau(\langle \mathbf{left}(i, \rho), \varphi, X \multimap (Y : \emptyset), \sigma \rangle)$,
   where Y has no (inclusion) index set.

(τ3) $\tau(\langle \rho, \varphi, X_1 \multimap (Y : \psi), \sigma \rangle) = \langle \rho, \varphi, X_2 \multimap (Y : \psi), \lambda x.t \rangle \cup \Gamma$,
   with Y atomic, x a fresh variable, and
   $\tau(\langle \rho, \phi, X_1, (sx) \rangle) = \langle \rho, \phi, X_2, t \rangle \uplus \Gamma$

(τ4) $\tau(\langle \rho, \phi, X \multimap ((Y \multimap Z) : \psi), \sigma \rangle) =$
   $= \tau(\langle \rho, \phi, X \multimap (Y : \pi), \lambda y.\sigma(\lambda z.y) \rangle) \cup \tau(\langle \epsilon, \iota, Z, \varsigma \rangle)$
   with ι a new index, y, z fresh variables, and $\pi = \iota \cup \psi$

Atomic Categories Without $\Box^{\downarrow}_i$

(τ5) $\tau(\langle \rho, \varphi, X, \sigma \rangle) = \langle \rho, \varphi, X, \sigma \rangle$, with X atomic.

Categories With $\square_i^\downarrow$

(τ6) $\tau(\langle \rho, \varphi, \square_i^\downarrow X, \sigma \rangle) = \langle \rho, \varphi, \square_i^\downarrow X, ^{\wedge i}(\sigma) \rangle$, with K atomic.

(τ7) $\tau(\langle \rho, \phi, \square_i^\downarrow X_1, \sigma \rangle) = \langle \rho', \phi, \square_i^\downarrow X_2, ^{\wedge i}(\sigma') \rangle$,

   with X_1 non-atomic, and

   $\langle \rho', \phi, X_2, \sigma' \rangle = \tau(\langle \rho, \phi, X_1, \sigma \rangle), \langle \rho', \phi, \square_i^\downarrow X_2, ^{\wedge i}(\sigma') \rangle \uplus \Gamma$

Definition 2 (Elimination Rule). *The elimination rule for the tabular deduction method is defined as follows. Δ represents a possibly empty sequence of directionality markers. If Δ is empty, then $\Delta[\Theta] \overset{definition}{\equiv} \Theta$.*

$$\frac{\langle \vartheta, \varphi, A \multimap (B : \alpha), \lambda v.a \rangle \quad \langle \theta, \psi, B, b \rangle}{\langle \Delta[\Theta], \pi, A, a[b//v] \rangle} E$$

with $\pi = \varphi \uplus \psi, \alpha \subseteq \psi$, and

$$\Theta = \begin{cases} (\omega \circ_i \theta) & \text{iff } \vartheta \text{ was of the form } \textbf{right}(i, \Delta[\omega]), \\ (\theta \circ_i \omega) & \text{iff } \vartheta \text{ was of the form } \textbf{left}(i, \Delta[\omega]) \end{cases} \quad (1)$$

As in [1], $\uplus$ (disjoint union) will enforce linear usage, whereas the requirement on α will ensure proper usage. In addition, the formation of the resource Θ out of ω and θ in an order depending on the (premisse) marking of ϑ will ensure proper sensitivity to directionality. Observe that the handling of directionality as presented here is general enough to handle *any* kind of directional implication definable in MMLG, notably in compar

ison to other proposals like [4], where compilation is restricted to the Lambek calculus L in a way that is not straightforwardly generalizable to a logic involving commutativity.

Thus, direction-sensitive implications ($\backslash, /$ or $\rightarrow, \leftarrow$) become available, as well as the entire range of structural rules one can normally formulate in MMLG. The method of [2] has been implemented in Prolog and in Perl, on the basis of a bottom-up Early parsing schema.

3   Robustness

In robust parsing, a distinction can be made between *shallow* (or *phrasal*) parsers, and *full* parsers. The former may be said to aim at providing analyses that segment a sentence into a minimal (bracketed) structure, whereas the latter intends to derive a structure in which the distinction between predicate(s) and argument(s) is elucidated. We aim at building a full parser, and we can conceive of at least the following issues that need to be addressed in the context of robustness. From the viewpoint of *composition/deduction*, we need to be able to provide at least partial analyses (proofs) for those parts recognized by the grammar – even if an overall analysis fails. Put differently, we need to be able to cope with *incompleteness of the grammar* (i.e. the set of structural rules). Similarly, incompleteness may arise within the lexicon: we might be facing *undergeneration*, meaning the failure to provide an analysis due to missing lexical items. On the other hand, robustness may also be increased by *generalization*: Generalization of lexical items in the sense of required complementations, and generalization of structural rules in terms of applicability.

Moortgat and Morrill (cf. [3]) have proposed to use additive logical operators $\sqcup, \sqcap$ in categories to express lexical generalizations: $\sqcup$ for disjunctive assignment, $\sqcap$ for multiple assignment. For example, $np\backslash s/(np\sqcup s)$ means a verb taking either a sentential or np complement (like En. *to know*). Or, we can use $(pp\sqcap(n\backslash n))/np$ for En. *at* to express that it can combine with an np to either form a pp or a nominal modifier. Another application of the additive operators (particularly $\sqcup$) is in *coordination of unlike categories*, like the coordination in "Pat is either stupid or a liar".

Using a "gensymmed" *goal* α (cf. also [1]), we can compile the category $np \backslash s/(np \sqcup s)$ into $s \circ\!\!-(np : \emptyset) \circ\!\!-(\alpha : \sqcup\{i,j\})$. The indexlist for the function's α argument contains the indices of the $\sqcup$'d formulas np and s. Naturally, the label of $s \circ\!\!-(np : \emptyset) \circ\!\!-(\alpha : \sqcup\{i,j\})$ will only have to indicate the directionality regarding np, in this case **left**, and α, being **right**[1].Consequently, we formulate the following compilation rules in addition to $(\tau 1) - (\tau 7)$ above.

$(\tau 1a)$ $\tau(\langle \rho, \varphi, X/_i(Y \sqcup Z), \sigma\rangle = \tau(\langle \mathbf{right}(i, \rho), \varphi, X \circ\!\!-(\alpha : \{\delta, \eta\}), \sigma\rangle) \cup$
$\qquad \tau(\langle \rho, \{\delta\}, Y, s\rangle) \cup \tau(\langle \rho, \{\eta\}, Z, s'\rangle)$
$\qquad$ with α a unique goal, and δ, η new indices.

$(\tau 2a)$ $\tau(\langle \rho, \varphi, X \backslash_i (Y \sqcup Z), \sigma\rangle = \tau(\langle \mathbf{left}(i, \rho), \varphi, X \circ\!\!-(\alpha : \{\delta, \eta\}), \sigma\rangle) \cup$
$\qquad \tau(\langle \rho, \{\delta\}, Y, s\rangle) \cup \tau(\langle \rho, \{\eta\}, Z, s'\rangle)$
$\qquad$ with α a unique goal, and δ, η new indices.

To extend the elimination rule of definition 2 so as to include (the effects of compiling) $\sqcup$, then, we only need to alter the conditions on α. Namely, whereas α used to be a simple list, we may now encounter either a simple list or a list prefixed by $\sqcup$. The Proper Use Condition therefore becomes as follows: $\alpha \subseteq \psi$ if α is a simple list, $\gamma \subseteq \psi$ if $\alpha = \sqcup \Gamma, \gamma \in \Gamma$.

We leave providing a solution to the problem of undergeneration for another paper; as for obtaining partial analyses, it is easy to see that a table (or chart) by its very nature provides us with all the partial analyses we might possibly be interested in.

References

1. Mark Hepple. Memoisation for glue language deduction and categorial parsing. In *Proceedings of the COLING-ACL'98 Joint Conference*, Montreal, Canada, 1998.
2. Geert-Jan M. Kruijff. Implementing tabular deduction for multimodal logical grammar. *Prague Bulletin of Mathematical Linguistics*, 71, to appear 1999.
3. Michael Moortgat. Categorial type logics. In Johan van Benthem and Alice ter Meulen, editors, *Handbook of Logic and Language*. Elsevier Science B.V., Amsterdam New York etc., 1997.
4. Glyn V. Morrill. Memoisation of categorial proof nets: parallelism in categorial processing. In V. Michele Abrusci and Claudia Casadio, editors, *Proofs and Linguistic Categories: Proceedings of the 1996 Roma Workshop*, pages 157–169, 1996.

---

[1] It is important to observe that in compilation, directionality markers are kept "local".

Classifying Visemes for Automatic Lipreading

Michiel Visser[2], Mannes Poel[1], and Anton Nijholt[1]

[1] University of Twente, Department of Computer Science,
P.O. Box 217 NL-7600 AE Enschede, The Netherlands
{anijholt, mpoel}@cs.utwente.nl
[2] Philips Medical Systems, The Netherlands

Abstract. Automatic lipreading is automatic speech recognition that uses only visual information. The relevant data in a video signal is isolated and features are extracted from it. From a sequence of feature vectors, where every vector represents one video image, a sequence of higher level semantic elements is formed. These semantic elements are "visemes" the visual equivalent of "phonemes" The developed prototype uses a Time Delayed Neural Network to classify the visemes.

1 Introduction

Automatic Speech Recognition (ASR) normally uses acoustic information as input. Often this information is too unreliable to get good recognition. This unreliability, indicated by the signal-to-noise ratio, is caused by noise of machinery, other people speaking, effects in the measurement, etc. In these cases visual information can be used to improve recognition, because speech production produces several visual side effects. This is normally referred to as lipreading, but is also called speech reading, to clarify that this information is not all located in the lips. Other relevant features for lipreading can be extracted from the jaw, tongue, teeth and skin, see for instance [2] and [1].

In this paper we will study the problem of mapping lip movements, contained in a digital video of the face, onto relevant semantic features, called visemes.

Section 2 describes the stages in the recognition process theoretically, while Section 3 describes a prototype that implements this decomposition. This prototype is evaluated in Section 4. The conclusions can be found in section 5.

2 Decomposition of the Lipreading System

The lipreading system is decomposed in three subsystems, First we have the liplocalization system, which localizes the lips in the digital input, then there is the feature extracting system, which computes the relevant features of the lips, and finally we have the classification system, which maps feature vectors to visemes.

Lip localization. The lip localizator must find the position of the lips in the digital image. This is a difficult task, due to the demands on robustness, accurateness, and the system should be person-independent.

F. Jelinek, E. Nöth (Eds.): TSD'99, LNCS 1692, pp. 349–352, 1999.

Most lip localization techniques are template- or edge-based. The template approach has as one of the disadvantages the lack of speed. A lot of templates must be tested on a lot of areas before an accurate position of the position of the mouth can be found. Lips can also be found using edge detection. Standard edge finders can find edges of the lips using a hard-crafted or learned model of the lip edges. One of the main disadvantages of the lip-edge approach is the lack of robustness.

Instead of finding the lips directly, the position of the lips can be calculated using the positions of other objects in the face, like the nose and the eyes. Using the nose and the eyes, the position of the mouth area can be predicted to a certain extent and therefore only a small piece of a video image has to be tested for lips.

Feature extraction. After the position, size and rotation of the mouth area are determined a set of features must be extracted using a parameter set of a lip edge model, pixel map of the mouth area, or a compressed pixel map. Using a lip edge model has the advantage that it gives a small set of well-defined parameters, that are more invariant in lighting and personal characteristics than pixel based parameters. The disadvantage is that it looses a lot of information and accuracy.

The set of parameters changes dynamically in time. Therefore often dynamical features are added to the feature set.

Translation to a semantic representation. To get a semantic representation of the spoken words the time sequence of the feature set must be translated into a sequence of semantic symbols. As semantic symbols we use visemes. Visemes are classes of phonemes which are indistinguishable from a lipreading point of view, see Figure 1.

viseme	phoneme class	viseme	phoneme class
0	<silence>	8	I e:
1	f v w	9	E E:
2	s z	10	A
3	S Z	11	@
4	p b m	12	i
5	g k x n N r j	13	O Y y u 2: o: 9 9: O:
6	t d	14	a:
7	l		

Fig. 1. Visemes as phoneme classes, using SAMPA notation.

However, the utterance of a phoneme does not generate exactly the same lip position and movement all the time. Various factors, such as person, situation and mood, causes these variances in phoneme production.

3   A Prototype of the Lip Reading System

For the above system a prototype is constructed, following the decomposition above.

Lip localization. For the liplocalization we used a non-standard approach. From a set of digital lip images (generated by hand) the principal components were computed. These principal components were used to compress the fixed regions of the picture. Since the principal components were fine tuned for lips, the lips will be in the region with the minimum loss of information.

Feature extraction. The feature extractor has as input a normalized rectangle containing the lips. Principal component analysis is used to generate a feature vector for the rectangler input frame. The principal components are computed using a single layer feedforward neural network and Sanger's learning algorithm [4].

Translation to visemes. The feature vectors computed by the feature extractor must be classified as (the produced) visemes. In order to incorporate temporal aspects in the classification process a Time-delay Neural Network (TDNN) [3,5], was used to classify the feature vectors.

In order to train the TDNN, a dataset was manually generated. Three different people uttered, under identical conditions, a list of non-existing words, which precisely contained all the viseme combinations. One testset was made, which consists of the utterances of a list of real words. These datasets were manually analyzed, and every feature vector was labeled with the corresponding viseme. These labeled sets were used to train and test the TDNN.

4   Evaluation of the Prototype

The succes rate of the prototype is given in figure 2. As can be seen from this table the average succes rate is 0.226. In order to gain more insight in the mistakes

viseme number	average succes rate	viseme number	average succes rate
0	0.260	8	0
1	0	9	0.667
2	0.167	10	0
3	0.389	11	0.419
4	0	12	0
5	0	13	0
6	0.867	14	0
7	0.600		

Fig. 2. Succes rate of the lipreading prototype.

made by the prototype a confusion matrix was constructed, figure 3. As one can see there is a lot of confusion between the first 7 visemes.

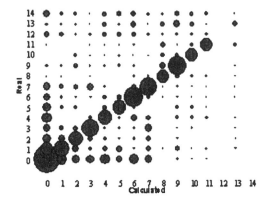

Fig. 3. Viseme confusion matrix

5   Conclusions

The prototype lacks speed and gives non-optimal results. The localization of the mouth is by far the slowest component in the prototype. More sophisticated techniques should be used to get a significant speedup. The following causes can be mentioned for the non-optimal classification results:

- The visemes, seen as a set of phonemes, are not correct, different visemes correspond to almost identical lip movements. An indication for this can be found in the confusion matrix in figure 3.
- Too little data: more instances of the trainings data should be generated and in the training sets natural words are preferable above synthetic ones.
- Differences in lighting: It turned out that differences in illuminations were larger than expected.

References

1. Benoît, Abry, Cathiard, Guiard-Marigny et all.: Read my lips: where? how? when? and so ... What? In: Proceedings of the 8th Int. Congress on Event perception and Action (1995).
2. Benoît, Guiard-Marigy, Le Goff, Adjoudani: Which components of the face do humans and machines best speechread? In: Stork, D. (eds.): Speechreading by Man and Machine. Springer-Verlag.
3. Lang, K.J., Hinton, G.E.: The development of the time-delay neural network architecture for speech recognition. Techical Report CMU-CS-88-152, Carnegie-Mellon University (1988).
4. Sanger, T.D.: Optimal unsupervised learning in a single-layer linear feedforward neural network. Neural Networks. **12** 459–473.
5. Waibel, A., Hanazawa, T., Hinton, G., Shikano, K., Lang, K.J.: Phonemen recognition using time-delay neural networks. IEEE Transactions on Acoustic, Speech, and Signal Processing, **37** (1989) 328–339.

Semantic Inference in the Human–Machine Communication*

Leo Hadacz

Faculty of Science
Masaryk University
Janáčkovo nám. 2a, CZ-662 95 Brno, Czech Republic
hadacz@math.muni.cz

Abstract. One of the recent trends in computer science is to create information retrieval systems which will be able to communicate with the user in natural language. Such a system requires certain abilities, especially semantic analysis of the natural language and semantic inference.

1 Introduction

Semantic inference means for us the following: a set of input facts and one question are given. The task is to infere the answer to this question using the set of input facts or to decide that the input data are unsatisfactory. The facts are expressed through propositions. Propositions and questions have to be represented in an appropriate form. Building this representation means in fact performing semantic analysis. For this purpose we use the Transparent Intensional Logic, or TIL. For the explanation see [1] and also [3,4]. The semantic analysis is realized through the so called Normal Translation Algorithm (see [3,4]). Its output are TIL constructions. One of our goals is to create a computer program which will do the semantic inference. We call it simply TIM, which means TIL Inference Machine.

2 Basic Definitions and Principles

The basic entity, which we are working with, is a so called *match*, which is a pair of two components. The first component in the pair is an object (that means a member of some type, a constant) or some variable. The entity which is a constant or a variable we call an *atom*. The latter component is a construction. Intuitivelly, the atom in a match tracks the entity, which the construction constructs. A match is designated a : C.

The input facts and question are given through a set of matches $\mathfrak{P}1, \ldots, \mathfrak{P}n$, $\mathfrak{Q}$. We assume that the appropriate propositions are true in the reference world

---

* This research has been funded by the Czech Ministry of Education under the grants
VS97028 and 0408/99.

F. Jelinek, E. Nöth (Eds.): TSD'99, LNCS 1692, pp. 353–356, 1999.

W and at the time of utterance T. This assumption is expressed through the constant T on the place of atoms in matches. The input facts are:

$$\mathsf{T}:\mathrm{Pl}_{\mathrm{WT}},\ldots,\mathsf{T}:\mathrm{Pn}_{\mathrm{WT}}$$

We restrict ourselves to those questions, where the answer is yes or no or some object or a class of objects.

We use inference system similar to Gentzen's Sequent Calculus, which is well known from the propositional or predicate logic. That is why the inference rules have the following form:

$$\frac{\Psi_1 \vdash \mathfrak{M}_1; \Psi_2 \vdash \mathfrak{M}_2; \ldots; \Psi_k \vdash \mathfrak{M}_k}{\Psi \vdash \mathfrak{M}}(*)$$

The inference behaves like operations over sequents. Above the line there could be also an empty set of sequents. For more explanation see [2,4].

To answer the given question involves infering the sequent:

$$\mathsf{T}:\mathrm{Pl}_{\mathrm{WT}},\ldots,\mathsf{T}:\mathrm{Pn}_{\mathrm{WT}} \vdash x:\mathrm{Wh}_{\mathrm{WT}}$$

where $x/o\xi$ or x/ξ or x/o, and determining the evaluation of this variable, which holds the answer to the question. It can happen that it will be impossible to infere such a sequent.

When performing the inference we use the characteristic function $\chi_{\mathbf{S}}:\mathbf{C} \longrightarrow \{0,1\}$, where $\mathbf{S}$, $\mathbf{C}$ are classes of objects, $\mathbf{S} \subset \mathbf{C}$. This function is defined as usual:

$$\chi_{\mathbf{S}}(c) = \begin{cases} 1 & \forall c \in \mathbf{S} \\ 0 & \forall c \notin \mathbf{S} \end{cases}$$

where c/α, $\mathbf{S}, \mathbf{C}, \chi_{\mathbf{S}}/o\alpha$.

The TIM system and implementation of NTA forms a system for knowledge representation. If we want this system to be practically usable, we need to augment it with a large set of facts. These facts can be obtained from the electronical lexical reference systems called wordnets. There are currently being processed at least two wordnets, for english language it is WordNet and for some european languages (including czech) it is EuroWordNet. Wordnets include words of the natural language, collocations and various relations between them. A typical lexical relation is for example the relation "ISA" (comes from "something is a something else") also called hypernymic or superordinate relation. For example ISA(oak,tree) ("oak is a tree"), ISA(tree,plant), ISA(plant,organism). These relations can on the other hand be directly represented as TIL constructions. So in the semantic inference it is sufficient to work with the inference over TIL constructions. For more information about wordnets see [7].

The TIM system is being implemented in the CAML language, which is a functional language based on the ML language. TIM is being developed simultaneously with the implementation of the Normal Translation Algorithm (NTA). NTA outputs TIL constructions of analyzed natural language expressions. These constructions form the basic data for the inference machine.

3   Demonstration of the Semantic Inference

We can demonstrate the semantic inference on the following simple example: given one proposition "Teddy is a bear." and one question "Is Teddy an animal?" We want the system for knowledge representation to answer "yes". The system should first semantically analyze the input. Then it probably detects, that the outcoming constructions do not carry enough information. The system tries to obtain some information about animal. It probably discovers, that ISA(bear,animal) (if this information is in the used wordnet). This is already enough to infer the answer, as shown in the example.

Example 1.

> P1 Teddy is a bear.
> P2 Each bear is an animal.
> Yn Is Teddy an animal?
> A Yes.

Of course, the expression P2 we get from the WordNet database, where is defined the relation ISA. For our case ISA(bear,animal) expresses, what exactly says the expression P2. More about possible relations in wordnets see [6,7].

The analysis of given input facts and the question is

1. P1: $\lambda w \lambda t.^0 animal_{wt} {}^0 Teddy$
2. P2: $\lambda w \lambda t. \forall x.^0 bear_{wt} x \supset {}^0 animal_{wt} x$
3. Yn: $\lambda w \lambda t.^0 animal_{wt} {}^0 Teddy$

The difference between the semantical analysis of a question and a fact cannot be seen before using matches. So on the input we have following sequents:

1. $\vdash \mathsf{T} : {}^0 bear_{\mathrm{WT}} {}^0 Teddy$
2. $\vdash \mathsf{T} : \forall x.^0 bear_{\mathrm{WT}} x \supset {}^0 animal_{\mathrm{WT}} x$
3. $\vdash y : {}^0 animal_{\mathrm{WT}} {}^0 Teddy$

At the first time it was nessessary to apply a concrete reference world and time. (*WT*). The constructions in matches are shown after the β-reduction.

The inference of the answer A (it means "yes", the truth) goes like that:

1. $\vdash \mathsf{T} : \forall x.^0 bear_{\mathrm{WT}} x \supset {}^0 animal_{\mathrm{WT}} x$
2. $\vdash \chi_\iota : \lambda x.^0 bear_{\mathrm{WT}} x \supset {}^0 člověk_{\mathrm{WT}} x$
3. $\mathsf{T} : {}^0 \chi_\iota(x) \vdash \mathsf{T} : {}^0 bear_{\mathrm{WT}} x \supset {}^0 animal_{\mathrm{WT}} x$
4. $\mathsf{T} : {}^0 \chi_\iota(x) ; \mathsf{T} : {}^0 bear_{\mathrm{WT}} x \vdash \mathsf{T} : {}^0 animal_{\mathrm{WT}} x$

The unification algorithm unifies the second match in the left-hand side of this sequent with P1 and finds the substitution $x = {}^0 Teddy$. So we get

$$\mathsf{T} : {}^0 \chi_\iota({}^0 Teddy) ; \mathsf{T} : {}^0 bear_{\mathrm{WT}} {}^0 Teddy \vdash \mathsf{T} : {}^0 animal_{\mathrm{WT}} {}^0 Teddy$$

The unification algorithm unifies the match on the right-hand side of the above sequent with Yn and the result is the substitution $y = \mathsf{T}$. Because *Teddy* $\in \iota$ and we assume the truth of P1, after double application of the rule (4) we get the sequent

$$\vdash \mathsf{T} : {}^0 animal_{\mathrm{WT}}{}^0 Teddy$$

and after double application of the rule (5) we get the sequent

$$\mathsf{T} : \forall x.{}^0 bear_{\mathrm{WT}} x \supset {}^0 animal_{\mathrm{WT}} x \,;\, \mathsf{T} : {}^0 bear_{\mathrm{WT}}{}^0 Teddy \vdash \mathsf{T} : {}^0 animal_{\mathrm{WT}}{}^0 Teddy$$

which is the sequent, we wanted to infere.

$$\frac{\Phi \vdash \mathsf{T} : \forall x.A}{\Phi \vdash \chi_\alpha : \lambda x.A}(1)$$

$$\frac{\Phi \vdash \mathsf{T} : C_1 \supset C_2}{\Phi \,;\, \mathsf{T} : C_1 \vdash \mathsf{T} : C_2}(3)$$

$$\frac{\Phi, \mathfrak{N} \vdash \mathfrak{M} \quad \Phi \vdash \mathfrak{N}}{\Phi \vdash \mathfrak{M}}(5)$$

$$\frac{\Phi \vdash a : A}{\Phi \,;\, \mathsf{T} : a\,x \vdash \mathsf{T} : A\,x}(2)$$

$$\frac{}{\Phi \vdash \mathfrak{M}}(4)$$

Φ is set of matches, $x/\alpha; A/\beta\alpha; a/\beta\alpha,\ C_1, C_2/o,\ \mathfrak{M} \in \Phi$

Table 1. Inference rules used in Example 1.

4   Conclusions

We expect that our system for knowledge representation enables a human to communicate with a computer in a natural language (or some part of it). A human can put his questions in natural language, the computer will make the semantic analysis using the NTA and then possibly infere the answer. The possibility to infere new facts could also improve the correctness of semantical analysis of text, text disambiguation or speech analysis.

References

1. Pavel Tichý: *The Foundations of Frege's Logic*, de Gruyter, Berlin, New York, 1988.
2. Pavel Tichý: Foundations of Partial Type Theory, *Reports on Mathematical Logic 14*, 1982, pp. 52-72.
3. Leo Hadacz: Using TIL for Semantic Analysis of Text, *Proceedings of TSD 1998*, Brno, 1998.
4. Leo Hadacz: *TIL a inference* (TIL and Inference), manuscript of PhD thesis.
5. Tomáš Chrz: *Reprezentace znalostí při komunikaci člověk-stroj* (The Knowledge Representation in the Human-Machine Communication), PhD thesis, Praha, 1984.
6. Karel Pala: Semantic Annotation of (Czech) Corpus Texts, *Proceedings of TSD 1999*, Pilsen, 1999
7. Miller G. A. et al.: *Five papers on WN*, electronic revised version, August 1993.

Playing with RST: Two Algorithms for the Automated Manipulation of Discourse Trees

Floriana Grasso

Department of Computer Science - University of Liverpool, UK
floriana@csc.liv.ac.uk

Abstract. This paper presents two algorithms for modifying RST based discourse trees in order to solve two given problems. By only exploiting syntactic properties of the trees, information originally presented is re-organized, to produce new, coherent text.

1 Introduction

This paper describes two algorithms for manipulating discourse trees produced by a text planner based on the Rhetorical Structure Theory (RST) principles [2]. Although many techniques exist to this purpose, the algorithms presented here are original insofar as they concentrate on purely syntactic manipulations, on the grounds that coherent text can be produced from existing text by only exploiting the property of nuclearity stated by the RST. The hypotheses behind them are general enough to ensure they can be applied to the output of any RST based text planner. We assume that a step of surface generation follows the algorithms application, and ignore practical problems of smooth phrasing of the text.

2 Definitions and Assumptions

We refer to the definitions and assumptions in [2,3], some of them slightly modified to add generality, and include a few more. These summarize the minimal set of characteristics that would be reasonably expected in any discourse tree adhering to the RST criteria.

Definition 1. *A **Rhetorical Structure tree (RS-tree)** is a tree whose nodes are defined by the triple <Name, Type, Content> where Name is an identifier, Type is either Root or the role (Nucleus or Satellite) that the node plays in the rhetorical relation (RR) associated with its parent, and Content is either the RR holding among the node's children (if the node is intermediate), or the informative unit (IU) associated with the node (if it is a leaf).*

We do not fix any limit on the number of nucleus and satellite children of a node, provided that each node has at least two children, with at least one nucleus.

Definition 2. *Given two sets of nodes $N = \{n_1, \ldots, n_j\}$ and $M = \{m_1, \ldots, m_k\}$ of T, then **N precedes M in T** ($N <_T M$) if each node in N is considered before every node in M when exploring T in a depth-first, left to right fashion.*

F. Jelinek, E. Nöth (Eds.): TSD'99, LNCS 1692, pp. 357–360, 1999.

Definition 3. *Given T a RS-tree, $L = \{l_1, \ldots, l_n\}$ a set of (not necessarily adjacent) leaves of T, and n a node (not leaf) of T, then:*
- **n generates L** *if L is contained in the set of leaves that n spans.*
- *The* **lowest generator of L** *(γ_L) is the unique node of T such that:*
  (i) γ_L generates L
  (ii) for all n_i, nodes of T generating L, γ_L is a descendant of n_i.
- *The* **context** *of L (χ_L) is the set of all leaves generated by γ_L.*
- *L is a* **span** *if $\chi_L = L$.*

Definition 4. *Two set of leaves L_1 and L_2 of a RS-tree are* **independent** *if their contexts do not overlap ($\chi_{L_1} \cap \chi_{L_2} = \emptyset$).*

Definition 5. *Given T a RS-tree, the* **most nuclear part** *of T (Nuc_T) is the set of T's leaves recursively defined as:*
(i) if T consists of a single node, then Nuc_T is T itself;
(ii) otherwise, if R_T is the root of T, Nuc_T is the union of the most nuclear parts of all R_T's children having a Nucleus role.
We define the most nuclear part of a node n as Nuc_{T_n}, where T_n is the sub-tree whose root is n, and the most nuclear part of a span S as Nuc_{γ_S}.

Definition 6. *Given T a RS-tree, the* **nuclear structure** *of T ($\mathcal{N}_T$) is the set of the most nuclear part of all T's nodes ($\mathcal{N}_T = \{N | N = Nuc_n, n \in T\}$).*

Assumption 1. *A rhetorical relation Rel holding between two spans S_1 and S_2 also holds between Nuc_{S_1} and Nuc_{S_2}. We say that Rel projects a* **deep-RR** *(Δ_{Rel}) between the two most nuclear parts.*

Assumption 2. *Two RS-trees having the same set of leaves (IUs), the same nuclear structure, and the same set of deep-RRs holding among the elements of their nuclear structures, have the same meaning (are* **equivalent**).

Definition 7. *A RS-tree manipulation operation is* **meaning preserving** *if the resulting RS-tree is equivalent to the original one.*

3   Algorithm 1: Extracting Sub-Trees

Problem 1. Given a RS-tree T and an arbitrary set $L = \{l_1, \ldots, l_n\}$ of leaves of T, extract T_1, the smallest sub-tree of T whose set of leaves contains L, such that $\mathcal{N}_{T_1} \subseteq \mathcal{N}_T$, and in which the deep-RRs defined by T in L are preserved.

Algorithm 1. Mark the nodes of T_L, the tree originating from γ_L, as follows:

1. let $N_k = \{n_1, \ldots, n_m\}$ be the set of nodes to mark at step k (where $N_0 = L$);
2. repeat until N_k is empty:
   a) mark in T_L each element belonging to N_k;
   b) for each element n_i of N_k, consider n_i's parent, p_{n_i}:
      i. if n_i has a satellite role, mark in T_L the most nuclear part of p_{n_i};
      ii. if $p_{n_i} \neq \gamma_L$ then add p_{n_i} to N_{k+1};
3. mark γ_L and prune out, from T_L, all unmarked nodes;
4. for each marked node n having only one child n_c, prune out n and connect n_c with n's parent, by also attributing n's role to n_c (see Fig. 1 for an example).

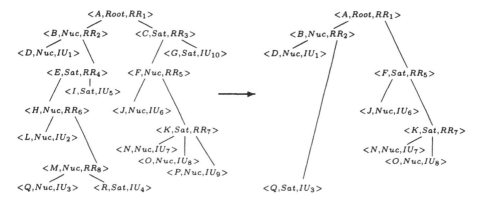

Fig. 1. Extraction of the smallest subtree generating $L = \{Q, N, O\}$.

4   Algorithm 2: Exchanging Text Spans

Problem 2. Given a RS-tree, T, and two independent sets $L_1 = \{l_j, ..., l_n\}$ and $L_2 = \{l_k, ..., l_m\}$ of leaves of T, such that $L_1 <_T L_2$, generate a RS-tree T_1, equivalent to T, such that $L_2 <_{T_1} L_1$.

We first describe two basic operations on the RS-tree, then the main algorithm:

Inversion of siblings: Let n be a node of a RS-tree T, and $N_i = \{n_{i_1}, ..., n_{i_k}\}$ and $N_j = \{n_{j_1}, ..., n_{j_h}\}$, two non overlapping subsets of n's children such that $N_i <_T N_j$. Then $\mathbf{Inv(n, N_i, N_j)}$ rearranges n's children so that $N_j <_T N_i$.
Exchange of satellite children: Let $< n_1, Role_1, RR_1 >$ and $< n_2, Role_2, RR_2 >$ be two nodes of a RS-tree T. Let Sa_1 and Sa_2 be the respective sets of T's subtrees originating from the children of n_1 and n_2 having a satellite role. An exchange of satellites between n_1 and n_2, $\mathbf{ExcSat(n_1, n_2)}$, consists of:
1. replacing $< n_1, Role_1, RR_1 >$ with $< n_1, Role_1, RR_2 >$;
2. replacing $< n_2, Role_2, RR_2 >$ with $< n_2, Role_2, RR_1 >$;
3. substituting the set Sa_1 with the set Sa_2 in n_1;
4. substituting the set Sa_2 with the set Sa_1 in n_2.

$\mathbf{Inv(n, N_i, N_j)}$ is always meaning preserving (we assume with [2] that a rhetorical relation application does not constrain the order of nuclei and satellites), whereas $\mathbf{ExcSat(n_1, n_2)}$ is meaning preserving only if $Nuc_{n_1} = Nuc_{n_2}$.

Algorithm 2. Let χ_1 and χ_2 be the contexts of L_1 and L_2 respectively, $L_{12} = \chi_1 \cup \chi_2$ and $\gamma_{L_{12}}$ the lowest generator of L_{12}. Two cases may occur:

1. $\gamma_{L_{12}}$ has at least one satellite child: let $\gamma_{Nuc_{12}}$ be the lowest generator of $Nuc_{\gamma_{L_{12}}}$, the most nuclear part of $\gamma_{L_{12}}$, and $L_{Nuc_{12}}$ the set of leaves generated by the nucleus children of $\gamma_{Nuc_{12}}$. Two cases may occur:
   a) $L_{Nuc_{12}}$ has empty intersection with L_{12}. Let γ_1 be the lowest generator of $\chi_1 \cup L_{Nuc_{12}}$ and γ_2 the lowest generator of $\chi_2 \cup L_{Nuc_{12}}$. Trivially, $Nuc_{\gamma_1} = Nuc_{\gamma_2} = Nuc_{\gamma_{L_{12}}}$. Then execute $\mathbf{ExcSat(\gamma_1, \gamma_2)}$ (Fig 2, left).

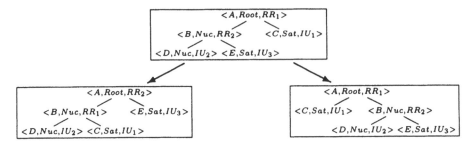

Fig. 2. Exchanging $\{C\}$ and $\{E\}$ (left) and $\{C\}$ and $\{D\}$ (right).

b) $L_{Nuc_{12}}$ has non-empty intersection with L_{12}. Because of the definition of context and the hypothesis of independence, it will have a non empty intersection with either χ_1 or χ_2 but not both. Then, if N_1 and N_2 are the sets of children of $\gamma_{L_{12}}$ generating χ_1 and χ_2 respectively, execute $\mathbf{Inv}(\gamma_{\mathbf{L_{12}}}, \mathbf{N_1}, \mathbf{N_2})$ (Fig 2, right).

2. $\gamma_{L_{12}}$ has no satellite children: treat as case 1(b).

Note that the algorithm can be applied only to two independent sets of leaves. If the independence hypothesis is relaxed, a purely syntactic exchange cannot be performed, and semantics has to be taken into account.

5 Conclusions

Modifying an existing text generally requires semantic knowledge, and can therefore be a difficult task. However, for a restricted class of modification problems, good results can be efficiently achieved by exploiting the syntactic properties of the RS-tree representing the text. Such problems concern the mere re-organization of existent text, with the only constraint that the deep rhetorical relations established among the informative units of the original text are preserved. For these problems, algorithms can be found that take in input a RS-tree and produce new, equivalent RS-trees satisfying a given requirement.

This paper described two of these text modification problems, and introduced two algorithms to solve them. The algorithms' results have been evaluated, and they were applied to two typical cases: re-structuring of one-shot documents [1] and generation of dynamic hypermedia. The two applications showed that these algorithms can be successfully used in practical contexts.

References

1. F. de Rosis, F. Grasso and D.C. Berry. Refining Instructional Text Generation After Evaluation. *Artificial Intelligence in Medicine*, to appear.
2. W. Mann and S. Thompson. Rhetorical Structure Theory: Toward a Functional Theory of Text Organization. *Text*, 8(3):243–281, 1988.
3. D. Marcu. Building up Rhetorical Structure Trees. In *Proceedings of the 13th National Conference on Artificial Intelligence (AAAI96)*, vol. 2, p. 1069–1074.

Another Step in the Modeling of Basque Intonation: Bermeo

Gorka Elordieta[1], Iñaki Gaminde[2], Inma Hernáez, Jasone Salaberria, and
Igor Martin de Vidales

[1] Dpto. Filología Inglesa, E.U.I.T.I. e I.T.T., Universidad del País Vasco,
Nieves Cano 12, 01006 Vitoria-Gasteiz, Spain
fipelalg@vc.ehu.es
[2] Dpto. Hizkuntza eta Literaturaren Didaktika Saila,
Irakaslegoaren Unibertsitate Eskola, Euskal Herriko Unibertsitatea,
Ramón y Cajal 72, 48014, Bilbo, Spain
tlpgatei@lg.ehu.es

Abstract. In this paper the basic features of the intonational structure of Bermeo Basque (BB) are analyzed. In BB there exists a lexical distinction between accented and unaccented words. Accented words
are always stressed, and unaccented words (not containing any accented
morphemes) only receive stress on their final syllable when they are immediately preceding the verb; otherwise they surface stressless. BB shows
a hierarchically organized intonational structure. Accentual Phrases are
identified by an initial boundary tone, a phrasal H-tone associated to the
second syllable, and a nuclear H* pitch accent. H-spreads rightwards to
other syllables until a H* pitch accent is met. A H* pitch accent triggers downstep on the following H*. An Intermediate Phrase contains one
or more APs, and is the domain of downstep. Finally, an Intonational
Phrase is signalled by a final L% in declarative.

1 Introduction

In this paper we provide an analysis of the intonational structure of a dialectal
variety of Basque spoken in Bermeo, on the coast of Bizkaia (the westernmost
Basque dialect). The importance of this study lies on the fact that it constitutes
another step towards modeling the intonation of Basque, a language about whose
intonation very little is known so far. Basque is substantially fragmented into
several dialects and local varieties within each of them. There are seven dialects,
according to modern commonly-accepted assumptions, and approximately fifty
dialectal varieties, according to some authors. The standard dialect cannot be
studied, because there is no commonly agreed-on prosody in this dialect (the
Academy of the Basque language has not provided any rules or guidelines in this
domain). Therefore, in order to know what the main properties of Basque intonation are, one must study as many varieties as possible. Only after a comparison
of the intonational characteristics of the different dialects will it be possible to
establish the general features that are common to all or most of the dialects,

F. Jelinek, E. Nöth (Eds.): TSD'99, LNCS 1692, pp. 361–364, 1999.

as well as the main differences among them. In the end, if as it is our hope the differences are not very dramatic, it will be possible to offer a model of Basque intonation. Such a model will have advantages both for the completion of the standardization process of the language and for technological applications. In the latter domain, our goal is to implement the model we obtain on a speech synthesizer for Basque.

In section 1, we present the basic features of the accentual system of Bermeo Basque. Section 2 describes the methodological aspect of the work. The results of our analysis are discussed in section 3. Finally, section 5 contains some concluding remarks.

2   Accentual System

Like all dialects spoken in Northern Bizkaia, the dialect of Bermeo is a pitch accent dialect, which could be classified into the accentual variety known as Getxo-Gernika. Previous work on the accentual system of Bermeo [3], [4] shows that pitch is lexically distinctive in this variety, as words can be lexically distinguished between being accented or unaccented. Lexically accented words are those that contain one or more accented morphemes, and display stress on the syllable which precedes the syllable that contains the accented morpheme. Affixes of the plural nominal declension are accented, as well as some affixes in the singular as well [3]. Some roots may be accented as well [4]. Lexically unaccented words, on the other hand, are those that do not contain any accented morphemes and only receive word-level stress when they occur in immediate preverbal position. In this type of contexts, stress falls on the last syllable of the word. In all other positions, unaccented words do not display stress. Thus, the lexically unaccented word *laguné* '(the) friend' occurs immediately to the left of the verb in (3), and receives stress on its final syllable. In (4), however, this word does not occur adjacent to the left of the verb, and therefore it does not receive stress. Unlike lexically unaccented words, lexically accented words always display surface stress, irrespective of their position in the sentence.

1. *gi.xó.nak*   (*gixon* 'man' + accented affix *-ak* 'determiner plural')
   mu.ti.llé.gas (*mutil* 'boy' + *-a* 'det.sg.' + accented affix *-gas* 'comitative')
2. *lé.ku* 'place'   *a.mú.ma* 'granny'
3. *laguné etorri da*   'the friend has come'
   friend come aux
4. *lagune gixonágas etorri da*   'the friend has come with the man'
   friend man-with come aux

3   Methodology

With these properties in mind, for this pilot study we recorded 365 utterances from a female speaker of the dialect of Bermeo, with different word orders and numbers of phrases. Thus, the utterances could be of the following kind: a) containing one phrase of two or three syllables, before and after the verb; b) containing two phrases of two or three syllables, before and after the verb; and

c) containing three phrases of two or three syllables, before and after the verb. The speaker had to read the utterances as naturally as possible. The recordings were made in a quiet room, with the use of a monodirectional microphone and a digital recorder. The data were analyzed by measuring F0 values for every syllable.

4 Experimental Results

The dialect of Bermeo presents a very similar intonational structure to that of the dialect of Lekeitio, previously described and analyzed in [2,6,7]. Couching our analysis within the framework known as the ToBI model of intonation analysis [1,8,9,10], we identified three intonational constituents: the Accentual Phrase (AP), the Intermediate Phrase (iP) and the Intonation Phrase (IP). All three constituents presented the same tonal/intonational cues as in Lekeitio.

An Accentual Phrase (AP) can be identified by three tonal cues. Its left edge is marked by a %L boundary tone which associates to the first syllable of the AP. Then, a phrasal H- is phonologically associated to the second syllable of the AP, and is usually phonetically realized on the second syllable. This tone spreads phonetically until the third tonal cue of an AP is met: the H* pitch accent, phonologically associated to the accented syllable, whether underlying (i.e., of an accented word) or derived (i.e., of a lexically unaccented word which occurs in immediate preverbal position). This pattern is illustrated in Fig. 1 which shows the F0 curve corresponding to the sentence "gixonak óllarra saldu dau" (*the man has sold the rooster*). After the high tone on the accented syllable there is a steep fall in pitch on the following syllable. The H* pitch accent marks the last word of the AP. That is, if after a H* accent there is another word in the utterance, it will display the same tonal pattern: %L on the first syllable, H- on the second syllable, and H* on the accented word.

As for issues of relative prominence among high tone peaks, it should be pointed out that as in Lekeitio, the accent in the word immediately preceding the verb is the most prominent in the utterance. After the preverbal word, the pitch level is reduced, and therefore all following accents present reduced peaks in the F0 curve. Another phenomenon affecting relations of prominence that can be observed in Bermeo Basque is downstep; a H* accent, derived or underived, causes downstep on a following H* within the same syntactic phrase. Thus, the presence of two or more APs in a syntactic phrase, each with its own H*, may give rise to a ladder-type configuration, as shown in Fig. 2, corresponding to "mallúkidxe lagunek jan dau" (*the friend has eaten the strawberry*).

An Intermediate Phrase (iP) has no tonal cues, but can be identified by the blocking of downstep. The left edge of an iP corresponds to the left edge of an AP that is the initial AP in a syntactic phrase. An iP may contain one or more APs, and is thus the domain of downstep.

An Intonational Phrase (IP) is signaled by L% boundary tones on the right edge of declarative utterances, and H% for list-type sentences and interrogative utterances. Another property that marks the right edge of an IP is lengthening of the final vowel.

5   Conclusions

To conclude, the model of intonation of Bermeo Basque (mid-northern Bizkaian) provided in this study presents striking similarities with the one of Lekeitio, a northwestern Bizkaian variety [2,7]. [5] reports similar intonational properties for the northeastern Bizkaian dialect of Gatika. This indicates that there is a common intonational structure for northern Bizkaian. This finding constitutes an important step towards the bigger goal of achieving a unified model of Basque intonation. Our next objective is to investigate the intonation of other dialects.

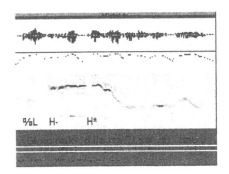 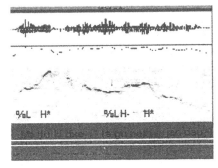

Fig. 1. Spreading of a H- upto H*   **Fig. 2.** Ladder-type configuration

References

1. Beckman, M.; and Ayers, G.: Guidelines for ToBI labelling, Ms., Ohio State University. (1994).
2. Elordieta, G.: Accent, tone and intonation in Lekeitio Basque. In: Martínez-Gil F., Morales A..(eds.): Issues in the Phonology and Morphology of the Major Iberian Languages., Washington, D.C.: Georgetown University Press. (1997) pp. 3–78.
3. Gaminde, I.: Bermeoko Azentu-Ereduaz. Uztaro 8, (1993) pp 105–117 (in Basque).
4. Gaminde, I.: Bizkaieraren Azentu-Moldeez. Labayru, (1995) Bilbo (in Basque).
5. Gaminde, I.: Gatikako Euskaraz. Gogoz Euskara Taldea, (1997) Gatika (in Basque).
6. Hualde, J.I.; Elordieta, G. and Elordieta, A.: The Basque Dialect of Lekeitio. Supplements of ASJU. UPV y Diputación Foral de Gipuzkoa. (1994).
7. Jun, Sun-Ah, Elordieta G.: Intonational structure of Lekeitio Basque. In: A. Botinis, G. Kouroupetroglou and G. Carayiannis (eds.): Intonation: Theory, Models and Applications. Proc. ESCA Workshop, (1997) pp. 193–196.
8. Pierrehumbert, J.: The Phonetics and Phonology of English Intonation. MIT doctoral dissertation. (1980).
9. Pierrehumbert, J.; Beckman, M.: Japanese Tone Structure. Cambridge, Mass.: MIT Press. (1988).
10. Silverman, K.; Beckman, M.; Pitrelli, J.; Ostendorf, M.; Wightman, C.; Price, P.; Pierrehumbert, J.; and Hirschberg, J.; ToBI: a standard for labeling English prosody. Proc. 2nd Int. Conference on Spoken Language Processing, Vol. 2 Banff, Canada. (1992) pp. 867–870.

Electronic Dictionaries: For Both Humans and Computers

Igor A. Bolshakov, Alexander F. Gelbukh, and Sofia N. Galicia-Haro

Natural Language Laboratory,
Center for Computing Research (CIC), National Polytechnic Institute (IPN),
Av. Juan de Dios Bátiz, CP 07738, Zacatenco, Mexico City, Mexico
{igor, gelbukh, sofia}@pollux.cic.ipn.mx

Abstract. The modern electronic dictionaries of natural languages should be universal. In the linguistic aspects, they should be a multi-linked database similar in their contents to the combinatorial dictionary by I. Mel'čuk, but with more stress on thesaurical links and word combinations. In interface aspects, they should have their data accessible to a text processing software, a human user and a lexicographer.

1 Introduction

During several decades, two different types of dictionaries of natural languages existed in parallel. In the paper form, they were oriented to various readers, in the electronic form, to needs of automatic text processing.

Recently electronic dictionaries have also appeared repeating a paper form and directly oriented to a human. All limitations on the size of dictionaries and on the complexity of demos on the screen were eliminated, with the tendency to minimize the role of the paper. Some computer scientists consider this as an ultimate solution of the problem of electronic dictionaries, but the situation is not so optimistic. The ex-paper dictionaries, even academically complete, do not contain all information necessary for text processing, and no automatic procedure can derive this information from human-oriented texts.

We argue for a universal dictionary, similar in its contents to [1], but with more stress on word combinations and thesaurical links. Three groups of possible use of the universal dictionary and some requirements oriented to various applications are described.

2 Some Deficiencies of Human-oriented Dictionaries

Trying to use the contents of two big electronic dictionaries of Spanish [2,3] for automatic processing of texts, the authors found a lack of information of graphical, morphological and syntactical nature. Indeed, no automata can calculate what lexemes in the pairs *lunes* vs. *mes* are invariable. All dictionaries give the labels of transitive verbs, adjoining pronominal clitics in accusative case. Meanwhile, for the group of dative verbs the number of agglutinated clitics that can be up to two, and without label of dativity to process such forms is impossible.

F. Jelinek, E. Nöth (Eds.): TSD'99, LNCS 1692, pp. 365–368, 1999.

In English, morphological peculiarities (nonstandard endings of plural for nouns like *phenomenon*, nonstandard paradigms for such verbs as *do, see, go,* etc.) are given in Merriam-Webster and other dictionaries. For Russian, a formal representation of its sophisticated morphology was given 20 years ago in [4] and immediately adopted by numerous software developers.

However, attempts to find in academic dictionaries combinatorial properties of words are usually in vain. The information about valences of nouns, adjectives, and especially verbs is scarse, even for English. Therefore, we cannot know how to express in Spanish or Russian combinations like *to pay attention.*

3   Some Deficiencies of Computer-Oriented Dictionaries

The main problem with computer-oriented dictionaries is the same: each of them contains only specific kind of information, so that several dictionaries are necessary to process the information on different language levels. On the first glance, computer dictionaries can be easily combined with each other, but it is so.

First, the sets of words in various dictionaries are different. Except for a small kernel, dictionaries tend to differ very significantly in their coverage. Second, combining dictionaries is not at all straightforward. The result is consistent only if the corresponding senses of the homonymous words are combined correctly. However, both the number and the sequence of the senses in different dictionaries are different, and there is no way to automatically recognize mutual correspondences. Computer dictionaries lack in human-oriented remarks, grammar reference, tables of abbreviations, etc.

4   Idea of the Universal Dictionary

Hence, the problems of the existing dictionaries are caused, apart from their natural incompleteness, by disruptive information spread across various sources. One needs to look up (and probably search for) many of them to see all about a particular word. Thus, our main idea is rather trivial:

- *A computer dictionary must present all the information about each word and the language as a whole.*
- *It must present all the possible ways of accessing and searching this information.*

We mean that the information should be accessible to both human users and other programs. A dictionary is so large and expensive database, that it is better to maintain, keep, and use its common version for all applications. It should be presented uniformly and be available in an integrated environment, such as a common browser (for users) or Application Program Interface (for programs).

The universal dictionary must also give all the available grammatical information, with all necessary cross-references. Since grammatical information may have a form of algorithms, the dictionary should not only show texts and describe algorithms, but also provide programs realizing them, e.g., various tools of checking and parsing. The new dictionary would not be a mechanical combination of different sources, though it is hardly possible to organize right now a great project on creation. All available sources should be merged by a program parsing various formats and compileing all data to a consistent whole.

The important problem is to avoid repetitions of in the common entries. Well-formalized information like morphological can be easily uniformed and merged. However, it is not the case for the explanations. At the same time, the merge should involve minimum of manual work. In any case, the number of possible updates should be great.

As the idea of the universal dictionary becomes more popular, it will be possible some standards, for unification of the formats of the sections. This will give lead to better procedures for further merges. If the source dictionaries cover different domains, it is enough to mechanically combine them, adding special marks to combined parts.

5   Contents of the Universal Dictionary

The universal dictionary should ideally contain the following types of information:

- Orthographic form of the keyword, including options and standard abbreviations
- Pronunciation including options
- Phonemic features, especially syllabic structure
- Morphological features: part of speech, inclination or conjugation class, etc.
- Syntactic features
- Explanations, maximally structured and consistent, including allusion and style features
- Semantic references, at least to synonyms and ideally to a thesaurs or a semantic network
- Combinatorial features, in style of a full combinatory dictionary and/or the dictionary of word combinations
- Equivalents in other languages
- Examples of usage

The necessity in the dictionary of the combinatorial information should be especially emphasized, since it can be currently found only in special dictionaries [5]. To compile lists of co-occrring words is much easier as compared with listing the lexical functions. Meanwhile, together with a thesaurus this facility proves to be very useful for both users (text compilation) and programs (syntactical analysis and disambiguation).

6   Needs of the End User

For a common user, the dictionary should provide a browser giving necessary data from the linguistic database in a convenient form. The interface developers should take into account, that: there is no need to save the space in the dictionary; the colors, fonts, etc., can be wider used; a nested hierarchy of paragraphs is much better for an entry than a single paragraph; the examples can be used more intensively. The data can be customized on the screen, with removal of unwanted parts. The data should be sorted by various categories. The request can combine logical means, such as AND, OR, NOT operators.

The dictionary should also give access to word-formation, agreement within word combinations, determining the syntactical structure of a phrase, language learning, etc.

7   Needs of the Text Processing System

For text processing software, the dictionary should have a library of procedures permitting to service any separate step of language processing or all of them. All the data are accessible from other programs in a formal way. The inner representation should not be just strings from a paper source, but members of well-defined sets. The dictionary should service various other programs, from spell-checker to text translators. There is no need to wait when these utilities are brought to perfection; they should be available right now.

8   Needs of the Lexicographer

The universal dictionary should be the environment to elaborate new data for this or other dictionaries. It should provide a way to modify the information, make temporal notes, etc. It should contain a specialized language to create the private programs for investigating lexical data. Pieces of data should have labels of its completion. Those without the labels are incomplete or accessible only to privileged users, for further elaboration.

References

1. Mel'čuk, I. A., Zholkovsky, A. K.: Explanatory Combinatorial Dictionary. In: Even, M.W. (ed.): Relational Models of the Lexicon. Cambridge University Press, (1988) 41–74.
2. Diccionario del Español contemporaneo. Grupo ANAYA, http://www.anaya.es.
3. Diccionario de la lengua Española. Real Acadêmia Española, Edición en CD-ROM (1996).
4. Zaliznyak, A.A.: Grammaticheskij Slovar' Russkogo Yazyka (Russian Grammar Dictionary). Russkij Yazyk, Moscow (1974).
5. Bolshakov, I.A.: Multifunction Thesaurus for Russian Word Processing. In: Proc. 4th Conf. on ANLP. Stuttgart (1994) 200–202.

Statistical Evaluation of Similarity Measures on Multi-lingual Text Corpora

R. Neumann and R. Schmidt

Institut für deutsche Sprache, D-68161 Mannheim, Germany

Abstract. Subject of this paper is the investigation of a similarity measure considering different large contexts of words. A multi-lingual bibel corpus is used to verify the results.

1 Introduction

There are several reasons to investigate similarity measures. The constantly increasing amount of documents available by computers requires procedures to index and classify these documents automatically and to supply an intelligent retrieval, e.g. content analysis.

2 Similarity Measures

According to [1] there are several similarity measures already applied in the fields mentioned above. We just picked up one of those already well known association measures that does not consider any context and modified it in order to verify if there can be obtained an improvement.

The measure we investigated is the cosine coefficient

$$\cos(doc_i, doc_j) = \frac{\sum_{k=1}^{t}(word_{ik}word_{jk})}{\sqrt{\sum_{k=1}^{t} word_{ik}^2 \sum_{k=1}^{t} word_{jk}^2}} \tag{1}$$

We applied it to the vocabulary (called context 0) of cleaned bible texts and to cumulated word pairs of context 5 resp. 20 words.

3 Experiments and Results

According to the theory of bible we expected from our experiments
- strong similarity between Mark, Luke and Mathew
- stronger similarity between Mark and John than between Mark and John 1.–3.
  What we got was
- all expected results can be verified when context is left out of consideration.
- the book of Luke shows always great similarity to the book of Mark (rank smaller 3)

F. Jelinek, E. Nöth (Eds.): TSD'99, LNCS 1692, pp. 369–371, 1999.

- the book of Mathew shows in most cases strong similarity to the book of Mark but there are some deviations, e.g. rank 10 for the English version with context 5 and even rank 44 for the German version with context 20
- the book of John is more similar to that of Mark than to those of John 1.–3. execept for the German version with context 5
- the results are in general language independent

Table 1. Comparions of the book of Mark. Context 0, 5 and 20 (German)

Rank Book	Coeff.	Rank Book	Coeff.	Rank Book	Coeff.
1 Luke	0.996	1 Luke	0.702	1 Ruth	0.803
2 Mathew	0.995	2 Mathew	0.670	2 Luke	0.774
4 John	0.986	... ...	...	... ...	...
... ...	...	... ...	...	11 John	0.697
... ...	...	24 1.John	0.489	... ...	...
... ...	...	... ...	...	26 1.John	0.621
... ...	...	... ...	...	40 2.John	0.522
... ...	...	42 2.John	0.364	... ...	...
... ...	...	... ...	...	44 Mathew	0.508
... ...	...	51 3.John	0.291	... ...	...
58 1.John	0.937	... ...	...	... ...	...
... ...	...	60 John	0.264	... ...	...
61 3.John	0.923	... ...	...	... ...	...
64 2.John	0.916	... ...	...	... ...	...
... ...	...	... ...	...	... ...	...

Table 2. Comparions of the book of Mark. Context 0 and 5 (English)

Rank Book	Coeff.	Rank Book	Coeff.
1 Luke	0.996	1 Daniel	0.788
2 Mathew	0.995	2 Luke	0.764
4 John	0.983	... ...	...
... ...	...	10 Mathew	0.692
... ...	...	25 1.John	0.578
... ...	...	31 John	0.513
... ...	...	48 2.John	0.391
52 1.John	0.930	... ...	...
59 2.John	0.915	... ...	...
60 3.John	0.911	... ...	...
... ...	...	64 3.John	0.330

Table 3. Comparions of the book of Mark. Context 0 and 5 (Latin)

Rank Book	Coeff.	Rank Book	Coeff.
1 Mathew	0.992	1 Luke	0.799
2 Luke	0.991	2 Mathew	0.787
4 John	0.978	... ...	...
... ...	...	20 John	0.625
... ...	...	49 1.John	0.369
59 1.John	0.900	... ...	...
... ...	...	61 John	0.175
62 2.John	0.893	62 3.John	0.129
63 3.John	0.884	... ...	...

4 Conclusion and Further Work

Most of the expected results could be verified but there are some phenomena we can not explain at the moment. For instance, the high similarity of the book of Mark to those of Ruth and 1. Samuel obtained for German considering a context of 20 words seems to be curious. Thus, further work has to be done to investigate the behaviour of the cosine coefficient of cumulated word pairs to verify whether it is able to detect more information or whether there are any random effects.

References

1. Salton, G.; McGill, M.J.: Introduction to Modern Information Retrieval. New York. McGraw-Hill (1983)

Document Title Patterns in Information Retrieval*

Manuel Montes-y-Gómez[1], Alexander F. Gelbukh[1], and Aurelio López-López[2]

[1] Natural Language Laboratory,
Center for Computing Research (CIC), National Polytechnic Institute (IPN),
Av. Juan de Dios Bátiz, CP 07738, Zacatenco, Mexico City, Mexico
`mmontesg@susu.inaoep.mx`, `gelbukh@pollux.cic.ipn.mx`
[2] INAOE, Electronics.
Luis Enrique Erro, No. 1, Tonantzintla, Puebla, 72840, Mexico
`allopez@giscl.inaoep.mx`

Abstract. The document titles give an important information about documents. This is why they are frequently used to obtain document keywords. We use them to determine document intentions. To obtain some textual details, we use special information extraction techniques for the construction of extra-topical representations of the documents. This representation reflects a document more completely. A possible use for the representation in the information retrieval is described. This improves the retrieval results.

1 Introduction

Unlike the structured information or formal representations, raw texts have a very complex form. This allows them to describe more completely all entities and facts, but at the same time provokes many of the difficulties in the analysis.

Nowadays, almost every raw text operation, for example, text classification, information retrieval, text indexing or description, is done on the basis of key-words or, in the best case, of topics obtained from entire texts of some their parts. This method can give text characteristics beyond topicality, such as intentions, proposes, plans, etc., which are usually ignored [1].

In this paper, we reveal the link between document title and its author intentions. We also describe a method of the automatic extraction of the intentions and finally propose a possible use of this information in IR systems.

2 Intention Structure

By intention, we mean determination to do something. In this sense, intentions are related with some acts fixed in the document text. They are grammatically

---

* This work was done while Manuel Montes y Gómez was supported by CONACYT, Mexico, through scholarship to pursue his Master Sc. Degree. The work was also partially supported by REDII-CONACYT and DEPI-IPN, Mexico.

F. Jelinek, E. Nöth (Eds.): TSD'99, LNCS 1692, pp. 372–375, 1999.

associated with some verbs having the main topic of the document as their subjects, such as *introduce, describe* or *propose.*

On the basis of these features, the task of determining document intentions consists of finding verbs which actions are performed by the document. For instance, the intention of some document is to *describe* something if there is some evidence in the document body that relate the document with the action *to describe.*

With this approach, the extraction of the document intention it is not simple. Intentions are more than mere actions reflected, they additionally include an object of the action and sometimes more pieces of related information. For instance, it is not sufficient to say that the intention of some document is to describe. It is also necessary to indicate what is to be described (the object), as well as how, when, or why this action is done.

3   Title Patterns

A title is the part of the document most heavily used for such tasks as indexing and classification. Just this prompts us to use titles for extraction of the intentions. We can note the following facts about the relation between titles and intentions [4]:

- Intentions are associated with a noun pattern:
    - A noun is followed by a preposition *of* or *to* in the beginning of the title (*An Introduction to a Machine-Independent Data Division*)
    - A substantive coordinated group is followed by a preposition *of* or *to* (*Implementation, evaluation, and refinement of manual SDI service*)
    - The case is similar to the previous, but with a dash instead of conjunction (*Computer simulation – discussion of the technique...*)
- Intentions are related to some gerund patterns:
    - The gerund is at the beginning of the title (*Proving theorems by recognition*)
    - The sequence adjective – gerund starts the title (*Automatic indexing and generation of classification systems*)
    - Prepositional group with gerund is anywhere except at the end (*A language for modeling and simulating dynamic systems*)

4   Intention Extraction Method

The system of intention extraction developed by us follows a information extraction scheme [2]. It contains a tagger, a filtering component, a parser, and a module of generation of the output data. As an example, let us process the title *Algebraic Formulation of Flow Diagrams.* First, each word is supplied with a syntactic-role tag[1]:

---

[1] The Tagger we are using is based on the Penn Treebank Tagset.

Algebraic|JJ formulation|NN of|IN flow|NN diagrams|NNS|

The next component selects only the titles containing some information about intentions. This filtering is based on the patterns previously described. Then the chosen titles are parsed and their structured representation[2] is formed [5]:

[[np,[n,[formulation,sg]],[adj,[algebraic]],
[of,[np,[n,[diagram,pl]],[n_pos,[np,[n,[flow,sg]]]]]]],'.']

This representation is entered to the last component, i.e., the output generator, where the structured representation is transformed into a semantic representation of the document (a conceptual graph) [3]:

[formulate] > *(manr)* > *[algebraically]*
[formulate] > *(obj)* > *[flow-diagram,{ * }]*

5   Experimental Results

Our system was tested on a collection of 4663 documents. Manual evaluations gave 802 useful documents (17.2 % of their total number), while our system 738 (15.7 %). The low percentage of the documents that can be processed by our method does not mean its low usefulness. The described method of intention extraction from titles is to be used together with our method of intention extraction from abstracts [1,4]. As it is shown below, the two methods work on nearly complementary distributed sets of documents and together cover up to 90 % of the collection.

Analyzing only those documents from the collection that have abstracts (normal situation), we obtained the following results:

Table 1. Abstracts and Titles

	Number	Procents
Documents with abstract	1587	100 %
Documents with graph from abstract	1207	76 %
Documents with graph from title	301	19 %
Documents with at least one graph	1438	**90 %**

Thus, for collections of documents with abstracts, the percentage of documents that are assigned at least one conceptual graph is close to 90 %, being sufficient for any retrieval application.

---

[2] The parser we are using was created in the New York University by Tomek Strzalkowski. It is based on "The Linguist String Project (LSP) Grammar" designed by Naomi Sager.

6   Future Infomation Retrieval Application

With the electronic information explosion caused by the Internet, more and more diverse information is on hand, so the need in better search engines is staggering. The more information about documents we have, the better we can evaluate the documents.

Basing on these ideas, we designed a new IR system. It will perform the document selection taking into account two different levels of document representation. The first is the keyword document representation. On this level, the documents are represented by means of keywords and the search is done by traditional retrieval method. The first level will select all documents related to the given general topics.

The second level is formed with semantic representations of the document intentions. This second level complements the topical information about documents and provides a new way to evaluate the relevance of a document. Intentions of documents are extracted from titles and abstracts [1,6].

7   Conclusions

With this article and with [1,4,6], we try to break down the keyword representation paradigm and begin to use other document characteristics. This paper shows the relations between document titles and their intentions and demonstrates how these intentions are reflected in titles. The method of intention extraction is domain independent, so that can be applied to documents on any topic.

References

1. López-López, A. and Myaeng, S. H.: Extending the Capabilities of Retrieval Systems by a Two Level Representation of Content. In: Proc. of the 1st Australian Doc. Comp. Symposium (1995).
2. Cowle, J. and Lehnert, W: Information Extraction. Com. of the ACM, 39 (1), (1996).
3. Sowa , J. F.: Conceptual Structures: Information Processing in Mind and Machine. Addison-Wesley (1983).
4. Montes y Gómez, M.: Extracción de Información de Títulos de Documentos. M.Sc. Thesis, Electronics, INAOE, México (1998).
5. Strzalkowski, T.: TTP: A fast and Robust Parser for Natural Language.PROTEUS. Project memorandum 43-A (1992).
6. López-López., A. and Montes y Gómez, M.: Nominalization in Titles: A Way to Extract Document Details. In: Memoria del Simposium Internacional de Computación CIC'98, México (1998).

Statistical Approach to the Automatic Synthesis of Czech Speech*

Jindřich Matoušek, Josef Psutka, and Zbyněk Tychtl

University of West Bohemia, Department of Cybernetics,
Univerzitní 8, 306 14 Plzeň, Czech Republic
{jmatouse, psutka, tychtl}@kky.zcu.cz

Abstract. The usage of multiple Hidden Markov Models (HMMs) to construct a Czech speech segment database (SSD) and a speech synthesis based on this inventory are presented in this paper. HMMs are used to model triphones. Binary decision trees are applied to automatically cluster the states of triphone HMMs. The clustered states are then employed to automatically segment the speech corpus and to create a SSD. The SSD constructed in this way is assumed to enable more precise context modeling than was previously possible. Several speech techniques are discussed to construct a concatenation-based synthesizer. Special attention is paid to an MFCC-based pitch-synchronous residually excited approach.

1 Introduction

Traditionally, formant synthesis was used in the context of TTS synthesis. Nowadays, concatenation-based speech synthesizers are widely used. Speech synthesis is performed by concatenating speech units (or segments) into a desired utterance.

The traditional speech units used in concatenation-based speech synthesis are diphones. They were traditionally obtained by manually doing segmentation of a speech corpus, which is a labor-intensive process. More contexts than diphones should be taken into account to achieve more natural synthetic voice quality. The suitable unit seems to be a triphone.

This paper presents a new approach to speech synthesis [1,2]. It uses triphones as the basic speech units. These units contain both the stationary parts of the phone and the transitional information of the adjacent phones. HMMs are used to model triphones. The states of triphone HMMs are clustered down using a set of automatically generated binary decision trees and then used to automatically segment the speech corpus and to create a SSD. Concatenation-based techniques can be then used to perform speech synthesis from these segments.

The paper is organized as follows. Section 2 describes the SSD construction process. Section 3 is dedicated to an MFCC-based pitch-synchronous residually

* This work was supported by the project No. VS97159 of the Ministry of Education of Czech Republic

F. Jelinek, E. Nöth (Eds.): TSD'99, LNCS 1692, pp. 376–379, 1999.

excited speech synthesis technique. Finally, Section 4 contains the conclusion, and outlines our future work.

2   Automatic Speech Segment Database Construction

The first step to create a SSD is to prepare a speech corpus. We use a number of utterances that represent approximately one hour of speech. These utterances were obtained by arbitrary text reading.

The next step involves word-level and phonetic transcription of the speech corpus. To do that, a set of Czech phonetic transcription rules is used, and phonetic transcription dictionary is then created. The dictionary is often extended by some mostly non-Czech words (so called exceptions), i.e. words, which cannot be transcribed by the rules. Presented approach to the SSD construction is based on modeling triphones, which were chosen as the basic speech units. HMMs are used to model triphones on the basis of the speech corpus. However, simply modeling triphones, problems with very big number of these speech units must be faced up to. Using a Czech 45-phone set, 91 125 triphones are theoretically possible. Moreover, there are too few occurrences of many triphones in the training data to properly estimate Gaussian distribution for each of the model states. In addition, there is a lot of triphones that do not occur even in the large speech corpus. To get rid of these problems, an effective clustering of a similar speech context in a model state level can be performed [1]. The clustering procedure results in clusters of model states (so called clustered states) and each cluster is represented by a single state. The process of clustering is realized by the automatically generated binary decision trees. They are built for each corresponding model states using a large list of questions concerning the immediate phonetic context and two stopping parameters. The trees make the models more robust and enable previously unseen triphones to be added to the most suitable state cluster. Changing clustering parameters, a trade-off between memory requirements and quality context modeling can be done.

Finally, the training data is aligned to state-clustered models (so called tied-state triphones) using the Viterbi search and label files of triphone and clustered state names against time are produced. The segmentation process could lead to some errors (like misplaced segment boundaries – mostly caused by incorrect transcriptions) when compared to manual segmentation. Nevertheless, good context coverage and consistent segmentation by HMMs typically overcome the drawback of an imperfect automatic segmentation [2].

The last step in the SSD construction process is a speech segment selection stage. Having the segmented corpus, one or more representative speech segments (i.e. acoustic realizations) per each speech unit can be selected and used later in speech synthesis.

Experiments were carried out on the basis of the one-hour speech corpus and resulted in approximately 4.600 word-internal and 7 800 cross-word triphones. Relatively high values of clustering parameters have been used so far and caused that tied-state triphones shared about 1 300 clustered states in case of word-internal triphones and 2 500 cross-word clustered states. No verification of the segmentation results was done since it is very laborious to manually segment such a large corpus. The SSD consists of one representative segment per each clustered state. The main criterion how to select a segment was an HMM score imposed by the Viterbi search.

3   Speech Synthesis

There are several techniques used in the context of a concatenation-based speech synthesis such as a speech synthesis based on waveform coding, an LP synthesis or variations of PSOLA technique.

In our approach we focus on an MFCC-based speech production system in the context of a TTS. We took into account the fact that we use the mel frequency cepstral coefficients (MFCCs) [3] as a basic parameterization for the purpose of the SSD construction. If a different kind of parameterization were used in this process, a different SSD would be obtained. So using the different parameterizations both in SSD construction and speech synthesis could impose inaccuracies, which we would not have under control. Being led by this reasoning, we decided to try to use the MFCCs instead of the LPC coding to synthesize speech. This approach should result in better intelligibility and naturalness of produced speech. The another advantage is that there is no need to compute any additional parameterization in that process.

Firstly, let's consider the algorithm used for an MFCC computing. The FFT-based algorithm is used to obtain the MFCCs. This parameterization has been refined mainly for the purpose of speech recognition and can be thought rather as a model of process of hearing than a model of speech production. MFCCs also benefit in their perceptually scaled frequency axis. The mel frequency scaling is achieved by applying the bank of triangular band-pass filters uniformly spaced in the mel frequency scale. In addition due to the cepstral nature of MFCCs, they offer abilities to model both poles and zeros (in contrast to e.g LPC).

MFCCs-based speech production system [4] is used for speech synthesis. The system is based on the straight approximation technique of log magnitude spectra for the linear filters. Since MFCCs represent smoothed a log magnitude spectra in the mel-scale, the basic requirement was to obtain linear (if possible) model, which would have an equal frequency response with spectrum represented by the vector of MFCCs. It means, apart from other things, that it was desired to find the linear filter with an exponential frequency response. For this purpose the Padde approximation of the exponential function by the rational function is employed. The mel frequency scale has to be approximated as well. It is performed using the All-pass transform applied in the form of the first order all-pass filter with one coefficient depending on the sampling frequency used. This approach results in a quite simple linear filter with a set of coefficients that can be easily and quickly computed from the stored MFCCs. Satisfying certain conditions speech production model we describe can be considered stable and even of a minimal phase. This fact is very important for our speech synthesis system, because it allows us to stay focused on achieving high quality pitch-synchronous residually excited speech synthesis system.

The excitation of the MFCC based production model cannot be straightforwardly declared because, as was mentioned, it does not model the process of speech production but rather the process of hearing. The hypothesis about a possibility of producing a real speech by means of a model of hearing results from a partial similarity of these two processes. A correct answer to the question how to excite the described model can be given by analyzing the residual signal obtained by the technique of an inverse filtering.

In our system we aspire to reach as high quality synthesis as possible so we are trying to take advantages of all above mentioned approaches to the speech synthesis. Concretely, the speech segments stored in the SSD are off-line scanned to track the pitches, windowed by Hanning window and filtered by inverse filter (inverse to the described MFCC-based speech production model) to obtain frames of pitch synchronous residual signals related to the respective speech segments. These residuals are then stored as well. During the speech synthesis the higher-level logic of our system supplies string of clustered states represented by vectors of MFCCs, the excitation as frames of residuals and prosodic parameters. The speech synthesis is then performed using the PSOLA approach on the residual signal (we call it Residual Excited Mel Cepstral PSOLA – REMC-PSOLA) and subsequently the MFCC-based speech production model is excited by that modified residuum to produce high quality synthetic speech.

4   Conclusion

This paper is dedicated to the problem of an automatic HMM-based SSD construction and a synthetic speech signal production. An MFCC-based concatenative speech synthesis technique is proposed to produce a resulting speech. The system is based on a single speaker SSD and should be able to produce intelligible speech that mimics the voice of the original speaker. The future work will be directed towards improving the naturalness and intelligibility of the synthesized speech by employing various prosodic parameters. We will focus on the REMC-PSOLA synthesis technique. A Czech TTS system will be also designed.

References

1. Donovan R.E., Eide M.: The IBM Trainable Speech Synthesis System. Proceedings of ICSLP'98, Sydney (1998).
2. Huang X., Acero A., Adcock J., Hon H-W., Goldsmith J., Liu J., and Plumpe M.: Whistler: A Trainable Text-to-Speech System; Proceedings of ICSLP'96, Philadelphia, (1996) 2387–2390.
3. Davis S., Mermelstein P.: Comparison of Parametric Representations for Monosyllabic Word Recognition in Continuously Spoken Sentences. IEEE Trans. ASSP, ASSP-28 (1980) 357–366.
4. Tychtl Z., Psutka J.: Speech Production Based on the Mel-Frequency Cepstral Coefficients. Proceedings of Eurospeech'99 (1999).

Language Model Representations for the GOPOLIS Database

Janez Žibert, Jerneja Gros, Simon Dobrišek, and France Mihelič

University of Ljubljana, Faculty of Electrical Engineering,
Tržaška 25, SI-1000 Ljubljana, Slovenia
janezz@luz.fe.uni-lj.si,
WWW home page: http://luz.fe.uni-lj.si

Abstract. The formation of a domain-oriented sentence corpus by sentence pattern rules is described. The same rules were transformed into word networks to serve as a language model within a HTK based speech recognition system. The performance of the word network language model was compared to the one of the bigram model.

1 Introduction

Nowadays, language models are mostly given by bigrams, 3-grams and generally n-grams [1]. They are statistical models and the transition probabilities have to be trained on large corpora.

An alternative approach is to use word networks which can be built directly or by transformation of generic rules given by a sentence pattern grammar.

In the paper we describe a domain-oriented sentence corpus acquisition procedure on the example of the Slovene GOPOLIS sentence corpus [2]. Standard domain-oriented sentences are transcribed by a sentence pattern grammar in form of rewrite rules. The same rules were transformed into word networks to serve as a language model within a HTK based speech recognition system. The performance of the word network language model was compared to the one of the bigram model.

2 Sentence Pattern Grammar

The GOPOLIS database is a large multi-speaker Slovene speech database, derived from real situation dialogs concerning airline timetable information services [2]. It was used within the SQEL multi-lingual speech recognition and understanding dialog system.

A sentence corpus was drawn from listening to recordings of real situation dialogs between anonymous clients inquiries and telephone operators. About 300 representative dialog sentences were drawn from the recordings, related only to flight timetables. They were compiled into the form of rewrite rules to obtain a generative sentence pattern grammar. Using this sentence pattern grammar,

F. Jelinek, E. Nöth (Eds.): TSD'99, LNCS 1692, pp. 380–383, 1999.
© Springer-Verlag Berlin Heidelberg 1999

we produced 22,500 different sentences for short introductory inquiries, long inquiries and short confirmations. These sentences have been already used to for training of a 3-gram language model [3] within the ISADORA speech recognition system.

The grammar rules consist of words, keywords and meta symbols and they have the following structure:

This (will be)/(is)/(was) a [really] nice * TIME_EXPRESSION.
* TIME_EXPRESSION=day, morning, evening, week.

The / sign indicates the exclusive OR relation: one of the items, given in round brackets must be selected. The [] brackets stand for an optional item, which can be selected or skipped during sentence generation. The keywords are marked with a *. Their values are randomly chosen from lists given below the rules.

3   Word Network

Further we constructed a word network based on the rules described in the previous section. We use the term word network according to the terminology in the HTK recognition engine [4].

The main advantage of the word network language model is in that it can be easily corrected or expanded while in other statistical models additional training is needed for every modification of the language model. Other advantages are: a word network built from rules represents a language model which defines all syntactically legal word sequences that can be recognised; a word network is integrated into the HMM framework within HTK; a word network can be easily visualized, so it can provide fine and detailed control over rules; in a word network we can simply transform one rule to another by insertion, deletion or substitution of words or word sequences.

Word network language models are extremely efficient in automatic dialog systems. We can easily insert a typical word sequence or phrase, which is often used but it has not be foreseen in the model and rebuild the word network again without additional training.

The transformation of a rule from a sequence of words and meta symbols into a list of nodes representing words and list of arcs representing transitions between words depending on the rule was performed automatically.

Each rule is represented by a subnetwork which is integrated into the word network. In the word network we define only one entry node as a *start node* marked as !NULL (according to HTK). Similarly, we define only one exit node as the *end node* with the same mark. The two basic building blocks of the subnetwork are shown in Figure 1. A word sequence $word_1 (word_2) / (word_3) word_4$ associated with an exclusive OR relation / is shown in Figure 1(a). A word sequence $word_1 [word_2] word_3$ associated with an optional item in [] brackets is represented in Figure 1(b). These basic elements are combined according to the sequence of meta symbols in the rule. Similarly, keywords marked with * are modelled using another subnetwork.

Fig. 1. Two basic building blocks of the word network.

Figure 2 shows an example of a word subnetwork built from the rule *(dober (dan) / (vecer)) / (dobro jutro) [zelim]* representing welcome greetings in the Slovene language.

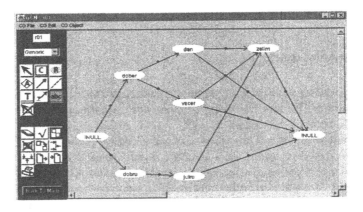

Fig. 2. Snapshot of the graphical visualisation tool interface representing a word subnetwork from the above rule.

The main network consists of subnetworks which are in parallel and they have the same start and end node. In the start node it is decided which subnetwork (rule) will accept the current word sequence. When this decision has been made then we can further choose only transitions (arcs) inside the current subnetwork (rule). Transitions between subnetworks are not allowed. By introducing an arc connecting the end and the start nodes a back-loop is established. In the given example we used unweighted arcs. However, in other applications certain arcs can be favourised with respect to the others by assigning them appropriate weights.

We developed a software that enables conversions of lists of arcs and nodes to the HTK standard lattice format and vice versa.

3.1   Visualisation of Word Networks

We used a conceptual graph editor CharGer [5] for visualization of word networks and subnetworks. CharGer is an editor to create visual display of graphs intended to support research projects and education. It was used for inspection and correction of the improper transition arcs between nodes. This tool when used as a word network visualisation interface (Figure 2) is very useful for ad hoc construction of simple language models for new application domains and for creation of new rules to obtain a generative sentence pattern grammar.

4   Evaluation of the Language Model Performance

Finally, we show how the presented language model improves speech recognition results in comparison to an unrestricted language model where all word transitions are allowed. A word recogniser was built using the HTK toolkit. All the acoustical model parameters were estimated on the training part of the GOPO-LIS speech database.

Table 1. *Word recognition scores in terms of correctness and accuracy (including insertions, deletions and substitutions) for three different language models.*

No Grammar		Bigram Model		Word Network Model	
Corr.	Acc.	Corr.	Acc.	Corr.	Acc.
83.0%	80.6%	97.4%	95.8%	98.4%	97.8%

Table 1 shows the final word recognition score for three different language models. The bigram model was trained on the extended GOPOLIS corpus [3]. The word network language model performed slightly better in comparison to the bigram model since all the test sentences were incorporated within the word network.

5   Conclusion

In the paper it was shown that an alternative approach to language modelling with word networks is efficient in systems with strictly domain oriented-sentences and mid-size vocabularies. Word networks can be modified without any additional training that is necessary in n-gram models. However, they can not recognise word sequences that are not expected in the word network. For this reason it is important to create rules that cover most syntactically correct word sequences in the given application domain. We showed that this way of language modelling for the GOPOLIS sentence corpus was successful.

References

1. Jelinek, F.: Continuous speech recognition by statistical methods. Proceedings IEEE **64** (1976) 532–556.
2. Dobrišek, S., Gros, J., Mihelič, F., Pavešić, N.: GOPOLIS: A mulitispeaker Slovenian speech database. Proceedings of the 1st LREC Conference. Granada Spain (1998).
3. Gros, J., Mihelič, F., Pavešić, N. . Sentence hypothesisation using ng-gram models. In Proceedings of the EUROSPEECH'95. Madrid Spain (1995) 1759–1762.
4. Young, S., Odell, J., Ollason, D., Vatchev, V., Woodland, P.: The HTK Book. Cambridge University. Entropic Cambridge Research Laboratory Ltd. (1997).
5. http://concept.cs.uah.edu/CharGer/.

Recognition of Alkohol Influence on Speech*

Richard Menšík

Institute of Radio electronics, Brno University of Technology
Purkyňova 118, 612 00 Brno, Czech Republic
Mensik@urel.fee.vutbr.cz

Abstract. This paper deals with two relevant questions: 1. Are the standard speech parameterizations able to detect alcohol influence to speech? 2. Which are the most appropriate phonemes, phonemes crossings and types of parameterization to detect alcohol influence? A small speech database for alcohol influence recognition was built for our purpose. All records in the database were parametrized using standard speech parameterizations and they were statistically evaluated. We found the display of alcohol is evident for all used parameterizations, but the localization in time or parametric domain was less evident.

1 Introduction

Research of alcohol recognition from speech signals was started by accident of tanker Exxon Valdez in March 1989. A suspicion arised the master was influenced by alcohol during the accident, but it was impossible to prove it, because blood alcohol tests were executed too late. A tape with recording of a dialogue between master and terrestrial radio communication station was the only material, which could clarify the situation. Therefore an intensive research of alcohol influence to speech followed and the suspicion was confirmed 2 years later [1]. Insurance offices and security organizations began to support next research, but the results were not publicized.

1.1 Influence of Alcohol on Human Organism

Alcohol causes changes in psychic (emotional) state and changes in psychomotorics in short-term point of view. It means the recognition of alcohol influence from speech will be superimposed by recognition of emotional state. It was proved that the emotion information in speech signal is mainly carried by the excitation rather than by the vocal tract in linear modeling of speech [2]. So if we want to separate influence of alcohol and influence of emotions, we must concentrate on vocal tract information. Vocal tract parameters and their changes can display the quality of psychomotorics in fact. Psychomotorical changes are noticeable on levels over 0.5 ‰ BAC. When we exceed level 1.5 ‰ BAC, the changes in

---

* This work was supported by the Ministry of Education of Czech Republic Project No.: VS97060

F. Jelinek, E. Nöth (Eds.): TSD'99, LNCS 1692, pp. 384–387, 1999.
© Springer-Verlag Berlin Heidelberg 1999

psychomotorics are so distinct that speech defects are audible by human ear. Higher BAC levels can be also detected by speech rate measuring. Our research is focused on levels from 0.5 to 1.5 ‰ BAC to separate the influence of alcohol as much is possible. Another reason for this demarcation is the unavailability of records on higher levels BAC.

2 Database

Our database contains records of sober speakers, records of speakers influenced by alcohol and the approximated values of blood alcohol concentration (BAC). Values of BAC were measured by respiration BAC meter *Drivesafe 2000*. The level of BAC are between 0.0 and 1.0 ‰. There are actually 25 speakers in the database, 13 males and 12 females, age between 18 and 50 years. The recorded utterances are words and short word conjunctions. The texts of utterances were chosen by empirical criterion, mostly they are words containing liquids ('r','l'), so they are relatively difficult to pronounce. Each speaker said 2 × 5 utterances for each given phrase, first 5 utterances on level 0.0 ‰ BAC and another 5 utterances on level greater than 0.5 ‰ BAC.

A portable DAT recorder was used for recording, the sampling frequency was 44.1 kHz, 16 bit quantisation. The records were converted into sound files (format '.wav') by soundcard SB AWE 64. The new sampling frequency was 22.05 kHz.

3 Statistical Evaluation and Results

All records from the database were parameterized – LPC, LPCC, PARCOR, log area ratio coefficients and delta parameters of each of previous ones. Standard segmentation 20 ms, 10 ms overlap. and windowing by Hanning window were used.

We simplified the classification task using 2 level categorization of speaker state. First state stands for sober speaker and second state for level over 0.5 ‰ BAC. The mean utterances were computed using DTW averaging for each speaker, each state and each type of parameterization. We observed the dispersion inside the state and dispersion between the states. At first these dispersions were summed across the time dimension and parameter dimension. The ratio of inter-state to intra-state dispersion represents a global gauge of suitability to distinguish the states.

The second part of our research was the localization of differences in time and parametric domain. For each time and each parameter index we have defined an objective function as a difference between those intra-state dispersion and inter-state dispersions. Higher value of objective function means that parameter located in this place (time, parameter index) is relevant for alcohol recognition.

Intra-state dispersion for speaker m, state s, parameterization q is defined as:

$$^I E_{m,s}^q = \frac{1}{K \cdot I \cdot J} \sum_{k=1}^{K} \sum_{l=1}^{L} \sum_{j=1}^{J} [P(m,k,s,DTW(l),j) - V(m,s,l,j)]^2 \quad (1)$$

P	- parameter matrix	l	- time (segmental) index
V	- mean parameter matrix	j	- parameter index
m	- speaker index	K	- number of utterances
k	- utterance index	L	- number of segments
s	- state index (logical variable)	J	- parametrization order

Corresponding inter-state dispersion:

$$^E E_{m,s}^q = \frac{1}{K \cdot I \cdot J} \sum_{k=1}^{K} \sum_{l=1}^{L} \sum_{j=1}^{J} [P(m,k,s,DTW(l),j) - V(m,\bar{s},l,j)]^2 \quad (2)$$

Objective function:

$$O_m^q(l,j) = {}^E E_m^q(l,j) - {}^I E_m^q(l,j) \quad (3)$$

Table 1. The intra-state and inter-state dispersions - word 'Laura'

	LPC	LPCC	Parcor	LAR	dLPC	dLPCC	dParcor	dLAR
^{I}E	33.71	5.13	5.33	15.65	22.98	3.07	3.15	10.14
^{E}E	107.40	20.34	17.61	58.75	49.00	6.74	6.96	22.78
^{E}E/^{I}E	3.19	3.97	3.3	2.24	2.13	2.19	2.21	2.24

When evaluating objective functions of each parameterization we can conclude the delta log area ratio (dLAR) coefficients appears to be the most suitable parameterization for alcohol recognition. The objective function of dLAR has minimal variations - about 75% speakers have the same location of objective function peaks. They were located around the consonant 'r', especially on the crossing between 'r' and vowels. Also crossing between nasals and consonants resulted in some peaks in objective function. In general, each swift change of spectra causes specific peak.

We assume that the inter-speaker variations of the objective function are probably caused by the heterogeneity of mood changes after alcohol consummation and heterogeneity of physical state - muscle tiredness and tiredness of vocal system. There is wide range of mood changes after alcohol consummation, the simplest classification being excitation-inhibition. The same alcohol level can cause excitation when drinking starts and inhibition when drinking ends. The another effect causing variability is that people moving during alcohol consumation are influenced less than people who do not encumber their muscles regardless they are on the same BAC level. (An encumbered muscle holds more blood so less alcohol intoxicates the brain).

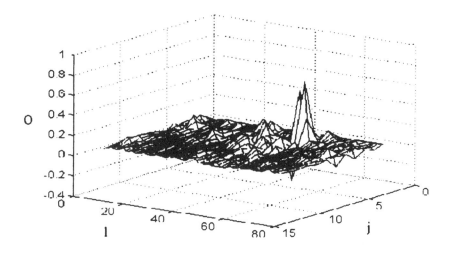

Fig. 1. Example of objective function (word 'Laura', dLAR parameterization)

4   Conclusions

We found the inter-state dispersion about 2-3 times higher than intra-state one. So it is evident the standard parameterizations are able to detect influence of alcohol on speech. The dLAR coefficients was evaluated to be the most suitable parametrization.

Further work will be conducted in enlarging of the database to investigate the factors described in last paragraph of chapter *Results*. We are also planing to search other parameterizations, more suitable for given task. Finally the speaker's BAC level should be classified, possibly using neural network classifier.

References

1. Brenner, M., Cash, J.R.: Speech Analysis as An Index of Alcohol Intoxication – The Exxon Valdez Accident, National Transportation Safety Board, Washington DC, 1991.
2. Cummings, K.E., Clements, M.: Analysis of Glottal Waveform Across Stress Styles, Proc. ICASSP-90, pp.369–372, 1990.
3. Kuenzel, H., Braun, A., Eysholdt, U.: Einfluss von Alkohol auf Sprache und Stimme, Kriminalistik Verlang, 1992.
4. Steven, B., Pisoni, D.: Alcohol and Speech, Acad. Press, San Diego 1996.
5. Rabiner, L.R.: Digital Processing of Speech Signals, Prentice Hall, New Jersey, 1978.
6. Riedel, O., Vondráček, V.: Klinická toxikologie, Avicenum, Praha 1971.

Recording of Czech and Slovak Telephone Databases within SpeechDat-E*

Jan Černocký[1], Petr Pollák[2], Milan Rusko[3], Václav Hanžl[2], and Marián Trnka[3]

[1] Brno Univ. of Technology, Inst. of Radioelectronics,
cernocky@urel.fee.vutbr.cz
[2] Czech Technical University in Prague, FEE K331,
{pollak,hanzl}@feld.cvut.cz
[3] Slovak Academy of Sciences, Inst. of Control Theory and Robotics,
{utrrrusk,trnka}@savba.sk

Abstract. The databases of 5 East-European languages: Czech, Slovak, Russian, Polish and Hungarian are being created within the SpeechDat-E project. This paper describes the overall design of SpeechDat-E databases and concentrates on the Czech (1000 speakers) and Slovak (1000 speakers). The item structure and recording specifications are presented. More detailed description is included for the language-specific items. Attention is paid also to the geographic and dialect distribution of speakers. The paper also presents the recruitment strategy.

1 Introduction

SpeechDat-E is the newcomer in the series of European projects aiming at the creation of large telephone speech databases (DB). Those projects started with SpeechDat(M), and last year, SpeechDat-II was successfully completed. The resulting databases are validated by SPEX and distributed by the European Language Resources Agency (ELRA). SpeechDat-E has started in December 1999 as Inco-Copernicus Project No. 977017. Participating institutions are Auditech (extending the existing Russian DB), CTU Prague and TU Brno (Czech), Inst. of Control Theory and Robotics of Slovak Academy of Sciences (Slovak), and the project is coordinated by Matra-Nortel (France). The consortium was joined by the University of Budapest and Philips (Hungarian), and by the University of Wroclaw and Siemens (Polish).

SpeechDat-E is a set of databases following the standard defined within SpeechDat-II [1]. The collection is performed over the telephone via an ISDN connection. The DB should be balanced in gender of speakers (45–55 % of males, 45–55 % of females), and in their age: 0–15 years 1% recommended, 16–30 min. 20 %, 31–45 min. 20 %, 46–60 min. 15 %, 61–∞ optional. The call structure is as follows: **2 isolated digits:** single isolated digit, sequence of 10 isolated digits, **4 digit/number strings:** prompt sheet #, telephone #, credit card #,

---

* This work is supported by EC programme INCO COPERNICUS under the project No. 977017: "Eastern European speech databases for creation of voice driven teleservices".

F. Jelinek, E. Nöth (Eds.): TSD'99, LNCS 1692, pp. 388–391, 1999.

PIN-code, **1 natural number**, **2 money amounts:** local currency, international currency (US dollar, euro), **2 yes/no questions** (spontaneous answer), **3 dates:** birth-date (spontaneous), prompted phrase, relative and general date expression, **2 times:** time of day (spontaneous), prompted time phrase, **6 application keywords/keyphrases**, **1 word spotting phrase** using embedded application word, **6 directory assistance names:** city of birth/growing up (spontaneous), city, company/agency, surname, forename and surname, own forename (spontaneous), **3 spellings:** artificial sequence, city name, own forename (spontaneous), **4 phonetically rich words** and **12 phonetically rich sentences**. The total number of items in a call is **48**.

2   Regional Coverage

The regional coverage (Table 1 and Fig. 1) should respect the major dialect groups. The repartition of speakers should be proportional to the population in regions, with 5 % tolerance, and with a minimum of 5 % speakers per region.

For the Czech, 5 regions were defined together with Assoc. Prof. Zdena Hladká from the Masaryk University Brno (Faculty of Arts). Slovakia still has many (about nine) dialect regions (the Slovak literary language is very young). We will try to cover them all in the database, but for the practical purposes it is better to have smaller number of bigger dialect regions. After a discussion with

Table 1. Dialect regions with the population.

Czech Republic		Slovakia	
region	population	region (including counties)	population
Central, West and North Bohemia	4378918	West (Bratislava,	
South Bohemia	2190396	Trnava, Trenčín)	1773000
Central Moravia	1887377	Central (Nitra, Žilina,	
East Moravia	710849	Banská Bystrica)	2065000
Silesia	760869	East (Prešov, Košice)	1518000
TOTAL	9928409	TOTAL	5356000

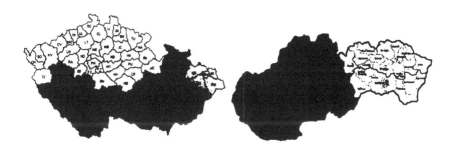

Fig. 1. Dialect regions for the recording of Czech and Slovak.

the phoneticians we decided to geographically divide Slovakia into three dialect regions. Their borders follow the borders of the recent administrative regions – counties.

3 Language-Specific Items and Phonetically Rich Phrases

Numerous items in the DB have to be generated with taking the given language into account. Due to a highly inflective nature of Czech and Slovak, we have encountered problems for the items, where the read form is different from the prompt. For example for natural numbers, it was necessary to check, if all forms of "hundred" appear with a sufficient number in the DB. Note, that in Czech, "hundred" can have 4 possible forms: "sto", "sta", "stě" and "set". The list of natural numbers had to be written so as to balance the coverage of all those forms. For the verification, a Perl-script was written to convert the numbers into words (for example: 9676200 → "devět miliónů šest set sedmdesát šest tisíc dvě stě"), and on the resulting list, a simple lexical analyzer counting the numbers of occurrences was run. A similar situation was encountered for local and international currency amounts, where all forms of crown ("koruna", "koruny", "korun"), dollar and euro had to be balanced. The creation of natural dates and times brought similar difficulties.

When creating the lists of Czech proper names, the telephone directory was analyzed and the lists of most frequent male and female first and last[1] names were created. The final lists have been generated in order to balance the two genders. For the spelled items, which are quite important in the DB, both Czech and Slovak have two deviations from the existing SpeechDat-II languages:

1. "Ch" is spelled as a unique letter. It is necessary to take this into account when analyzing the number of occurrences of different letters in items to spell (cities and proper names) and in the generation of artificial sequences.
2. in both languages, two ways of spelling coexist: the official one, spelling for example "B" as "bé", and the unofficial (but widely used also by educated people), where simply the phonetic form is read ("b", followed by a brief schwa). Those two systems have to be taken into account in the annotation of calls.

Each caller should provide several **phonetically rich sentences** read from the prompt sheets. The intention here is to obtain adequate training material for continuous speech modeling using sub-word units. To achieve this goal, sentences should be carefully designed to contain enough examples of rare phones - the most rare ones could be easily missed altogether if sentences were selected at random. Sentences were selected from a large corpus of daily newspaper texts containing more than two million sentences. First we applied several filters to exclude sentences which: 1) were too long or too short, 2) contained digits,

---

[1] In Czech and Slovak, most last names have their male and female form (Navrátil – Navrátilová).

abbreviations, parenthesis etc. Then we transcribed the sentences to sequences of phones using our program transc. This transcription reflects the most probable pronunciation. Each sentence was scored by number expressing its contents of rare units. The scoring function used weight for every phone and these weights were adapted iteratively. (following steps were repeated several times and their outcome influenced the weights). Sentences were sorted by descending score and sub-corpus was selected from sentences with the highest score. Word frequencies were taken into account in this selection to suppress too many repetitions of words. Phone frequencies were computed in the sub-corpus and phone weights adapted (iteration goes till this point). Resulting 8.000 sentences were hand-checked for hard-to-pronounce words, offensive contents, grammatical errors etc. and possibly corrected or changed. 5.300 sentences resulted from this step. All the previous automatic processing was repeated on this small clean corpus to receive the final set of sentences.

4 Recruitment Strategy

Due to bad experiences of Czech and Slovak people with unserious publicity campaigns, and to lack of special agencies (and of funding to pay them), the *snowball recruitment* strategy has been adopted. Students, relatives and friends are asked to recruit speakers, which may themselves become recruiters. In Slovakia, the members of Acoustic society of Slovakia are willing to recruit speakers, too. For the motivation of speakers, a small present (a camera film) will be offered for each completed call, as well as for a certain number of recruited speakers. The idea of a lottery was rejected due to a huge number of unserious ones.

5 Conclusion

For the first time, large telephone speech corpora are being collected for Czech and Slovak. After initial difficulties: technical (ISDN lines, green numbers, installation of recording platforms), linguistic (definition of items, regions, etc) and administrative (communication within the consortium and with the EC), the recording enters its main phase. We hope the DB will be useful not only for the Universities, but primarily for companies in the telecommunication industry and for telecommunication operators. In this field, the liberalization of Czech (in 2001) and Slovak (2002) telecommunication markets will bring increased competition among operators and one of credits of each operator will be the availability, robustness and user friendliness of its voice-driven teleservices. Our DBs will promote the creation of such services.

References

1. R. Winski. Definition of corpus,scripts and standards for fixed networks. Technical report, SpeechDat-II, January SD.1.1, workpackage WP1, http://www.speechdat.org.

Pragmatic Features of the Electronic Discourse in the Internet

Irina Potashova

Simferopol State University, Department of Russian Language,
Jaltinskaya 2, Simferopol Crimea Ukraine 333036
tvweek@pop.cris.net

Abstract. The paper is dedicated to the problem of pragmatic analysis of specific text represented on Web pages. The Internet, being now one of the most widespread electronic means of communication, uses its special structure and language which have number of features. This paper represents the attempt of revealing of these features on the example of different home pages wrote in English and Russian languages. Two classes of speech acts – assertives and directives – are revealed and described.

1 Introduction

The poster considers a part of Simferopol State University research project for elaborating cognitive and speech strategies of interactive discourse and constructing systems supporting computer-assisted human interaction. The research is connected with Incocopernicus project called Larflast started 01.11.98.

Three types of interactive discourse are analyzed:
- spoken communication
- written dialogue (exchange of usual letters and papers)
- electronic dialogue (Internet communication)

Nowadays different forms of communication by electronic means become more and more widespread, and especially popular now is the Internet. It's no more privilege of only professionals and is easy of access almost to everyone. According to G. Jones, at the end of 1995 there were about 35 millions people hooked up to the Internet [2]. And now we can even confirm that a new kind of speech develops – electronic, with oral and written speech.

World Wide Web (WWW, or W3) is the most important part of the Internet. It is an incredible amount of information compiled in special Web pages containing hypertext (see Sect. 2).

First, the interaction between a person and a computer has dialogical form, so it follows special rules of interactive communication; second, it has a number of pragmatic, lexical, syntax and other features, which are present on Web pages.

There are different kinds of electronic communication. These are E-mail, IRC (Chat) by which people can communicate with each other. It's a computer-mediated communication. It may have a dialogical form (one-to-one) or a form

F. Jelinek, E. Nöth (Eds.): TSD'99, LNCS 1692, pp. 392–394, 1999.

of mass communication (one-to-many or one-to-everyone). Strictly, browsing a Web page is not a dialogue itself, because there is no turn taking: an active communication position belongs to a user during his interaction with the Web page text. But according to M. Bachtin, the dialogue with text is possible [1]. In this case a user takes certain sense position relating to the text: he may have certain emotions, certain attitude to what he reads, and such position may be positive, negative, or neutral. So, following the Bachtin's idea, we can consider such communication as a dialogue.

2 Web Page: Structure, Language, Pragmatics

2.1 Structure

The dialogue between a person and the Internet is provided with a special language – HTML, or *hypertext*. It is represented on Web pages and includes headlines, words, and graphics being the references (or *hot links*) to other pages elsewhere on the Internet. Differing from printed book or magazine, Web pages are arbitrary connected, and to some extent a user don't need to use rigid linear formats.

2.2 Language

The natural language on Internet is English, the absolute majority of Web pages are wrote in it. There are pages in other languages, of course, but their number is comparatively small. Home pages wrote in Russian are bilingual: both English and Russian languages are used in time. On the Russian entertainment site *Weekend* home page (*www.weekend.ru*) such words as *Help, Webmaster* are used inside the main text. Also the forms of objective cases of the word *weekend* with Russian endings are used – *Weekend'a, na Weekend'e*. The world stem is English and the endings are Russian.

2.3 Pragmatics: the Speech Acts

The theory of speech acts considers the language communication as a purposeful action [4]. The unit of communication is the speech act. By definition, the dialogue is a sequence of the speech acts of the participants of dialogue. The set of such acts is a speech move of the participant. It represents the choice of required item from all ones represented on the home page by a user. Since the illocution of the Internet speech operations is a message or a command, it is possible to select two classes of speech acts used on home pages:

1. *assertives* (assertions, statements, reminders etc.). Nouns, adjectives and narrative sentences are used (*Exposition. Excursion. You will see the new exposition, exhibitions, and collections of the museum.* – State Darwin Museum home page.).

2. *directives* (requests, allowances, advices, recommendations etc.). The verbs in the form of imperative mood are frequently used for expression of directives (*search, enter, browse, click here; see* the new exposition).

The perlocution of these speech acts is the user's choice of one item from various alternate variants represented on home pages.

When browsing the home pages a user meets such features as underlining and coloring of the words being the hot links. We consider these features as a metalanguage and as a paralinguistic means, because these are communicative signs, and they are used specially for the delivery of the information.

3   Conclusion

There are the rules of interactive communication, which includes P. Grice's maxims and specific dialogue rules. We communicate with the Internet by means of written language, so it is necessary to take into account its space character, as well as rigidly limited format of display size. Therefore the home page information should be represented in precise and concise form. The Webmasters should be short, clear and avoid the ambiguity. The following to such demands provides the best communication and the receiving of the best results.

References

1. Bakhtin, M. M.: Problema teksta v lingvistike, filologiyi i drugich gumanitarnych naukach: Opyt filosofskogo analyza. In: Literaturno-krititcheskye statyi (in Russian). Moscow, Hudozch. Lit. (1986)
2. Jones, G.: How to Use the Internet. How To Books, United Kingdom (1996)
3. Ronginsky, V., Dikareva, S., Ilovayskaya, H.: Interactive Knowledge Base: Research Project of Simferopol State University. In: Dikareva, S., Ronginsky, V. (eds.): Cognitive Processes in Spoken and Written Communication: Theories and Applications. Proceedings. Ukraine, Crimea, 18–23 September (1995)
4. Searle, J.: Taxonomy of Illocutive Acts (in Russian). In: Novoye v zarubezchnoy lingvistike, Vyp. XVII. Moscow, Progress (1986)

Author Index

Aretoulaki, M. 1
Arruti Illarramendi, A. 181

Batliner, A. 193
Batůšek, R. 209
Baudoin, G. 262
Ben Hamadou, A. 127
Bieliková, M. 66
Böhmová, A. 34
Bolshakov, I. A. 219, 365
Buckow, J. 193
Byrne, W. 235

Čermák, F 39
Černocký, J. 262, 388
Chollet, G. 262
Cikhart, O. 109
Coughlin, D. A. 95

Dobrišek, S. 248, 380
Dvořák, J. 209

Ellouze, M. 127
Elordieta, G. 361

Ferencz, A. 333
Filasová, A. 258
Fink, G. A. 229

Galicia-Haro, S. N. 219, 365
Gallwitz, F. 1
Gaminde, I. 361
Gavat, I. 171
Gelbukh, A. F. 133, 219, 365, 372
Grasso, F. 357
Grigoryan, A. S. 160
Gros, J. 223, 241, 380
Guzman-Arénas, G. 133

Haas, J. 1
Hadacz, L. 353
Hajič, J. 109, 235
Hajičová, E. 20, 44
Hanžl, V. 388
Harbeck, S. 1, 187
Hejda, P. 290

Hermansky, H. 10
Hernáez, I. 361
Hlaváčová, J. 321
Holada, M. 296
Horák, A. 101, 105
Huber, R. 1, 193

Ipšić, I. 315
Ircing, P. 235
Ivanov, A. V. 215

Jelinek, F. 235
Jo, Ch.-W. 333
Julinek, R. 101

Khudanpur, S. 235
Kim, D.-H. 333
Klečková, J. 268
Klein, M. 274
Klíma, M. 280
Kopeček, I. 203, 262, 302
Král, J. 121
Krauwer, S. 19
Krokavec, D. 252, 258
Kruijff, G.-J. M. 83, 89, 345
Kruijff-Korbayová, I. 44, 83, 89
Kučera, K. 62

López-López, M. 372

Machovikov, A. 152
Machovikova, I. 152
Matoušek, J. 376
Matoušek, V. 308
McDonough, J. 235
Menšík, R. 384
Mihelič, F. 223, 241, 248, 315, 380
Mikovec, Z. 280
Montes-y-Gómez, M. 372

Nakagawa, S. 286
Nenadić, G. 115
Neumann, R. 72, 369
Niemann, H. 1, 187, 193, 195
Nijholt, A. 349
Nishizaki, H. 286

Nollen, D.   146
Nöth, E.   1, 187, 193, 199

Oceliková, J.   308
Ohler, U.   187

Pala, K.   56, 325
Panevová, J.   34, 50
Pavešić, N.   223, 241, 248, 315
Peterek, N.   235
Petkevič, V.   77
Petrovsky, A. A.   215
Poel, M.   349
Pollák, P.   388
Postigo Gardón, A.   181
Potashova, I.   392
Psutka, J.   235, 376

Radová, V.   165, 341
Rothkrantz, L.J.M.   146
Ruiz Vázquez, C.   181
Rusko, M.   388
Rychlý, P.   321

Sabac, B.   171
Salaberria, J.   361
Say, B.   337
Sazhok, M. M.   175
Schmidt, R.   72, 369

Sgall, P.   34, 44, 139
Shalonova, K. B.   329
Sidorov, G.   133
Skrelin, P. A.   156
Slavík, P.   280
Smrž, P.   101, 105
Spasić, I.   115
Stemmer, G.   199
Straňáková, M.   50
Švenda, Z.   341

Trabalka, M.   66
Trnka, M.   388
Tufiş, D.   28
Tychtl, Z.   376

Veber, M.   101
de Vidales, I. M.   361
Vintsiuk, T. K.   175
Visser, M.   349
Volskaya, N. B.   162
Vopálka, P.   166

Warnke, V.   193, 199

Žáčková, E.   325
Žemlička, M.   121
Žibert, J.   380

Lecture Notes in Artificial Intelligence (LNAI)

Vol. 1535: S. Ossowski, Co-ordination in Artificial Agent Societies. XVI, 221 pages. 1999.

Vol. 1537: N. Magnenat-Thalmann, D. Thalmann (Eds.), Modelling and Motion Capture Techniques for Virtual Environments. Proceedings, 1998. IX, 273 pages. 1998.

Vol. 1544: C. Zhang, D. Lukose (Eds.), Multi-Agent Systems. Proceedings, 1998. VII, 195 pages. 1998.

Vol. 1545: A. Birk, J. Demiris (Eds.), Learning Robots. Proceedings, 1996. IX, 188 pages. 1998.

Vol. 1555: J.P. Müller, M.P. Singh, A.S. Rao (Eds.), Intelligent Agents V. Proceedings, 1998. XXIV, 455 pages. 1999.

Vol. 1562: C.L. Nehaniv (Ed.), Computation for Metaphors, Analogy, and Agents. X, 389 pages. 1999.

Vol. 1566: A.L. Ralescu, J.G. Shanahan (Eds.), Fuzzy Logic in Artificial Intelliegence. Proceedings, 1997. X, 245 pages. 1999.

Vol. 1570: F. Puppe (Ed.), XPS-99: Knowledge-Based Systems. VIII, 227 pages. 1999.

Vol. 1571: P. Noriega, C. Sierra (Eds.), Agent Mediated Electronic Commerce. Proceedings, 1998. IX, 207 pages. 1999.

Vol. 1572: P. Fischer, H.U. Simon (Eds.), Computational Learning Theory. Proceedings, 1999. X, 301 pages. 1999.

Vol. 1574: N. Zhong, L. Zhou (Eds.), Methodologies for Knowledge Discovery and Data Mining. Proceedings, 1999. XV, 533 pages. 1999.

Vol. 1582: A. Lecomte, F. Lamarche, G. Perrier (Eds.), Logical Aspects of Computational Linguistics. Proceedings, 1997. XI, 251 pages. 1999.

Vol. 1585: B. McKay, X. Yao, C.S. Newton, J.-H. Kim, T. Furuhashi (Eds.), Simulated Evolution and Learning. Proceedings, 1998. XIII, 472 pages. 1999.

Vol. 1599: T. Ishida (Ed.), Multiagent Platforms. Proceedings, 1998. VIII, 187 pages. 1999.

Vol. 1600: M. J. Wooldridge, M. Veloso (Eds.), Artificial Intelligence Today. VIII, 489 pages. 1999.

Vol. 1604: M. Asada, H. Kitano (Eds.), RoboCup-98: Robot Soccer World Cup II. XI, 509 pages. 1999.

Vol. 1609: Z. W. Ras, A. Skowron (Eds.), Foundations of Intelligent Systems. Proceedings, 1999. XII, 676 pages. 1999.

Vol. 1611: I. Imam, Y. Kodratoff, A. El-Dessouki, M. Ali (Eds.), Multiple Approaches to Intelligent Systems. Proceedings, 1999. XIX, 899 pages. 1999.

Vol. 1612: R. Bergmann, S. Breen, M. Göker, M. Manago, S. Wess, Developing Industrial Case-Based Reasoning Applications. XX, 188 pages. 1999.

Vol. 1617: N.V. Murray (Ed.), Automated Reasoning with Analytic Tableaux and Related Methods. Proceedings, 1999. X, 325 pages. 1999.

Vol. 1620: W. Horn, Y. Shahar, G. Lindberg, S. Andreassen, J. Wyatt (Eds.), Artificial Intelligence in Medicine. Proceedings, 1999. XIII, 454 pages. 1999.

Vol. 1621: D. Fensel, R. Studer (Eds.), Knowledge Acquisition Modeling and Management. Proceedings, 1999. XI, 404 pages. 1999.

Vol. 1623: T. Reinartz, Focusing Solutions for Data Mining. XV, 309 pages. 1999.

Vol. 1632: H. Ganzinger (Ed.), Automated Deduction – CADE-16. Proceedings, 1999. XIV, 429 pages. 1999.

Vol. 1634: S. Džeroski, P. Flach (Eds.), Inductive Logic Programming. Proceedings, 1999. VIII, 303 pages. 1999.

Vol. 1637: J.P. Walser, Integer Optimization by Local Search. XIX, 137 pages. 1999.

Vol. 1638: A. Hunter, S. Parsons (Eds.), Symbolic and Quantitative Approaches to Reasoning and Uncertainty. Proceedings, 1999. IX, 397 pages. 1999.

Vol. 1640: W. Tepfenhart, W. Cyre (Eds.), Conceptual Structures: Standards and Practices. Proceedings, 1999. XII, 515 pages. 1999.

Vol. 1647: F.J. Garijo, M. Boman (Eds.), Multi-Agent System Engineering. Proceedings, 1999. X, 233 pages. 1999.

Vol. 1650: K.-D. Althoff, R. Bergmann, L.K. Branting (Eds.), Case-Based Reasoning Research and Development. Proceedings, 1999. XII, 598 pages. 1999.

Vol. 1652: M. Klusch, O.M. Shehory, G. Weiss (Eds.), Cooperative Information Agents III. Proceedings, 1999. XI, 404 pages. 1999.

Vol. 1674: D. Floreano, J.-D. Nicoud, F. Mondada (Eds.), Advances in Artificial Life. Proceedings, 1999. XVI, 737 pages. 1999.

Vol. 1688: P. Bouquet, L. Serafini, P. Brézillon, M. Benerecetti, F. Castellani (Eds.), Modeling and Using Context. Proceedings, 1999. XII, 528 pages. 1999.

Vol. 1692: V. Matoušek, P. Mautner, J. Ocelíková, P. Sojka (Eds.), Text, Speech, and Dialogue. Proceedings, 1999. XI, 396 pages. 1999.

Vol. 1701: W. Burgard, T. Christaller, A.B. Cremers (Eds.), KI-99: Advances in Artificial Intelligence. Proceedings, 1999. XI, 311 pages. 1999.

Vol. 1704: Jan M. Żytkow, J. Rauch (Eds.), Principles of Data Mining and Knowledge Discovery. Proceedings, 1999. XIV, 593 pages. 1999.

Vol. 1705: H. Ganzinger, D. McAllester, A. Voronkov (Eds.), Logic for Programming and Automated Reasoning. Proceedings, 1999. XII, 397 pages. 1999.

Lecture Notes in Computer Science

Vol. 1652: M. Klusch, O.M. Shehory, G. Weiss (Eds.), Cooperative Information Agents III. Proceedings, 1999. XI, 404 pages. 1999. (Subseries LNAI).

Vol. 1653: S. Covaci (Ed.), Active Networks. Proceedings, 1999. XIII, 346 pages. 1999.

Vol. 1654: E.R. Hancock, M. Pelillo (Eds.), Energy Minimization Methods in Computer Vision and Pattern Recognition. Proceedings, 1999. IX, 331 pages. 1999.

Vol. 1655: S.-W. Lee, Y. Nakano (Eds.), Document Analysis Systems: Theory and Practice. Proceedings, 1998. XI, 377 pages. 1999.

Vol. 1656: S. Chatterjee, J.F. Prins, L. Carter, J. Ferrante, Z. Li, D. Sehr, P.-C. Yew (Eds.), Languages and Compilers for Parallel Computing. Proceedings, 1998. XI, 384 pages. 1999.

Vol. 1661: C. Freksa, D.M. Mark (Eds.), Spatial Information Theory. Proceedings, 1999. XIII, 477 pages. 1999.

Vol. 1662: V. Malyshkin (Ed.), Parallel Computing Technologies. Proceedings, 1999. XIX, 510 pages. 1999.

Vol. 1663: F. Dehne, A. Gupta. J.-R. Sack, R. Tamassia (Eds.), Algorithms and Data Structures. Proceedings, 1999. IX, 366 pages. 1999.

Vol. 1664: J.C.M. Baeten, S. Mauw (Eds.), CONCUR'99. Concurrency Theory. Proceedings, 1999. XI, 573 pages. 1999.

Vol. 1666: M. Wiener (Ed.), Advances in Cryptology – CRYPTO '99. Proceedings, 1999. XII, 639 pages. 1999.

Vol. 1667: J. Hlavička, E. Maehle, A. Pataricza (Eds.), Dependable Computing – EDCC-3. Proceedings, 1999. XVIII, 455 pages. 1999.

Vol. 1668: J.S. Vitter, C.D. Zaroliagis (Eds.), Algorithm Engineering. Proceedings, 1999. VIII, 361 pages. 1999.

Vol. 1671: D. Hochbaum, K. Jansen, J.D.P. Rolim, A. Sinclair (Eds.), Randomization, Approximation, and Combinatorial Optimization. Proceedings, 1999. IX, 289 pages. 1999.

Vol. 1672: M. Kutylowski, L. Pacholski, T. Wierzbicki (Eds.), Mathematical Foundations of Computer Science 1999. Proceedings, 1999. XII, 455 pages. 1999.

Vol. 1673: P. Lysaght, J. Irvine, R. Hartenstein (Eds.), Field Programmable Logic and Applications. Proceedings, 1999. XI, 541 pages. 1999.

Vol. 1674: D. Floreano, J.-D. Nicoud, F. Mondada (Eds.), Advances in Artificial Life. Proceedings, 1999. XVI, 737 pages. 1999. (Subseries LNAI).

Vol. 1675: J. Estublier (Ed.), System Configuration Management. Proceedings, 1999. VIII, 255 pages. 1999.

Vol. 1976: M. Mohania, A M. Tjoa (Eds.), Data Warehousing and Knowledge Discovery. Proceedings, 1999. XII, 400 pages. 1999.

Vol. 1677: T. Bench-Capon, G. Soda, A M. Tjoa (Eds.), Database and Expert Systems Applications. Proceedings, 1999. XVIII, 1105 pages. 1999.

Vol. 1678: M.H. Böhlen, C.S. Jensen, M.O. Scholl (Eds.), Spatio-Temporal Database Management. Proceedings, 1999. X, 243 pages. 1999.

Vol. 1679: C. Taylor, A. Colchester (Eds.), Medical Image Computing and Computer-Assisted Intervention – MICCAI'99. Proceedings, 1999. XXI, 1240 pages. 1999.

Vol. 1680: D. Dams, R. Gerth, S. Leue, M. Massink (Eds.), Practical Aspects of SPIN Model-Checking. Proceedings, 1999. X, 277 pages. 1999.

Vol. 1682: M. Nielsen, P. Johansen, O.F. Olsen, J. Weickert (Eds.), Scale-Space Theories in Computer Vision. Proceedings, 1999. XII, 532 pages. 1999.

Vol. 1684: G. Ciobanu, G. Păun (Eds.), Fundamentals of Computation Theory. Proceedings, 1999. XI, 570 pages. 1999.

Vol. 1685: P. Amestoy, P. Berger, M. Daydé, I. Duff, V. Frayssé, L. Giraud, D. Ruiz (Eds.), Euro-Par'99. Parallel Processing. Proceedings, 1999. XXXII, 1503 pages. 1999.

Vol. 1688: P. Bouquet, L. Serafini, P. Brézillon, M. Benerecetti, F. Castellani (Eds.), Modeling and Using Context. Proceedings, 1999. XII, 528 pages. 1999. (Subseries LNAI).

Vol. 1689: F. Solina, A. Leonardis (Eds.), Computer Analysis of Images and Patterns. Proceedings, 1999. XIV, 650 pages. 1999.

Vol. 1690: Y. Bertot, G. Dowek, A. Hirschowitz, C. Paulin, L. Théry (Eds.), Theorem Proving in Higher Order Logics. Proceedings, 1999. VIII, 359 pages. 1999.

Vol. 1691: J. Eder, I. Rozman, T. Welzer (Eds.), Advances in Databases and Information Systems. Proceedings, 1999. XIII, 383 pages. 1999.

Vol. 1692: V. Matoušek, P. Mautner, J. Ocelíková, P. Sojka (Eds.), Text, Speech, and Dialogue. Proceedings, 1999. XI, 396 pages. 1999. (Subseries LNAI).

Vol. 1694: A. Cortesi, G. Filé (Eds.), Static Analysis. Proceedings, 1999. VIII, 357 pages. 1999.

Vol. 1701: W. Burgard, T. Christaller, A.B. Cremers (Eds.), KI-99: Advances in Artificial Intelligence. Proceedings, 1999. XI, 311 pages. 1999. (Subseries LNAI).

Vol. 1704: Jan M. Żytkow, J. Rauch (Eds.), Principles of Data Mining and Knowledge Discovery. Proceedings, 1999. XIV, 593 pages. 1999. (Subseries LNAI).

Vol. 1705: H. Ganzinger, D. McAllester, A. Voronkov (Eds.), Logic for Programming and Automated Reasoning. Proceedings, 1999. XII, 397 pages. 1999. (Subseries LNAI).